Electrical Circuit Analysis and Design

Macmillan New Electronics Series
Series Editor: Paul A. Lynn

G. J. Awcock and R. Thomas, *Applied Image Processing*
Rodney F. W. Coates, *Underwater Acoustic Systems*
Martyn D. Edwards, *Automatic Logic Synthesis Techniques for Digital Systems*
Peter J. Fish, *Electronic Noise and Low Noise Design*
W. Forsythe and R. M. Goodall, *Digital Control – Fundamentals, Theory and Practice*
C. G. Guy, *Data Communications for Engineers*
Paul A. Lynn, *Digital Signals, Processors and Noise*
Paul A. Lynn, *Radar Systems*
R. C. V. Macario, *Cellular Radio – Principles and Design*
A. F. Murray and H. M. Reekie, *Integrated Circuit Design*
F. J. Owens, *Signal Processing of Speech*
Dennis N. Pim, *Television and Teletext*
M. Richharia, *Satellite Communications Systems – Design Principles*
M. J. N. Sibley, *Optical Communications, second edition*
Martin S. Smith, *Introduction to Antennas*
P. M. Taylor, *Robotic Control*
G. S. Virk, *Digital Computer Control Systems*
Allan Waters, *Active Filter Design*

Electrical Circuit Analysis and Design

Noel M. Morris

formerly Principal Lecturer
Staffordshire University

First published 1993 by
THE MACMILLAN PRESS LTD
Houndmills, Basingstoke, Hampshire RG21 2XS
and London
Companies and representatives
throughout the world

ISBN 0–333–55482–5 hardcover
ISBN 0–333–55483–3 paperback

A catalogue record for this book is available from the British Library.

Reprinted 1994

Printed in Hong Kong

Contents

Preface

Electrical Circuit Analysis and Design is intended for use with the early years of a first degree course in Electrical, Electronic and Control Engineering, and for Higher National Diploma and Certificate courses in Electrical and Electronic Engineering.

The main prerequisite to its use is a knowledge of the basic concepts of electricity, magnetism and mathematics; an introduction to calculus is more in the nature of a corequisite than a prerequisite.

The book has primarily been written for the student, and it is intended that readers should be able to teach themselves the analytical techniques involved. To this end, many fully worked examples are included in the body of the text, and a large number of unworked problems (with solutions) are included at the end of chapters. Throughout the book, both 'power' and 'electronic' circuit examples and problems have been included.

A 'plus' feature of the book is a chapter on the use of SPICE software (Simulated Program with Integrated Circuit Emphasis) for circuit analysis. Examples in this chapter range from resistive d.c. networks to a.c. solutions and transient analysis, and illustrate the practical advantages of this software, which is pre-eminent in the field of circuit analysis.

When writing the book, I decided that it should be written from a logical teaching viewpoint. That is, as with a conventional course, the more understandable parts of circuit theory are treated first, after which the less easy but, technically, more interesting topics are covered.

Chapter 1 covers d.c. circuits and introduces the concept of basic elements and laws, including Kirchhoff's laws together with simple circuit analysis, and described dependent and independent sources.

In chapter 2, we take a first look at network analysis using mesh, nodal and loop analysis. In undergraduate and some HND courses, the latter usually involves a knowledge of network topology, which is also introduced. Finally, an introduction to the duality between circuits having similar mesh and nodal equations is given.

In order to understand circuit analysis fully, the reader should have a grasp of a number of circuit theorems and this, for d.c. circuits, is provided in chapter 3.

To move on to alternating current theory, the reader needs to understand the basis of circuits containing energy storage elements, this information being provided in chapter 4. Here we deal with capacitors, inductors and mutual inductance. Engineers have devised the 'dot' notation to deal with the latter, and this is fully explained in this chapter.

In chapter 5, we look at some of the many interesting aspects of alternating current theory, including phasors and phasor diagrams, complex impedance and admittance, together with series and parallel combinations of elements and circuits. Also covered are power and power factor, together with complex power. Next, in chapter 6, we apply a range of circuit theorems to a.c. networks.

Power-based electrical engineers have a particular interest in polyphase circuits, and this topic is comprehensively covered in chapter 7. This chapter describes and analyses many aspects of three-phase systems, including power measurement and symmetrical components.

Two-port networks are of great significance to electronics and telecommunications engineers and, in chapter 8, the reader is introduced to y, z, h and transmission parameters, together with the relationship between them.

In chapter 9 we meet the transformer, both 'ideal' and 'linear'. A knowledge of these is vital to both electrical and electronic engineers alike.

In chapter 10, we deal comprehensively with the transient analysis of circuits. A practice in many courses is to deal with this topic using two or sometimes three different techniques, each time covering very similar ground! In this chapter we look, initially, at the process of solving first- and second-order circuits by classical methods. These methods generally have a number of disadvantages, which are overcome by the use of the Laplace transform method; the latter is used throughout the remainder of the chapter.

While the Laplace transform method has the minor drawback that we need to spend a little time looking at the development of Laplace transforms before moving on to circuit analysis, it has the great advantage that the solution of circuits (both without and with initial conditions) becomes relatively straightforward. This chapter covers step function (d.c.) and a.c. analysis of first- and second-order circuits, together with transients in magnetically coupled circuits.

A feature of many electrical and electronic courses is the treatment of the frequency response of circuits, and this is described in chapter 11. Additionally, an introduction to complex frequency and the s-plane is provided and, equally importantly, the transformation of time-domain impedance into its equivalent *s*-domain impedance is covered. Frequency response is described in terms of Bode diagrams, and the method of drawing the diagrams is outlined in a straigthforward manner for both first- and second-order circuits.

Resonance occurs both in electronic and power circuits, and comprehensive coverage of series and parallel resonance is provided in chapter 12. Additional features in this chapter include frequency scaling, selective resonance and tuned coupled circuits.

In chapter 13 the attention of the reader is directed to harmonics and Fourier analysis. A knowledge of Fourier analysis is vital for all engineers, and the chapter includes such topics as waveform symmetry, line spectra, circuit response and the effect of harmonics in a.c. systems. Also included is a section on harmonic analysis.

In chapter 14 we meet one of the most powerful software packages available for analysis of electrical and electronic circuits, namely SPICE (Simulated Program with Integrated Circuit Emphasis). This software, which is both fast and versatile, is widely available both in full and in educational versions, and can be used to solve almost any electrical problem. The solution of a wide range of problems is included in this chapter.

Chapter 15 is devoted to a number of mathematical 'tools' needed by engineers and technicians, namely complex numbers, matrices, determinants and partial fractions.

Noel M. Morris

Acknowledgements

I acknowledge with thanks the assistance I have received from Derek Hopewell of Nottingham Polytechnic, Mr F. W. Senior, M.Sc., former Senior Lecturer at the Staffordshire University and, in particular for his not inconsiderable assistance, Lionel Warnes of Loughborough University. I would also like to thank Mr A. Lewis, Director of the Open Terminal Computer Centre, Stevenage.

Finally, grateful thanks are due to my wife, without whose assistance and support the book would not have been possible.

Acknowledgements

1

Elements and Laws

1.1 Introduction

In this chapter the basic relations in electric circuits are reviewed, including current, voltage, resistance, Ohm's law, electric power, etc. Additionally, other topics including the application of Kirchhoff's laws to circuits are described.

The concepts of independent and dependent voltage and current sources, are introduced, some of which may be new to some readers. One application of dependent sources is in the operational amplifier, which is used in every sphere of electrical and electronic engineering. So vital is this to all engineers it is introduced in this, the first chapter of the book.

1.2 Electric current

We are all familiar with electrostatic charge – and its most dramatic effect, namely lightning discharge. It is the latter effect, namely the electrical charge in motion or *electric current* which attracts our attention here.

The current in a circuit is a measure of the rate at which electric charge passes through the circuit, and we define the *instantaneous value* of the current, i, as

$$i = \frac{dq}{dt}$$

The charge in motion is, generally, carried by electrons which move from a low (that is, negative) potential to a higher potential (that is, a positive potential). In this book, however, we adopt the more usual convention that *current flows from a point of positive potential to a point of negative potential.*

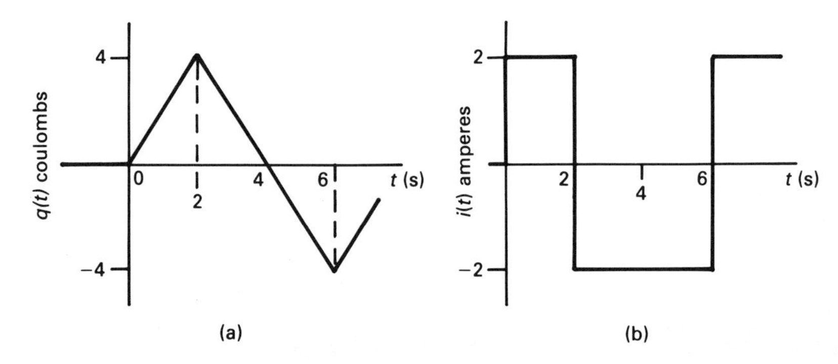

Figure 1.1 *(a) The charge passing a point in a circuit, and (b) the corresponding current in the circuit.*

The diagrams in figure 1.1 illustrate the concept of movement of charge and current. The waveform of movement of charge is shown in figure 1.1(a), and the corresponding current waveform is in diagram (b).

Lower-case symbols (such as i) are used in electrical engineering to represent an *instantaneous value* which varies with time. When *steady-state operating conditions* are reached, we use a capital symbol (such as I) to represent the quantity. We also use capital symbols to deal with the steady-state (root mean square) value of alternating quantities (see also chapter 5).

In the UK, current is represented by an arrow *drawn on the wire in which the current flows*, as shown in figure 1.2. If it is an instantaneous value of current, it is shown either as $i(t)$ or simply as i. Depending on the direction in which it is flowing, the current may be assigned either a positive or a negative value, as indicated by the 5 A current in figure 1.2. In the figure, a current of 5 A flows from left to right along the top wire; alternatively, −5 A flows from right to left. Engineers refer to a current as flowing either 'in' or 'through' a circuit.

In many US texts, the current is often shown as an arrow by the side of the wire or element in which it flows.

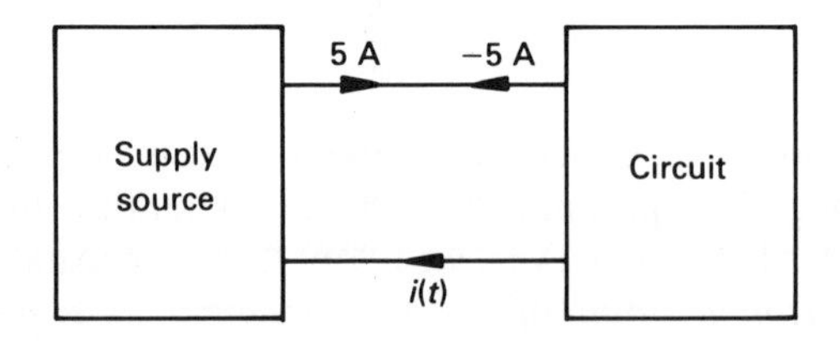

Figure 1.2 *Representation of current on a circuit diagram.*

1.3 Voltage, e.m.f. and p.d.

The voltage 'between' or 'across' a pair of terminals is a measure of the work required to move charge through the element or circuit connected between the terminals. The voltage across an element is, in fact, the work required to move a charge of 1 C from one terminal to the other. The SI unit is the *volt* (V) – named after the Italian physicist Giuseppe Antonio Anastasio Volta.

The instantaneous voltage is represented by the symbol v, and steady-state voltage (or r.m.s. voltage if alternating) by V.

Since voltage represents the potential needed to move a charge between a pair of terminals, a voltage can exist even if no current flows.

The energy converted per unit charge in an electrical source is known as the *electromotive force* (e.m.f.) of the source, and the electrical *potential difference* (p.d.) between two points in a circuit is a measure of the work required to move charge through the element. The general name given to both e.m.f. and p.d. is *voltage*.

Voltage may be represented on a circuit diagram either by a '+' and '−' pair of symbols (see figure 1.3(a)), or by an arrow pointing from one terminal to another (see diagram (b)). Where + and − symbols are used, the + terminal assumes the voltage written between the terminals with respect to the − terminal. For example, in figure 1.3(a), terminal A is +8 V with respect to terminal B, and in diagram (c) terminal B is −8 V with respect to terminal A.

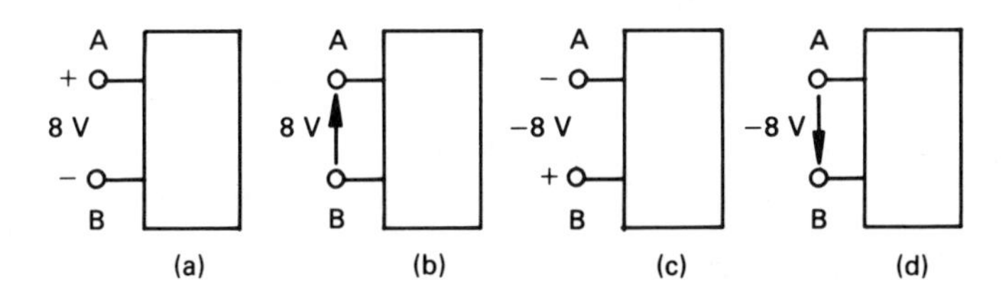

Figure 1.3 *Representation of voltage. In each case, terminal A is 8 V positive with respect to terminal B.*

When an arrow is used, the terminal which the arrowhead points to assumes the voltage written by the arrow with respect to the tail of the arrow. For example, in figure 1.3(b), terminal A is +8 V with respect to terminal B; in diagram (d) terminal B is −8 V with respect to terminal A. In some cases both sets of symbols are used.

1.4 Power in d.c. circuits

Power is the rate of transfer of energy; the unit of power is the watt (W) or joule per second (J s^{-1}). Since voltage has the dimensions of joules per coulomb, and current has the dimensions of coulombs per second, then

$$\text{power}, p = vi$$

Once again, lower-case p is reserved for the instantaneous value of power in the circuit, and upper-case P is reserved for *average power*. That is

$$P = VI$$

Figure 1.4 illustrates various combinations of electrical source and power absorbing element (or *load*). In general, *current flows out of the positive terminal of a power source, and into the positive terminal of a load* (see figure 1.4(a)).

In diagram (b), current flows out of the positive terminal of the left-hand block and into the positive terminal of the right-hand block. Consequently, the left-hand block is the power source and the right-hand block is the load. The power transferred from the source to the load is 10 × 5 = 50 W.

In figure 1.4(c) the lower line is −7 V with respect to the upper line, and the power transferred from the source (the right-hand block) to the load is 7 × 3 = 21 W. In diagram (d) the power absorbed by the load is 8 × 4 = 32 W.

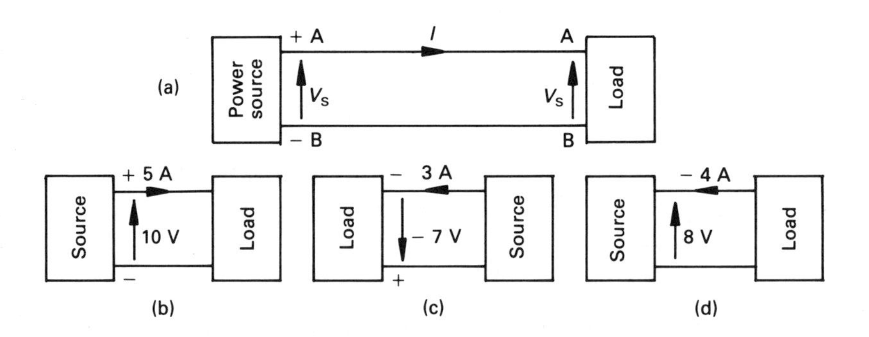

Figure 1.4 *Block diagrams showing a power source and its load.*

1.5 Linear passive circuit elements

The major part of electric circuits comprises *passive elements*, which can only dissipate or store electrical energy. A *linear* circuit element is one in which the voltage across the element varies linearly with the current through it. A number of elements are *non-linear*, in which case the voltage across the element does not vary linearly with the current through it.

There are two broad categories of passive element, namely those which dissipate energy and those which store energy. The former include *resistance*, which is the subject matter of this chapter, and the latter include *inductance* and *capacitance* which are described in chapter 4.

An *ideal circuit element* is one which is both linear and either dissipates energy or stores it. Unfortunately, most *practical circuit elements* have some degree non-linearity.

1.6 Resistance, conductance and Ohm's law

The outer electrons in the atoms of metallic conductors are loosely bound to the parent atom, and are relatively 'free' to move around the lattice structure at normal room temperature. When a p.d. is applied to the conductor, free electrons drift towards the positive pole of the supply, resulting in current flow in the conductor.

As the electrons move through the crystal lattice, they collide with other electrons and lose some of their energy. That is, there is some 'resistance' to current flow, and the resulting loss of energy is usually converted to heat. The resistance of the element enables us to broadly classify materials as follows:

Conductors: these have a low resistance, and include metals such as copper, brass, manganin, etc.
Insulators: these have a high resistance to current flow, and include wood, plastic and glass.
Semiconductors: these have resistance between that of conductors and insulators but, as students of electronics will know, it is not quite that simple! These include germanium, silicon, cadmium sulphide, etc.
Superconductors: when the temperature is within a few degrees of absolute zero, the resistance of these materials falls to zero (or nearly so). These materials include tin, lead, thallium, etc.

In 1827 George Simon Ohm published a pamphlet describing the result of experiments in electrical circuits, one relationship in the pamphlet being what we know as *Ohm's law*, and is

$$v = iR \quad \text{or} \quad V = IR$$

where v is the p.d. in volts across the linear element, i is the current in amperes through the element and R is a constant of proportionality called the *resistance*. The unit of resistance is the *ohm* (symbol Ω). An alternative form of Ohm's law is

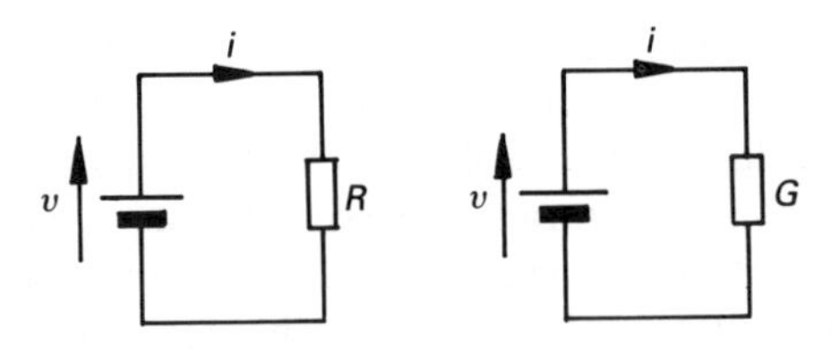

Figure 1.5 *Ohm's law.*

$$i = vG \quad \text{or} \quad I = VG$$

where G is the *conductance* of the element or circuit, and has the unit of the *siemens* (symbol S).

A resistance (or a conductance) is represented in a circuit by means of a rectangular box as shown in figure 1.5.

1.7 Power in a resistive circuit

In a linear resistive circuit the voltage is given by $v = iR$, and the power consumed is

$$p = vi = i^2R = v^2/R \quad \text{W}$$

or, alternatively

$$p = vi = v^2G = i^2/G \quad \text{W}$$

In many installations the power is measured in *kilowatts* (1 kW = 1000 W), and in large systems it is measured in *megawatts* (1 MW = 10^6 W = 1000 kW). In electronic circuits the power may be measured in *milliwatts* (1 mW = 10^{-3} W) or in *microwatts* (1 μW = 10^{-6} W).

Worked example 1.7.1

If $v = 10$ V and $i = -15$ mA, calculate the power consumed.

Solution

$$p = 10 \times (-15 \times 10^{-3}) = -0.15 \text{ W}$$

that is, the element can either be regarded as *consuming* −0.15 W, or as supplying or *generating* 0.15 W.

Worked example 1.7.2

If $P = 200$ mW and $I = 10$ mA, calculate the resistance of the circuit.

Solution

$$R = P/I^2 = 200 \times 10^{-3}/(10 \times 10^{-3})^2 = 2000\ \Omega$$

Worked example 1.7.3

A current of 0.1 A produces 5 W of power in a circuit. Calculate the conductance of the circuit.

Solution

$$G = I^2/P = 0.1^2/5 = 0.002\ \text{S}$$

1.8 Energy consumed in a resistive circuit

Power is the rate of energy transfer, that is

$$p = \frac{\mathrm{d}w}{\mathrm{d}t}$$

and energy is given by

$$W = \int_{t_1}^{t_2} p\ \mathrm{d}t$$

where W is the energy transferred in the interval t_1 to t_2. If the power supply is constant for time t, then

$$W = Pt$$

In a d.c. circuit where steady-state conditions exist

$$P = VI$$

hence

$$W = VIt$$

and since $V = IR$ in a d.c. circuit, then

$$W = I^2Rt = V^2t/R$$

In most systems, energy is measured in *kilowatt-hours* (1 kWh = $10^3 \times 60 \times 60 = 3.6 \times 10^6$ J), and in large systems it is measured in *megawatt-hours* (1 MWh = 10^3 kWh).

1.9 Independent and dependent supply sources

A source is an *active element*, that is, it can deliver energy to an external device; sources include batteries, alternators, oscillators, etc. Sources can be classified into 'ideal' or 'practical' as follows.

An *ideal source* is one which does not represent a 'real' device (other than special electronic circuits which have been engineered to provide an 'ideal' characteristic over a limited operating range); such sources can, theoretically, deliver an infinite amount of power. A *practical source* is one having limitations on its output voltage, current and power.

There are two types of source (ideal or otherwise), namely *voltage sources* and *current sources*. An ideal voltage source could, theoretically, maintain a constant voltage across a load of any resistance. An ideal current source could, also theoretically, maintain a constant current in a load of any resistance. An *independent source* is one whose source quantity (voltage or current) is independent of the remainder of the circuit.

The symbols in figure 1.6 depict typical independent sources. In diagram (a), the 6 V source supplies a current of 1.5 A to its load while, in diagram (b), the independent source (a battery) receives a current of 3 A, that is, it is being charged. Diagram (c) shows an independent current source, the voltage across its terminals depending on the resistance of the connected load.

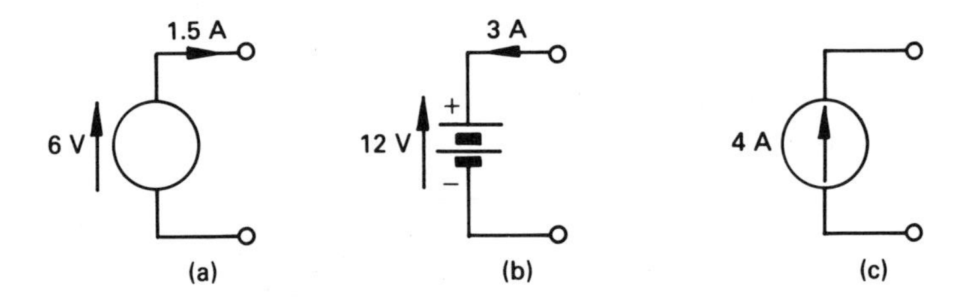

Figure 1.6 *An ideal or independent voltage source (a) delivering a power of 9 W, (b) absorbing 36 W. (c) An ideal or independent 4 A current source.*

There is another group of voltage and current sources, namely *dependent sources* or *controlled sources*, which are characterised by the diamond-shaped symbols in figure 1.7. The output voltage or current produced by the controlled source is dependent on a voltage or current at some other point in the circuit.

The dependent source in figure 1.7(a) produces a voltage which is dependent on the voltage V_1 existing at some other point in the circuit. We would describe this source as a *voltage-controlled voltage source*. The dependent voltage source in diagram (b) is controlled by a current I_2 which

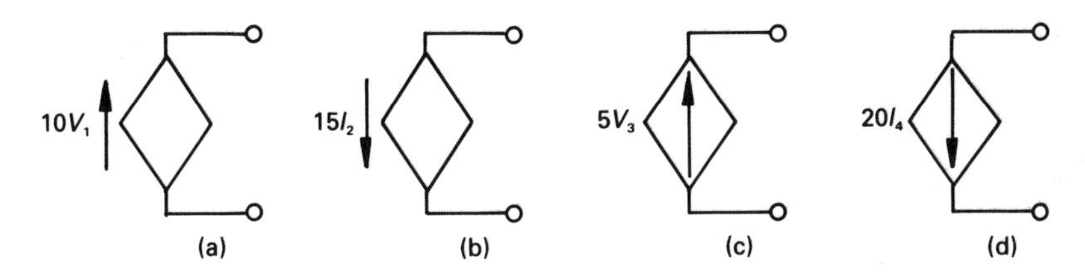

Figure 1.7 *Diagrams (a) and (b) represent dependent or controlled voltage sources, (c) and (d) represent dependent current sources.*

exists at another point in the circuit. Diagrams (c) and (d) respectively show a *voltage-controlled current source* and a *current-controlled current source*.

The reader should note, once again, that dependent sources are ideal sources.

1.9.1 The operational amplifier

The *operational amplifier* (often abbreviated to *op-amp*) is of fundamental importance in practically all forms of electrical circuit. The elements which comprise an *ideal* operational amplifier are shown in figure 1.8(a).

The amplifier has two input terminals marked v_1 and v_2, and one output terminal, v_o. We refer to the terminal marked '+' (to which v_2 is connected) as the *non-inverting input*, and to the terminal marked '−' as the *inverting input*; this is because the output voltage, v_o, is *in phase with* v_2

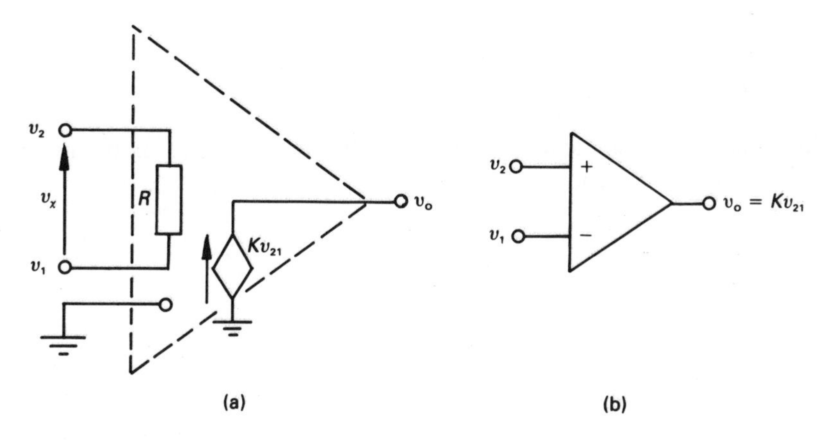

Figure 1.8 *(a) Idealised operational amplifier, (b) circuit representation.*

and *antiphase with* v_1. We can, in fact, connect a signal of any potential (either positive or negative) to either input terminal.

The *voltage gain* of the amplifier is K, and the output voltage is $v_o = K(v_2 - v_1) = Kv_x$. The value of K approaches infinity in an ideal op-amp and, if v_o is to be finite then the value of $v_x = v_2 - v_1$ must approach zero. That is, in an ideal op-amp, $v_2 \approx v_1$ which implies that the input current must be practically zero! Also, in an ideal op-amp, the input resistance R is infinite, so that the input current can be regarded as being zero (this also confirms that v_x must be zero!). The reader will also observe that the output from the op-amp is provided by a voltage-controlled voltage source, whose output resistance is zero (or nearly so!).

Worked example 1.9.1

Deduce an expression for the voltage gain of the operational amplifier circuit in figure 1.9. The op-amp is ideal.

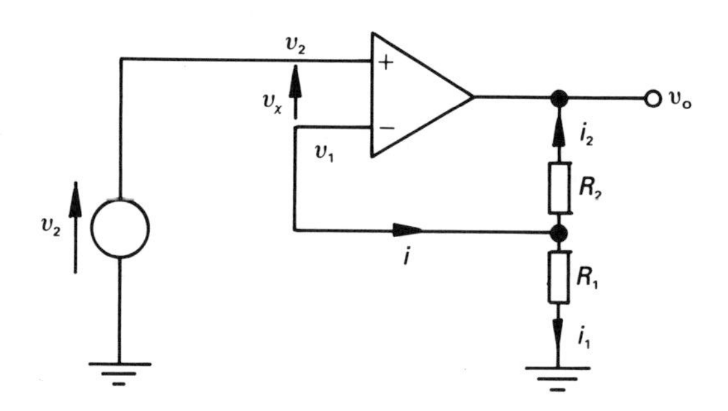

Figure 1.9 *Circuit for worked example 1.9.1.*

Solution

Since the op-amp is ideal, we can make the simplifying assumptions that $v_x = 0$ and $i = 0$, hence

$$v_1 = v_2$$

and

$$i_1 + i_2 = i = 0$$

that is

$$\frac{v_1}{R_1} + \frac{v_1 - v_o}{R_2} = 0$$

or, since $v_1 = v_2$, then

$$\frac{v_2}{R_1} + \frac{v_2 - v_o}{R_2} = 0$$

that is

$$v_o = \frac{R_1 + R_2}{R_1} v_2 = \left[1 + \frac{R_2}{R_1}\right] v_2$$

and the overall voltage gain is

$$\frac{v_o}{v_2} = 1 + \frac{R_2}{R_1}$$

1.10 Kirchhoff's laws

At about the same time as Ohm was carrying out his experimental work, Gustav Robert Kirchhoff was born in Germany; he was to revolutionise the work on electric circuit theory.

It is a simple fact that the rate of flow of charge into any point or *node* (a formal definition of a node is given in section 2.2) in a circuit is equal to the rate of flow of charge out of it. That is, charge cannot accumulate at a given point in the circuit. This is not a mathematical proof, but is the basis of *Kirchhoff's current law*, or KCL. That is

the algebraic sum of currents entering any node is zero

At the node in figure 1.10, KCL states that

$$i_1 - i_2 - i_3 + i_4 = 0$$

A simple way of expressing this is $\Sigma i = 0$ or, more precisely

$$\sum_{n=1}^{N} i_n = 0$$

where N is the number of wires meeting at the node. Alternatively, we can re-write KCL as follows

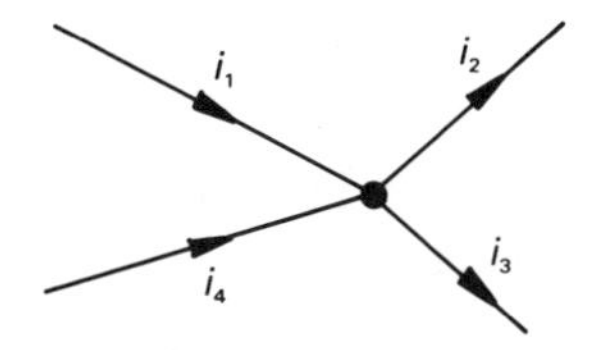

Figure 1.10 *Kirchhoff's current law (KCL).*

the algebraic sum of the current entering any node is equal to the algebraic sum of the current leaving the node

which, in figure 1.10 gives

$$i_1 + i_4 = i_2 + i_3$$

Also, if we proceed around the complete circuit, and return to the starting point, there is no change in electrical potential. This is the basis of *Kirchhoff's voltage law* or KVL, and can be stated in the form

the algebraic sum of the e.m.f.s. and p.d.s around any closed circuit is zero

If, for example, we proceed around the closed path ABCA in figure 1.11 we get

$$v_1 - v_2 - v_3 = 0$$

or, if we take the path ACBA we get

$$v_3 + v_2 - v_1 = 0$$

which, effectively, gives the same equation. A simple way of writing this is $\Sigma v = 0$ or, expressed in mathematical form, KVL says

$$\sum_{n=1}^{N} v_n = 0$$

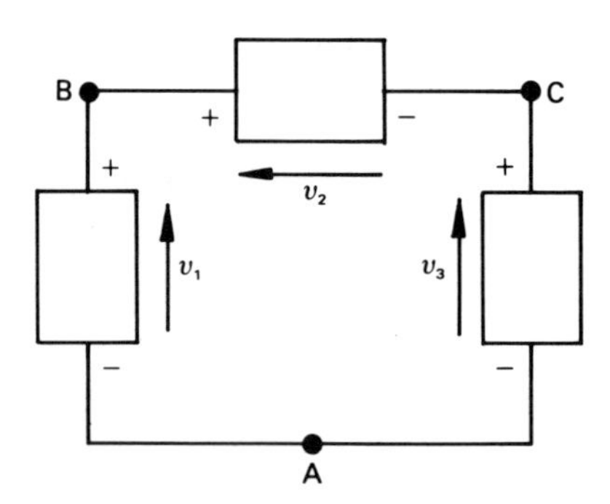

Figure 1.11 *Kirchhoff's voltage law (KVL).*

where N is the number of elements in the closed loop. Alternatively, KVL may be written in the following form

in any closed path the algebraic sum of the e.m.f.s is equal to the algebraic sum of the p.d.s

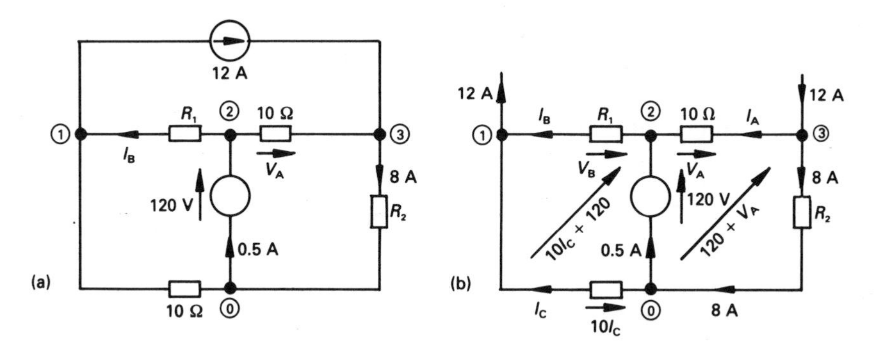

Figure 1.12 *Figure for worked example 1.10.1.*

Worked example 1.10.1

Calculate V_A, I_B, R_1 and R_2 in the circuit in figure 1.12(a).

Solution

The relevant section of the drawing is shown in diagram (b). Applying KCL to node 3 gives

$$12 = I_A + 8$$

or

$$I_A = 4 \text{ A}$$

hence, from Ohm's law

$$V_A = 10I_A = 40 \text{ V}$$

Applying KCL to node 2 yields

$$I_A + 0.5 = I_B$$

or

$$I_B = 4.5 \text{ A}$$

At node 0, KCL shows that

$$8 = 0.5 + I_C$$

or

$$I_C = 7.5 \text{ A}$$

Applying KVL to the closed-loop on the left of figure 1.12(b) shows that the voltage at node 2 with respect to node 1 is $10I_C + 120$, which is equal to V_B. That is

$$V_B = (10 \times 7.5) + 120 = 195 \text{ V}$$

Hence, from Ohm's law

$$R_1 = \frac{V_B}{I_B} = \frac{195}{4.5} = 43.33\ \Omega$$

Also, the voltage at node 3 with respect to node 0 is

$$120 + V_A = 160\ \text{V}$$

Hence $$R_2 = 160/8 = 20\ \Omega$$

1.11 The double-suffix voltage notation

Either when specifying or calculating the voltage of one node in a circuit with respect to another node, a *double-suffix voltage notation* is very useful.

Referring to the circuit diagram in figure 1.13, the voltage V_{AN} is read as 'the voltage of node A with respect to node N', and may be written in the form

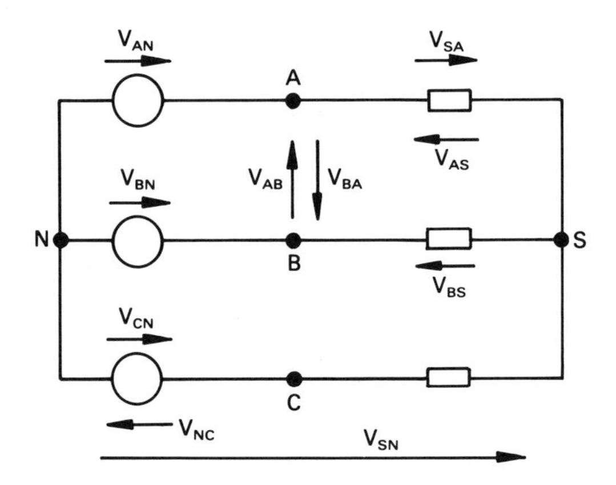

Figure 1.13 *The double-suffix voltage notation.*

$$V_{AN} = V_A - V_N$$

If node N is the *reference node* (whose potential may be regarded as zero), we can define V_{AN} simply as V_A. Thus

$$V_{AS} = V_{AN} - V_{SN} = (V_A - V_N) - (V_S - V_N) = V_A - V_S$$

also

$$V_{SA} = V_{SN} - V_{AN} = (V_S - V_N) - (V_A - V_N) = -V_{AS}$$

If in figure 1.13 $V_{AN} = 10$ V, $V_{BN} = 20$ V, $V_{CN} = 30$ V and $V_{AS} = -10$ V, then

$$V_{SN} = V_{AN} - V_{AS} = 10 - (-10) = 20 \text{ V}$$

Alternatively, we may say $V_{SA} = -V_{AS} = 10$ V, and

$$V_{SN} = V_{AN} + V_{SA} = 10 + 10 = 20 \text{ V}$$

If we need to calculate the voltage between two nodes, say node A and node B, then

$$\begin{aligned} V_{AB} &= \text{voltage of node A with respect to node B} \\ &= V_{AN} - V_{BN} = 10 - 20 = -10 \text{ V} \end{aligned}$$

or

$$V_{BA} = V_{BN} - V_{AN} = 20 - 10 = 10 \text{ V}$$

that is

$$V_{AB} = -V_{BA}$$

Also

$$\begin{aligned} V_{AB} &= V_{AS} - V_{BS} = V_{AS} - (V_{BN} - V_{SN}) \\ &= -10 - (20 - 20) = -10 \text{ V} \end{aligned}$$

Worked example 1.11.1

Calculate the voltage V_{AB} in figure 1.14.

Solution

In this case there are two closed meshes (a formal definition of a mesh is given in section 2.2) linked by a 4 Ω resistor. Since there is no return path for the current through the 4 Ω resistor, no current can flow in it, and $V_2 = 0$. The current I_1 circulating around the left-hand mesh is

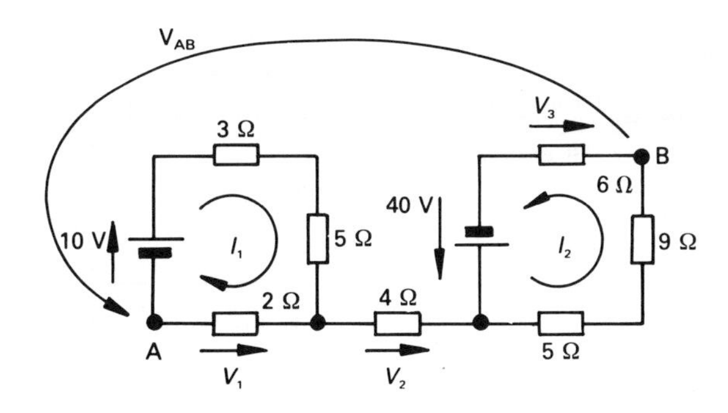

Figure 1.14 *Figure for worked example 1.11.1.*

$$I_1 = \frac{10}{2 + 3 + 5} = 1 \text{ A}$$

hence $$V_1 = 2I_1 = 2 \text{ V}$$

The current I_2 in the right-hand mesh is

$$I_2 = \frac{40}{5 + 6 + 9} = 2 \text{ A}$$

and $$V_3 = 6I_2 = 12 \text{ V}$$

To evaluate V_{AB}, we start at node B and proceed to node A (using any path), and we simply add (or subtract) voltages as we meet them. Selecting one of the available paths we get

$$V_{AB} = -V_3 + 40 - V_2 - V_1 = -12 + 40 - 0 - 2$$
$$= 26 \text{ V}$$

1.12 Practical (non-ideal) sources

An ideal voltage source can, theoretically, supply any current to a load, giving the flat *v–i* characteristic (shown dotted) in figure 1.15(a). However, a *practical voltage source* cannot do this, and the current it can deliver is limited. We account for this fact by introducing an *internal resistance*, or *source resistance*, or *output resistance*, R, in series with the ideal voltage source, e_S, as shown in figure 1.15(b).

The terminal voltage of the source is equal to e_S only when $i = 0$, that is, when the terminals of the practical source are open-circuited.

An ideal current source can, theoretically, supply any load with a constant current; the corresponding *v–i* characteristic is shown by the dotted line in figure 1.16(a). A practical current source cannot do this; the

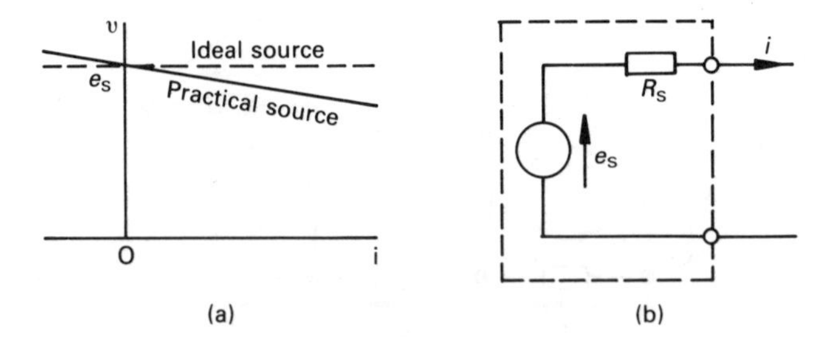

Figure 1.15 *(a)* v–i *characteristic of ideal and practical voltage sources, (b) an equivalent circuit of a practical voltage source.*

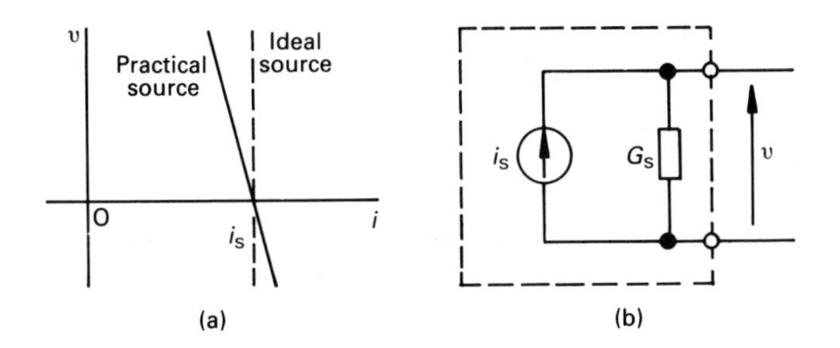

Figure 1.16 *(a)* v–i *characteristic of ideal and practical current sources, (b) an equivalent circuit of a practical current source.*

v–i characteristic of a *practical current source* is shown in full line in diagram (a), and an equivalent circuit for this type of source is shown in diagram (b). The change in characteristic is accounted for by the *internal conductance*, G_S, which is also known as the *source conductance*, or the *output conductance*.

1.13 Transformation of practical sources

When analysing circuits, it is often convenient to convert all the sources in the network either to voltage sources or to current sources. In the following we will consider the process of *source transformation*.

To simplify the process of analysis, let the internal conductance G_s of the practical current source be replaced by its equivalent resistance r_S, where $r_S = 1/G_S$. Consider the practical voltage and current sources in figure 1.17(a) and (b), respectively, which are connected to an identical load resistor.

The current i_L, in the load connected to the voltage source is (figure 1.17(a))

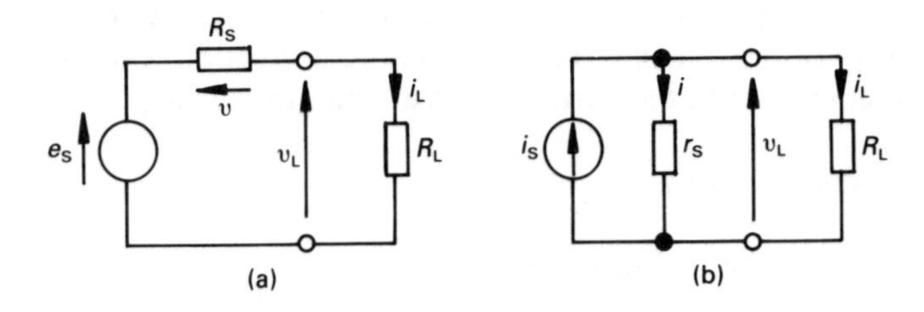

Figure 1.17 *Source transformation.*

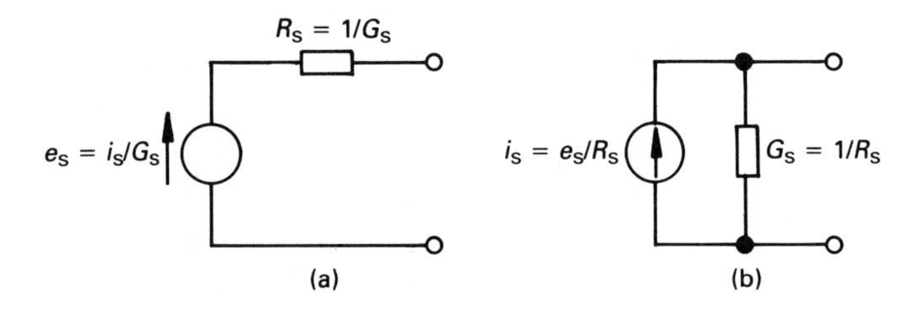

Figure 1.18 *Equivalent sources.*

$$i_L = \frac{v}{R_S} = \frac{e_S - v_L}{R_S}$$

or $$v_L = e_S - i_L R_S$$

The current flowing in the load connected to the current source is (figure 1.17(b))

$$i_L = i_S - i = i_S - \frac{v_L}{r_S} = \frac{i_S r_S - v_L}{r_S}$$

or $$v_L = i_S r_S - i_L r_S$$

Since the two sources are identical, we can equate the voltage across the load in the two cases as follows

$$v_L = e_S - i_L R_S = i_S r_S - i_L r_S$$

Equating terms in i_L shows that

$$R_S = r_S = 1/G_S$$

Also, for equality, it follows that

$$e_S = i_S r_S = i_S/G_S$$

That is, the voltage source and the current source in diagrams (a) and (b) are equivalent as far as measurements of voltage and current at the terminals are concerned.

When converting sources, the reader should be careful to ensure that the polarity of the voltage source acts in the correct direction to produce the correct current in the current source, and vice versa.

Worked example 1.13.1

Calculate the value of I in figure 1.19(a).

Solution

Initially, we will convert the 3 A practical current source into its equivalent voltage source. The result is shown in diagram (b). The equivalent voltage source has an internal resistance of 1/0.2 = 5 Ω, and a source voltage of I_S/G_S = 3/0.2 = 15 V.

The net voltage in figure 1.19(b) acting in the direction of I is 10 − 15 = −5 V, so that

$$I = -5/(10 + 5) = -0.333 \text{ A}$$

that is, its value is 0.333 A flowing in the *opposite direction* to that shown in figure 1.19(a).

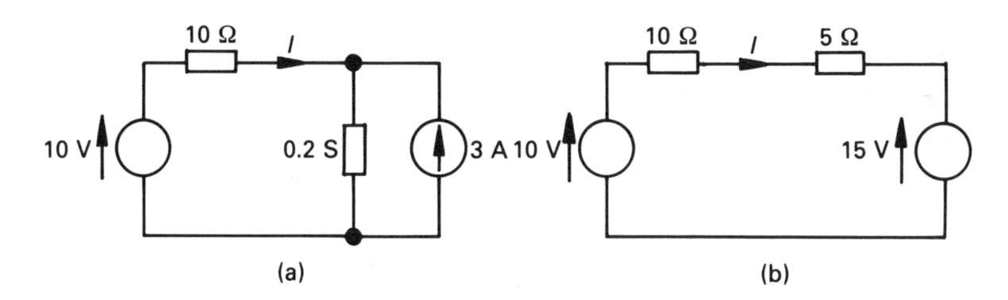

Figure 1.19 *Figure for worked example 1.13.1.*

1.14 Resistance of a series circuit

Elements which carry the same current are connected in series; the three resistors in figure 1.20 are connected in series.

Since the current, I, is common to each resistor, then

$$V_1 = IR_1, \quad V_2 = IR_2, \quad V_3 = IR_3$$

Applying KVL to the circuit gives

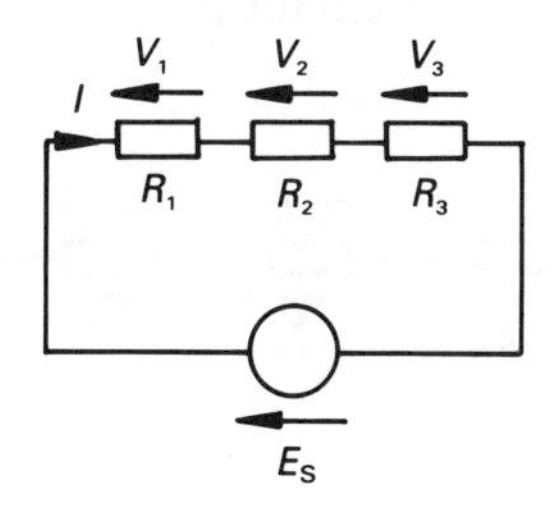

Figure 1.20 *Series-connected resistors.*

$$E_S = V_1 + V_2 + V_3 = I(R_1 + R_2 + R_3) = IR_E$$

where
$$R_E = R_1 + R_2 + R_3$$

and is the *equivalent resistance* of the series circuit. If there are n resistors in series, the equivalent resistance of the circuit is

$$R_E = R_1 + R_2 + \ldots + R_n$$

To summarise, *the equivalent resistance of a series-connected circuit is always greater than the largest value of individual resistance in the circuit.*

Resistors in series are often described as a *string* of resistors.

1.15 Voltage division in series-connected resistors

If there are n resistors in series, the voltage across the nth resistor is $V_n = IR_n$, where I is the current flowing through the resistors. If the voltage across the series circuit is V_S, then $V_S = IR_E$, where R_E is the equivalent resistance of the circuit. That is

$$\frac{V_n}{V_S} = \frac{IR_n}{IR_E} = \frac{R_n}{R_E}$$

or

$$V_n = V_S \frac{R_n}{R_E} = V_S \frac{R_n}{(R_1 + R_2 + \ldots + R_n)}$$

Worked example 1.15.1

Calculate the equivalent resistance of the series circuit in figure 1.21, and determine I, V_1, V_2 and V_3.

Solution

The equivalent resistance of the circuit is

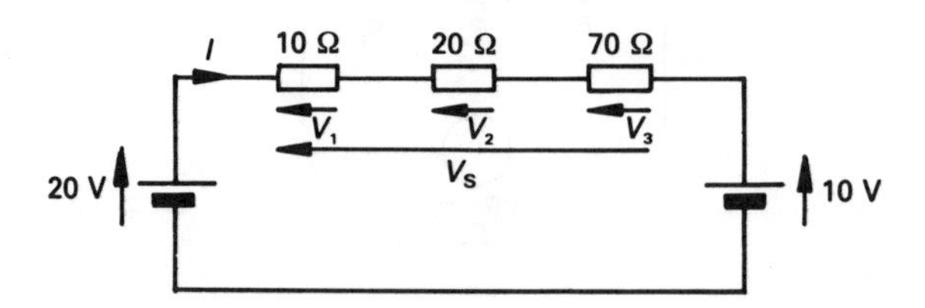

Figure 1.21 *Figure for worked example 1.15.1.*

$$R_E = R_1 + R_2 + R_3 = 10 + 20 + 70 = 100\ \Omega$$

Applying KVL to the circuit shows that

$$V_S = 20 - 10 = 10\ \text{V}$$

where V_S produces I in the direction shown in the figure, hence

$$I = V_S/R_E = 10/100 = 0.1\ \text{A}$$

hence

$$V_1 = IR_1 = 0.1 \times 10 = 1\ \text{V}$$

$$V_2 = IR_2 = 0.1 \times 20 = 2\ \text{V}$$

$$V_3 = IR_3 = 0.1 \times 70 = 7\ \text{V}$$

and from KVL, $V_S = V_1 + V_2 + V_3 = 1 + 2 + 7 = 10$ V or, alternatively, from the theory developed above we can say

$$V_1 = V_S R_1/R_E = 10 \times 10/100 = 1\ \text{V}$$

$$V_2 = V_S R_2/R_E = 10 \times 20/100 = 2\ \text{V}$$

$$V_3 = V_S R_3/R_E = 10 \times 70/100 = 7\ \text{V}$$

It can be seen from the above results that *the value of the voltage across any resistance in a series circuit is proportional to the value of the resistance.*

1.16 Resistance and conductance of a parallel circuit

Elements are said to be connected in parallel with one another *when they are connected between the same pair of terminals* in a circuit. The resistors in figure 1.22 are connected in parallel.

The solution of parallel circuits can be approached either by considering resistances in parallel or conductances in parallel. Since the former is the most popular approach, we will adopt it here.

By definition, each branch in a parallel circuit supports the same voltage, and the current in the nth branch is

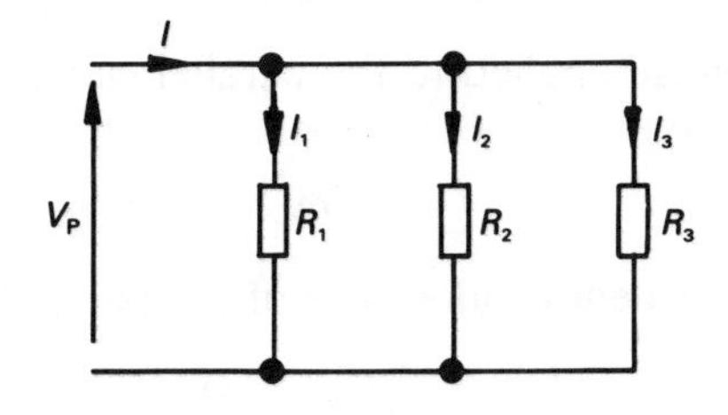

Figure 1.22 *Parallel-connected resistors.*

$$I_n = V_p/R_n$$

that is $I_1 = V_p/R_1$, $V_2 = V_p/R_2$, etc. Applying KCL to the circuit gives

$$I = I_1 + I_2 + I_3 = \frac{V_p}{R_1} + \frac{V_p}{R_2} + \frac{V_p}{R_3}$$

$$= V_p\left[\frac{1}{R_1} + \frac{1}{R_2} + \frac{1}{R_3}\right]$$

but $I = V_p/R_E$, where R_E is the *equivalent resistance* of the circuit. That is

$$\frac{1}{R_E} = \frac{1}{R_1} + \frac{1}{R_2} + \frac{1}{R_3}$$

If there are n resistors in parallel then

$$\frac{1}{R_E} = \frac{1}{R_1} + \frac{1}{R_2} + \ldots + \frac{1}{R_n}$$

In the special case of *two resistors in parallel*

$$R_E = \frac{R_1R_2}{R_1 + R_2}$$

Had we considered the case of parallel-connected conductances, and allowed $G_n = 1/R_n$ and $G_E = 1/R_E$, then for the general case of n parallel-connected conductances

$$G_E = G_1 + G_2 + \ldots + G_n$$

1.17 Current division in a parallel circuit

Possibly the simplest method of calculating the way in which current divides in a parallel circuit is to consider the case of a parallel circuit containing conductances. The current in branch n is

$$I_n = V_pG_n$$

where V_p is the voltage applied to the parallel circuit. The total current drawn by the circuit is

$$I = V_pG_E$$

where G_E is the equivalent conductance of the parallel circuit, hence

$$\frac{I_n}{I} = \frac{V_pG_n}{V_pG_E} = \frac{G_n}{G_E}$$

or

$$I_n = I\,\frac{G_n}{G_E}$$

Alternatively, considering parallel-connected resistors, it can be shown that

$$I_n = I\,\frac{R_E}{R_n}$$

where R_E is the equivalent resistance of the parallel circuit. In the special case of a *two-branch parallel circuit*

$$I_1 = I\,\frac{G_1}{G_1 + G_2} = I\,\frac{R_2}{R_1 + R_2}$$

$$I_2 = I\,\frac{G_2}{G_1 + G_2} = I\,\frac{R_1}{R_1 + R_2}$$

Worked example 1.17.1

Calculate the value of I_1, I_2, I_3 and V_P in figure 1.23.

Solution

The total current entering the top node of the parallel circuit is (14 − 4) = 10 A, and this divides between the three branches. The total conductance of the parallel circuit is

$$G_E = G_1 + G_2 + G_3 = \frac{1}{R_1} + \frac{1}{R_2} + \frac{1}{R_3} = \frac{1}{2} + \frac{1}{4} + \frac{1}{5}$$

$$= 0.5 + 0.25 + 0.2 = 0.95 \text{ S}$$

hence

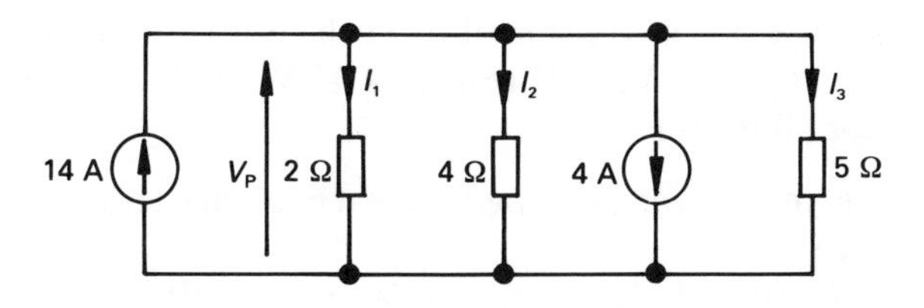

Figure 1.23 *Figure for worked example 1.17.1.*

$$I_1 = I\,G_1/G_E = 10 \times 0.5/0.95 = 5.263 \text{ A}$$

$$I_2 = I\,G_2/G_E = 10 \times 0.25/0.95 = 2.632 \text{ A}$$

$$I_3 = I\,G_3/G_E = 10 \times 0.2/0.95 = 2.105 \text{ A}$$

It can be seen from the result that *the current in each branch is proportional to the conductance of the branch* (or is inversely proportional to the resistance of the branch).
Also

$$V_P = (14 - 4)/G_E = 10/0.95 = 10.562 \text{ V}$$

Unworked problems

1.1. The repetitive waveform of current entering a circuit is shown in figure 1.24. (a) What is the value of the current at t = 0.01 s? (b) What charge enters the circuit between t = 80 ms and t = 150 ms? (c) What total charge has entered the circuit at t = 210 ms?
[(a) 0.05 A; (b) 15.25 mC; (c) 0.05025 C]

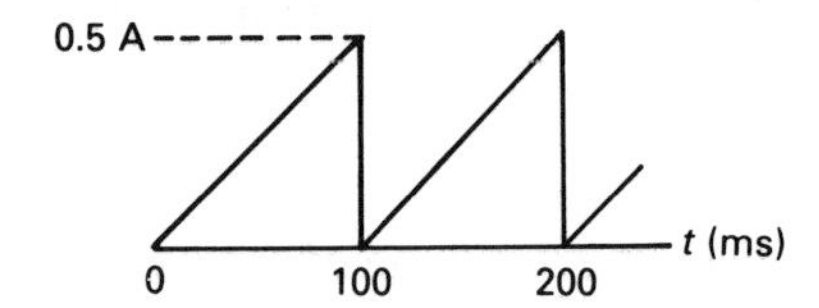

Figure 1.24

1.2. The electrical charge entering a terminal in a circuit is 10 sin $100\pi t$ mC. (a) What charge has entered between −3 ms and 3 ms? (b) Calculate the current at t = 2 ms.
[(a) 16.18 mC; (b) 2.54 A]

1.3. In figure 1.25 calculate V_{BC}, V_{CA}, V_{BD}, V_{EB} and V_{DC}.
[30.93 V; −20.93 V; −20 V; −110 V; −9.07 V]

1.4. Calculate the power consumed by each resistor in figure 1.25.
[10 Ω resistor 95.7 W; 20 Ω resistor 4.11 W; 25 Ω resistor 250 W; 40 Ω resistor 10.95 W]

1.5. Calculate the power *absorbed* by each of the circuit elements in figure 1.26.
[(a) 2.4 W; (b) $-180e^{-5t}$; (c) −20 W; (d) 30 W]

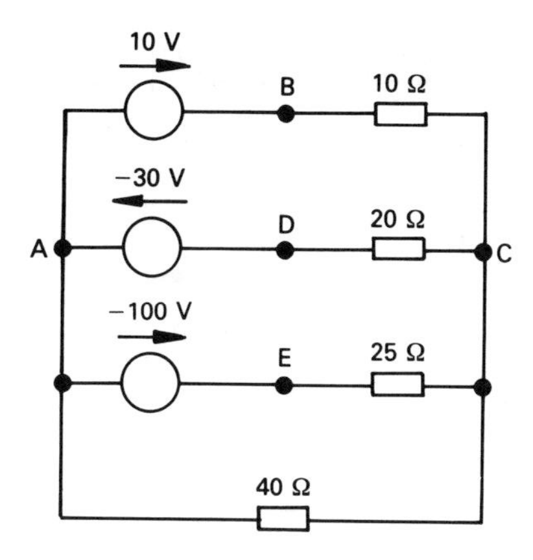

Figure 1.25

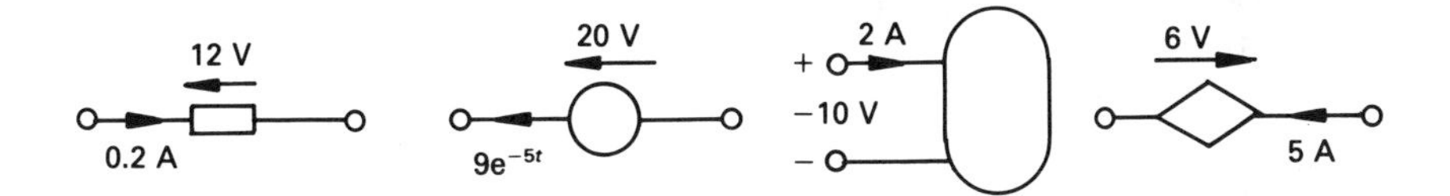

Figure 1.26

1.6. The resistance, R, of a conductor in ohms is given by the equation $R = \rho L/A$, where ρ is the *resistivity* (or *volume resistivity*) of the conductor in Ω m, L is the length of the conductor in m, and A is the cross-sectional area of the conductor in m^2. If a conductor of resistivity 0.027 μΩ m, which is 100 km long and of diameter 1 cm carries a current of 20 A, calculate the p.d. across the length of the wire, the power consumed in the wire and the energy consumed in 20 mins.
[687.5 V; 13.75 kW; 4.58 kWh]

1.7. For the circuit in figure 1.27. calculate (a) I, (b) V and (c) the power absorbed by each element. The calculation should verify the principle of conservation of energy.
[(a) 0.75 A; (b) 3.75 V; (c) 5 Ω resistor 90.31 W, 10 Ω resistor 5.625 W, resistor R 2.8125 W, 20 V source 85 W, 30 V source 22.5 W, 5 A source −206.25 W]

1.8. Calculate I in figure 1.28, and determine the power supplied by the 10 V independent source.
[−0.4 A; −4 W]

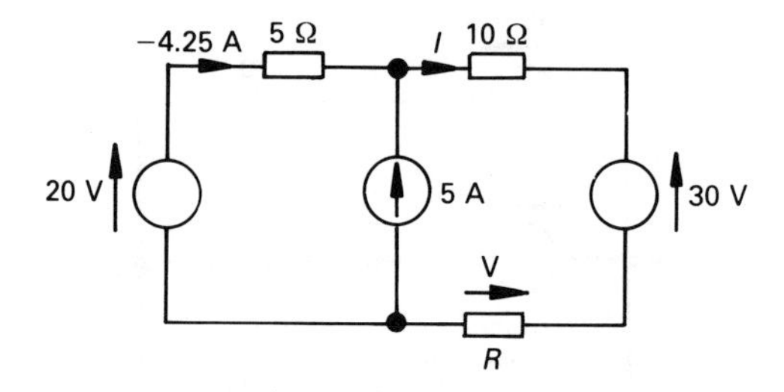

Figure 1.27

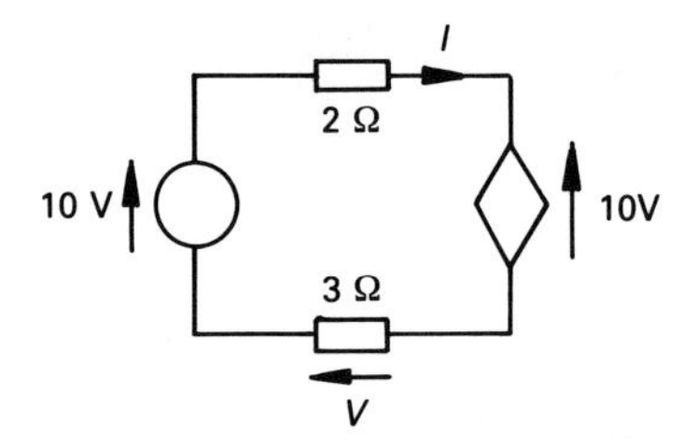

Figure 1.28

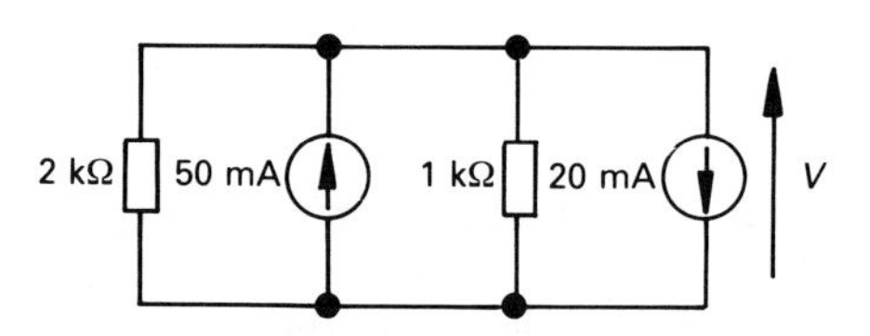

Figure 1.29

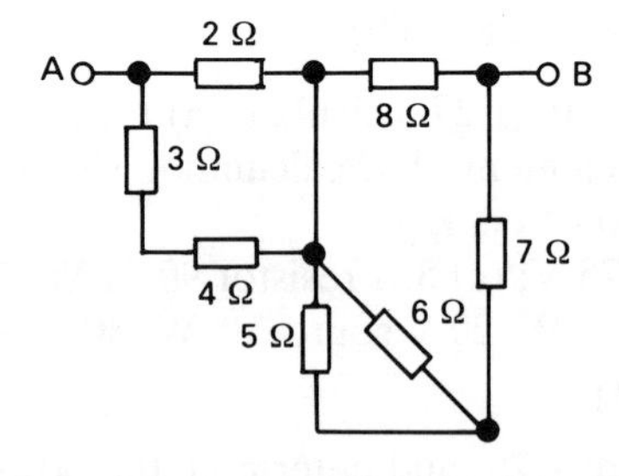

Figure 1.30

1.9. In figure 1.29, calculate V and the power supplied by the 50 mA source.
[20 V; 1 W]

1.10. Calculate the resistance between A and B in figure 1.30.
[5.95 Ω]

1.11. IF 10 V is applied between A and B in figure 1.30, calculate the voltage across and the current in each element.
[2 Ω resistor 2.617 V, 1.308 A; 3 Ω resistor 1.121 V, 0.374 A; 4 Ω resistor 1.495 V, 0.374 A; 5 Ω resistor 2.07 V, 0.414 A; 6 Ω resistor 2.07 V, 0.345 A; 7 Ω resistor 5.313 V, 0.76 A; 8 Ω resistor 7.384 V, 0.923 A]

2

Circuit Analysis

2.1 Introduction

By now the reader is familiar with the use of Ohm's and Kirchhoff's laws in the analysis of simple d.c. circuits. One of the primary goals of circuit theory is the attainment of knowledge and experience of analysing more practical systems. Among the methods adopted are mesh analysis, nodal analysis and loop analysis.

Any one of these methods can be applied to almost any circuit and, in many cases there is no simple way of saying which is the 'best' method of solution; we cannot lay down simple rules to determine the 'best' approach. A knowledge of each method can only by gained by acquiring a sound understanding of the features of each type of solution.

2.2 Definitions and terminology

An *electrical network* is a system of interconnected *circuit elements* containing, for example, resistors, inductors, capacitors, voltage and current sources, transformers, amplifiers, etc. If the network contains at least one closed path or mesh it is an *electrical circuit*; every circuit is a network, but not all networks are circuits (see figure 2.1). Despite this academic difference, engineers use the term network and circuit without differentiating between them.

Terminals A and B in figure 2.1(a) are not electrically connected, and an *open-circuit* is said to exist between them. Terminals C and D in figure 2.1(b) are connected (ideally) by a resistanceless piece of wire, so that current can pass from C to D without power loss; a *short-circuit* is said to exist between C and D. Strictly speaking, figure 2.1(a) is a network, and figure 2.1(b) is a circuit.

An *ideal* circuit element is one which does not, strictly speaking, rep-

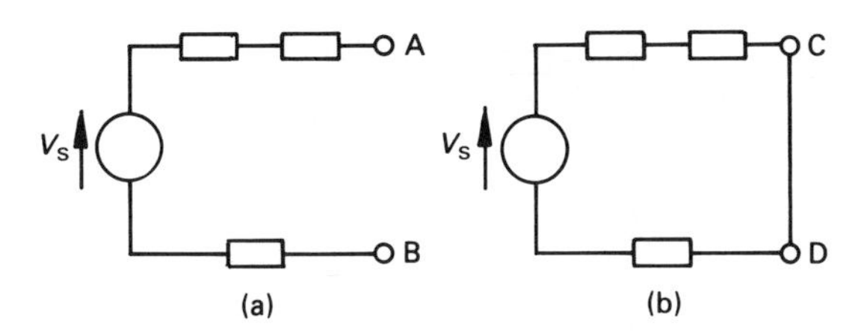

Figure 2.1 *(a) A network and (b) a circuit.*

resent a *practical* element. For example, a practical resistor has an ideal resistance element as part of its make-up, but it also has some inductance because the current passing through it produces a magnetic field, and it has some capacitance because it is insulated from, say, earth (the insulation acting as a dielectric). Fortunately it is usually possible either to neglect subsidiary effects or to compensate for them, but not in every case.

It is fortunate that the major feature of an element (such as the resistance of a resistor, or the capacitance of a capacitor) can be thought of as being at one point within the element; if this is the case, then we say we are dealing with a *lumped-constant* element. In other cases, such as a transmission line, we need to think of the element as an infinite number of infinitely small interconnected elements; in this case we say that we are dealing with a *distributed-constant* element. This latter type of network is dealt with in more advanced texts.

A *node* is a point in a circuit which is common to two or more circuit elements; the circuit in figure 2.2 has four nodes (numbered 0, 1, 2 and 3). A *junction* or *principal node* is a point in the circuit where three or more elements are connected together; nodes 0 and 2 in figure 2.2 are principal nodes. In many cases we drop the term 'principal', and refer to them as 'nodes'. In the majority of cases of circuit analysis, we choose one node (usually a principal node) to be a *reference node* or *datum node*, so that the voltage of other nodes can be defined with respect to it.

For example if, in figure 2.2, we define node 0 as the reference node, then the voltage at node 3 with respect to node 0 is V_{30}; however, since node 0 is the reference node, we simply say that the voltage of node 3 with respect to node 0 is V_3.

A *branch* in a circuit is a path containing one circuit element, and which connects one node to another. Thus R_1 in figure 2.2 is the element in the branch connecting node 1 to node 2. A *path* in a network is a set of connected elements that may be traversed without passing through the same node twice. For example, elements R_2 and R_4 in figure 2.2 are in the path which commences at node 0 and, after passing through node 2, arrives at node 3. A *loop* in a circuit is a *closed path* within the network; the closed path containing the current source I_S, and the resistors R_1, R_4 and R_3 in

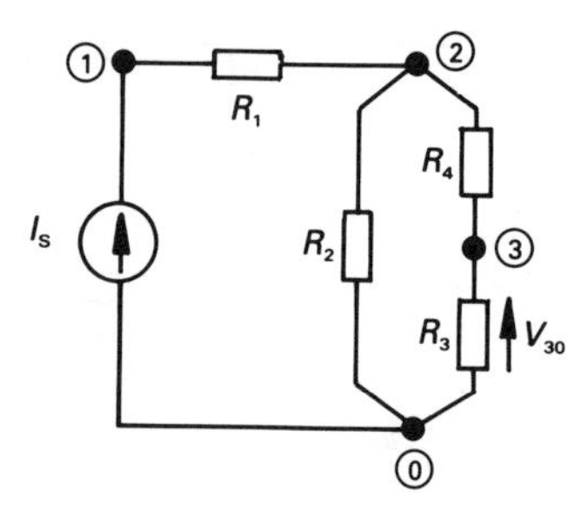

Figure 2.2 *A circuit with four nodes and five branches.*

figure 2.2 is a loop (that is, the loop passing through nodes 0, 1, 2, 3 and back to node 0)

An element which can be connected in a circuit in either direction (this assumes that it is a two-terminal element) without changing the electrical performance of the circuit is known as a *bilateral element*. Examples, include resistors, inductors and capacitors. The majority of networks are *bilateral networks*, that is, they contain only bilateral elements. Certain network theorems, such as the reciprocity theorem (see chapter 3 for details), are only applicable to bilateral networks.

A *planar network* is one which may be drawn on a flat surface, so that none of the branches passes over or under any other branch. When this cannot be done, the network is *non-planar*. The circuit drawn in full line in figure 2.3 is a planar network but, if the branch containing R_6 is introduced (shown broken), it becomes a non-planar circuit.

A *mesh* is a loop *which does not contain any other loops within it*. For example, the circuit in full line in figure 2.3 contains the meshes ABCDA (R_1, R_2, R_3, V_S), BCDB (R_2, R_3, R_4) and ABDA (R_1, R_4, V_S). The loop containing, for example, R_1 R_2 R_3 V_S in figure 2.3 *is not a mesh* because it contains two loops. However, the reader should note that, in some cases

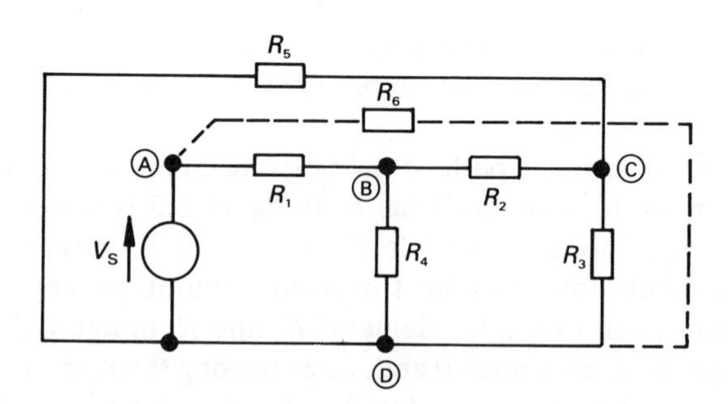

Figure 2.3 *The circuit drawn in full line is a planar network.*

(referring to figure 2.3), it is possible to re-draw the circuit so that a loop which is not a mesh in one version of the circuit can become a mesh in another version.

2.3 Mesh analysis

Mesh analysis involves the concept of *mesh current*, which is introduced via the two-mesh, four-node, five-branch planar circuit in figure 2.4; the reader should note that the four nodes comprise two 'nodes' and two 'principal nodes'. The circuit is shown with a mesh current circulating around the periphery of each mesh *in a clockwise direction* (I_1 and I_2 in the figure). The relationship between the mesh current and the *branch currents* (I_A, I_B and I_C) are

$$I_A = I_1, \qquad I_B = I_2, \qquad I_C = I_1 - I_2$$

Let us apply KVL to each mesh in turn. For mesh 1 (the left-hand mesh), the equation is (proceeding in the direction of I_1)

$$20 - 10I_1 - 30I_1 + 30I_2 = 0$$

or

$$20 = 10I_1 + 30I_1 - 30I_2 = 40I_1 - 30I_2$$

and for mesh 2 the equation is (proceeding in the direction of I_2)

$$-10 + 30I_1 - 30I_2 - 20I_2 = 0$$

or

$$-10 = -30I_1 + 30I_2 + 20I_2 = -30I_1 + 50I_2$$

The equations to be solved are, therefore

$$\begin{aligned} 20 &= 40I_1 - 30I_2 \\ -10 &= -30I_1 + 50I_2 \end{aligned}$$

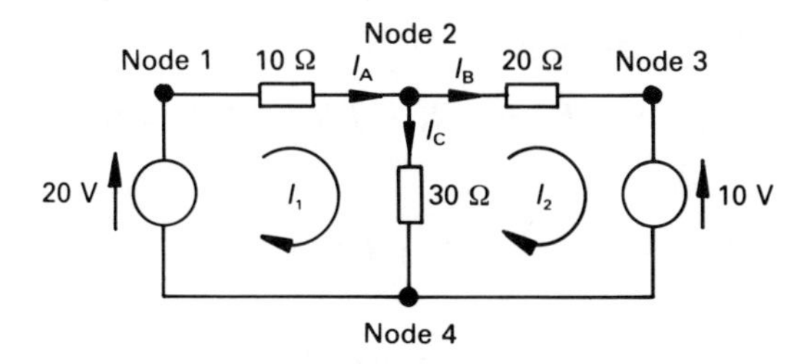

Figure 2.4 *Mesh analysis.*

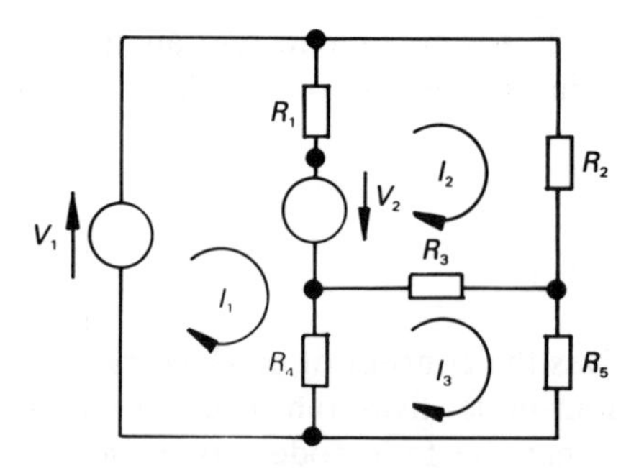

Figure 2.5 *A three-mesh, five-node, seven-branch circuit.*

Solving the two simultaneous equations is fairly straightforward, and the answers are $I_1 = 0.636$ A and $I_2 = 0.182$ A. While the above problem is useful in illustrating a simple application, it does not help us to 'look' and simply write down the mesh equations.

Before we can do this, we need to determine the equations of a more complex circuit, such as that shown in figure 2.5. The equations of this circuit will enable us to draw some general conclusions, which allow us to write down the mesh current equations *by observation*. The reader should carefully study the following passages.

Applying KVL to mesh 1 (in which I_1 circulates), the mesh equation is seen to be

$$V_1 - I_1R_1 + I_2R_1 + V_2 - I_1R_4 + I_3R_4 = 0$$

or

$$V_1 + V_2 = (R_1 + R_4)I_1 - R_1I_2 - R_4I_3 \tag{2.1}$$

Similarly for mesh 2 (in which I_2 circulates), the equation is

$$-V_2 + I_1R_1 - I_2R_1 - I_2R_2 - I_2R_3 + I_3R_3 = 0$$

or

$$-V_2 = -R_1I_1 + (R_1 + R_2 + R_3)I_2 - R_3I_3 \tag{2.2}$$

and for mesh 3 the equation is

$$I_1R_4 - I_3R_4 + I_2R_3 - I_3R_3 - I_3R_5 = 0$$

that is

$$0 = - R_4I_1 - R_3I_2 + (R_3 + R_4 + R_5)I_3 \tag{2.3}$$

The equations (2.1) to (2.3), inclusive, allow us to solve the circuit, and are grouped below

$$\begin{aligned} V_1 + V_2 &= (R_1 + R_4)I_1 - R_1I_2 - R_4I_3 \\ -V_2 &= -R_1I_1 + (R_1 + R_2 + R_3)I_2 - R_3I_3 \\ 0 &= -R_4I_1 - R_3I_2 + (R_3 + R_4 + R_5)I_3 \end{aligned}$$

The equations can be written in the following generalised form

$$\begin{aligned} E_1 &= R_{11}I_1 + R_{12}I_2 + R_{13}I_3 \\ E_2 &= R_{21}I_1 + R_{22}I_2 + R_{23}I_3 \\ E_3 &= R_{31}I_1 + R_{32}I_2 + R_{33}I_3 \end{aligned}$$

where E_1 is the sum of the source voltages driving I_1 in a *clockwise direction*, that is, $E_1 = V_1 + V_2$; E_2 is the sum of the source voltages driving I_2 in a clockwise direction, that is, $E_2 = -V_2$ (that is, V_2 tries to drive I_2 in a counterclockwise direction); E_3 is the sum of the source voltages driving I_3 in a clockwise direction, that is, $E_3 = 0$.

R_{11}, R_{22} and R_{33} are, respectively, the self-resistances of the mesh in which I_1, I_2 and I_3 flow. That is, R_{11} is the sum of all the resistances around the perimeter of mesh 1, that is $R_{11} = R_1 + R_4$; R_{22} is the sum of the resistances around the perimeter of mesh 2, that is $R_{22} = R_1 + R_2 + R_3$, etc.

If we write down R_{ij} as the resistance in row i and column j of the matrix containing the resistance elements, we see that R_{ij} is (for $\neq j$)

$(-1) \times$ the resistance in the branch
which is mutual to the meshes in
which I_i and I_j flow

that is, $R_{13} = -R_4$, $R_{23} = -R_3$, etc.

We can write down the resistance values in the above equation in what is known as *matrix* form as follows; the square brackets around the symbols tell us that we are dealing with a matrix (see chapter 15 for details).

$$\begin{bmatrix} R_{11} & R_{12} & R_{13} \\ R_{21} & R_{22} & R_{23} \\ R_{31} & R_{32} & R_{33} \end{bmatrix}$$

The *resistance matrix* is a *square matrix*, that is it has as many rows as it has columns. For a bilateral network it is symmetrical about the major diagonal, that is $R_{ij} = R_{ji}$, for example $R_{23} = R_{32}$, $R_{13} = R_{31}$, etc. In the case of a bilateral network, that is, a resistive network, all elements on the major diagonal of the resistance matrix are positive; elements not on the major diagonal are either zero or negative.

These comments do not always apply to a non-bilateral network or to networks containing sources other than independent voltage sources, for example, current sources (see worked example 2.6.2).

2.4 General rules for writing mesh equations

Depending on the circuit, there are either three or four general rules which need to be followed:

1. Draw a carefully labelled circuit diagram.
2. Assign mesh currents $I_1, I_2, \ldots, I_m$ to each mesh flowing in a *clockwise direction* in the circuit.
3. If the circuit contains only voltage sources, apply KVL to each mesh and solve the resulting simultaneous equations for the unknown mesh currents (if there are *m* meshes, there are *m* equations). If the circuit contains dependent voltage sources, relate the dependent source voltages to the unknown mesh currents.
4. If the circuit contains one or more current sources, we cannot deal with it in the normal way because the internal resistance of these sources is infinity. The following rule explains how to deal with it. However, since it is fairly technical, the reader should study worked example 2.6.2 in association with the following.

It is first necessary to replace each such source by an open-circuit (*note*: the mesh currents assigned in step 2 must not be changed). Each source current should then be related to the mesh currents assigned in step 2. The resulting simultaneous equations should then be solved

Where a circuit contains practical current sources (see the work on Thévenin's and Norton's theorems in chapter 3), each can be converted into its equivalent practical voltage source, and the problem solved as outlined in step 3 above.

2.5 Solution of three simultaneous equations

Solution of circuits containing two unknowns is relatively simple because we are dealing with only two simultaneous equations. Unfortunately, many practical circuits contain three or more unknowns. We look here at the principles involved in solving for three unknowns. Let us suppose that the three simultaneous equations representing the circuit are of the form

$$V_1 = AX + BY + CZ$$

$$V_2 = DX + EY + FZ$$

$$V_3 = GX + HY + JZ$$

where V_1, V_2 and V_3 are three numerical values, A to H and J ('I' is omitted for the reason that it may be confused with current) are coefficients, and X, Y and Z are the unknown variables.

The solution of three simultaneous equations is no more difficult than solving two simultaneous equations, but it simply involves a few more steps and takes a little longer. We will list the general procedure here, and then solve a set of three simultaneous equations.

1. Eliminate one of the variables (say X) from, say, the 1st and 2nd equations.
2. Eliminate the same variable (X) from two other equations (say the 1st and the 3rd). This leaves two simultaneous equations with two unknowns (Y and Z).
3. Solve for Y and Z from these two equations.
4. Insert the values of Y and Z into one of the original equations to determine the value of X.

Consider the equations

$$-20 = 3X + 2Y - 4Z \tag{2.4}$$

$$32 = 1.5X - 3Y + 4Z \tag{2.5}$$

$$-11 = X + Y - 2Z \tag{2.6}$$

Step 1: Eliminate X from equations (2.4) and (2.5) by multiplying equation (2.5) by 2 and subtracting it from equation (2.4).

$$\begin{array}{lrl} & -20 = 3X + 2Y - 4Z & ((2.4)\text{ re-written}) \\ & 64 = 3X - 6Y + 8Z & ((2.5) \times 2) \\ \hline \text{SUBTRACT} & -84 = \quad 8Y - 12Z & \end{array} \tag{2.7}$$

Step 2: Eliminate X from equations (2.4) and (2.6) by multiplying equation (2.6) by 3 and subtracting it from equation (2.4).

$$\begin{array}{lrl} & -20 = 3X + 2Y - 4Z & ((2.4)\text{ re-written}) \\ & -33 = 3X + 3Y - 6Z & ((2.6) \times 3) \\ \hline \text{SUBTRACT} & 13 = \quad - Y + 2Z & \end{array} \tag{2.8}$$

Step 3: Solving between equations (2.7) and (2.8) for Y and Z gives $Y = -3$ and $Z = 5$.
Step 4: Re-writing equation (2.4) in terms of X gives

$$X = (-20 - 2Y + 4Z)/3 = (-20 - 2(-3) + 4(5))/3$$
$$= 2$$

Alternatively, we can solve the simultaneous equations by *determinants* (which are fully described in chapter 15).

Yet another method is to solve three simultaneous equations by means of the BASIC language program given in listing 2.1. This also uses determinants to solve the equations. Some versions of BASIC do not use line numbers but, generally speaking, they are advanced forms of the language and will accept this program directly (including the line numbers).

Listing 2.1
BASIC program for the solution of three simultaneous equations.

```
10  CLS
20  PRINT TAB(3); "Solution of three simultaneous equations"
30  PRINT TAB(15); "of the form": PRINT
40  PRINT TAB(11); "V1 = A*X + B*Y + C*Z"
45  PRINT TAB(11); "V2 = D*X + E*Y + F*Z"
50  PRINT TAB(11); "V3 = G*X + H*Y + J*Z": PRINT
60  PRINT TAB(3); "Where V1, V2 and V3 are numerical values,"
70  PRINT TAB(3); "A to H and J are numerical coefficients,"
80  PRINT TAB(3); "and X, Y and Z are the variables."
90  PRINT
100 INPUT "V1 = ", V1
110 INPUT "A = ", A
120 INPUT "B = ", B
130 INPUT "C = ", C
140 PRINT
150 INPUT "V2 = ", V2
160 INPUT "D = ", D
170 INPUT "E = ", E
180 INPUT "F = ", F
190 PRINT
200 INPUT "V3 = ", V3
210 INPUT "G = ", G
220 INPUT "H = ", H
230 INPUT "J = ", J: PRINT
240 D1 = (A * E * J) + (B * F * G) + (C * D * H)
250 D2 = (G * E * C) + (H * F * A) + (J * D * B)
260 Det = D1 - D2
270 REM ** There is no solution if Det = 0 **
280 IF Det = 0 THEN PRINT TAB(3); "The equations cannot be solved.": END
290 REM ** Calculate Det X, Det Y and Det Z **
300 D1 = (V1 * E * J) + (B * F * V3) + (C * V2 * H)
310 D2 = (V3 * E * C) + (H * F * V1) + (J * V2 * B)
320 DetX = D1 - D2
330 D1 = (A * V2 * J) + (V1 * F * G) + (C * D * V3)
340 D2 = (G * V2 * C) + (V3 * F * A) + (J * D * V1)
350 DetY = D1 - D2
360 D1 = (A * E * V3) + (B * V2 * G) + (V1 * D * H)
370 D2 = (G * E * V1) + (H * V2 * A) + (V3 * D * B)
380 DetZ = D1 - D2
390 REM ** Calculate the value of the variables **
400 PRINT TAB(3); "X = "; DetX / Det
410 PRINT TAB(3); "Y = "; DetY / Det
420 PRINT TAB(3); "Z = "; DetZ / Det
430 END
```

2.6 Worked examples using mesh analysis

In this section we look at four examples – three of them using the same basic network – respectively involving independent voltage sources only, mixed voltage and current sources, and independent and dependent voltage sources. The fourth involves an operational amplifier circuit.

Worked example 2.6.1

Using mesh analysis, analyse the circuit in figure 2.6.

Solution

Since the circuit contains independent voltage sources only, a solution can be obtained using the first three steps outlined in section 2.4. The first two steps are already performed in figure 2.6. Next, we need to write down the three mesh equations, which we may do by observation in the manner outlined in section 2.3. The three equations are

$$
\begin{aligned}
10 &= 13I_1 - 6I_2 - 5I_3 \\
-8 &= -6I_1 + 16I_2 - 7I_3 \\
0 &= -5I_1 - 7I_2 + 16I_3
\end{aligned}
$$

We can solve for the three unknown mesh currents by any of the methods outlined in section 2.5, and the answers are

$$
\begin{aligned}
I_1 &= 0.789 \text{ A} \\
I_2 &= -0.119 \text{ A} \\
I_3 &= 0.195 \text{ A}
\end{aligned}
$$

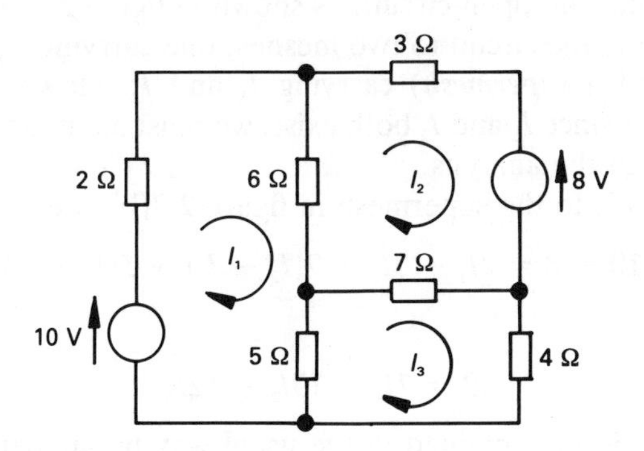

Figure 2.6 *Circuit for worked example 2.6.1.*

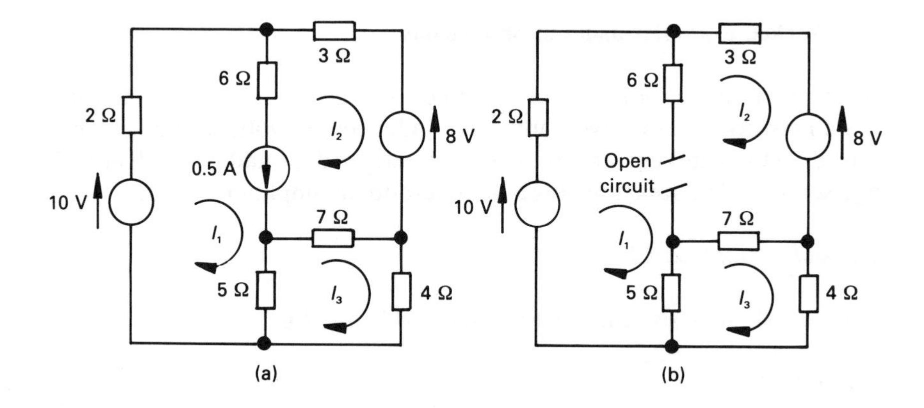

Figure 2.7 *Mesh analysis of a circuit containing an independent current source.*

The reader will note that I_2 is negative. This merely means that the current in mesh 2 circulates in a counterclockwise direction.

Worked example 2.6.2

In this case we use the basic circuit in figure 2.6, but have inserted a 0.5 A independent source in the branch containing the 6 ohm resistor, as shown in figure 2.7(a). The mesh currents can be evaluated as shown in the solution.

Solution

Since we employ KVL in mesh analysis, it is necessary to remove the current source from the network and use the method outlined in step 4 in section 2.4. That is, the independent current is replaced by its internal resistance, namely an open-circuit, as shown in figure 2.7(b). This has the effect of reducing the circuit to two meshes, one carrying I_3, and the other (which is called a *supermesh*) carrying I_1 and I_2. However, the reader should note that since I_1 and I_2 both exist, we must maintain their independent identities in the analysis.

Applying KVL to the supermesh in figure 2.7(b), we get

$$10 - 8 = 2I_1 + 3I_2 + 7(I_2 - I_3) + 5(I_1 - I_3)$$

or

$$2 = 7I_1 + 10I_2 - 12I_3$$

A second equation is obtained in the usual way by applying KVL to the mesh in which I_3 flows as follows

$$0 = -5I_1 - 7I_2 + 16I_3$$

Finally, the third equation is obtained by relating the current in the 0.5 A independent current source to the unknown mesh currents as follows

$$0.5 = I_1 - I_2$$

The three mesh current equations for the circuit in figure 2.7(a) are, therefore

$$\begin{aligned} 2 &= 7I_1 + 10I_2 - 12I_3 \\ 0 &= -5I_1 - 7I_2 + 16I_3 \\ 0.5 &= I_1 - I_2 \end{aligned}$$

Solving these equations by any of the methods described earlier gives

$$\begin{aligned} I_1 &= 0.547 \text{ A} \\ I_2 &= 0.047 \text{ A} \\ I_3 &= 0.191 \text{ A} \end{aligned}$$

Worked example 2.6.3

This example illustrates a method of solving a circuit which includes both independent and dependent voltage sources. In this case, the current source in figure 2.7 is replaced by a voltage-controlled voltage source. The circuit is shown in figure 2.8.

Solution

The dependent voltage source, $3V_X$ is regarded as a normal voltage source, with the exception that the voltage V_X across the 5 ohm resistor must be

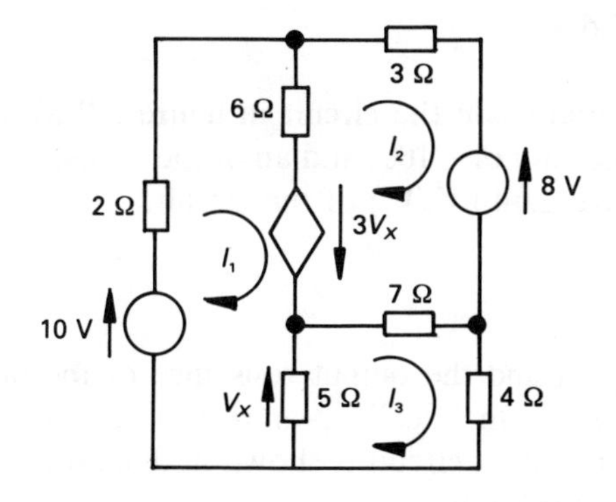

Figure 2.8 *Mesh analysis of a circuit containing independent and dependent voltage sources.*

related to the unknown mesh currents. By inspection, the voltage V_X is

$$V_X = 5(I_1 - I_3)$$

Applying KVL to the mesh in which I_1 flows gives

$$10 + 3V_X = 13I_1 - 6I_2 - 5I_3$$

or

$$10 = -2I_1 - 6I_2 + 10I_3$$

Applying KVL to mesh 2 gives

$$-3V_X - 8 = -6I_1 + 16I_2 - 7I_3$$

that is

$$-8 = 9I_1 + 16I_2 - 22I_3$$

and the equation for mesh 3 is

$$0 = -5I_1 - 7I_2 + 16I_3$$

That is, the three simultaneous equations are

$$\begin{aligned} 10 &= -2I_1 - 6I_2 + 10I_3 \\ -8 &= 9I_1 + 16I_2 - 22I_3 \\ 0 &= -5I_1 - 7I_2 + 16I_3 \end{aligned}$$

Solving gives the results

$$\begin{aligned} I_1 &= 4.776 \text{ A} \\ I_2 &= -2.847 \text{ A} \\ I_3 &= 0.247 \text{ A} \end{aligned}$$

Worked example 2.6.4

The operational amplifier in the circuit in figure 2.9(a) has infinite input resistance, a voltage gain of −100, and an output resistance of 20 kilohms. Calculate the voltage gain (V_2/V_1) of the circuit.

Solution

Since the voltage gain and the output resistance of the operational amplifier differ from that of an 'ideal' op-amp, we need to solve the circuit completely. The equivalent circuit is shown in diagram 2.9(b); since there is only one mesh, its solution is

$$1 + 100V_X = (1000 + 10\,000 + 20\,000)I$$

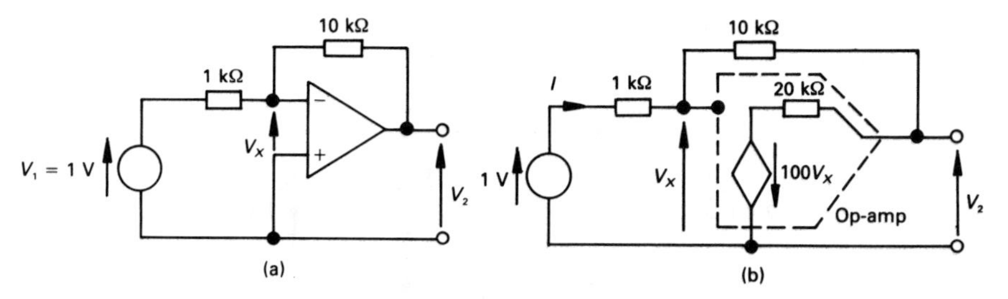

Figure 2.9 *Diagram for worked example 2.6.4.*

but $V_X = 1 - 1000I$, that is

$$1 + 100(1 - 1000I) = 31\,000I$$

or

$$I = 7.71 \times 10^{-4} A$$

The output voltage is

$$\begin{aligned} V_2 &= -100V_X + 20\,000I = -100(1 - 1000I) + 20\,000I \\ &= -100 + 120\,000I = -7.48 \text{ V} \end{aligned}$$

and the overall voltage gain is

$$V_2/V_1 = -7.48/1 = -7.48$$

This value contrasts with the gain of −10, which would prevail if an ideal operational amplifier was used. *Note*: if a load resistance of 10 kilohms is connected to the output of the circuit in figure 2.9(a), the overall gain falls to −6.4! The reader would find it an interesting exercise to verify this using mesh analysis.

2.7 Nodal analysis

Nodal analysis uses KCL to evaluate the voltage at each principal node in the circuit, and is valid for all circuits, both planar and non-planar. In this case we write down and solve a set of simultaneous equations in terms of the unknown voltage at each node. We will illustrate this initially by means of the simple three-node example in figure 2.10.

Of the three principal nodes in the circuit, we must choose one to be a reference or datum node; we select node 0. Generally speaking, if the circuit has n principal nodes, we need $(n - 1)$ simultaneous equations to solve the circuit. Applying KCL to each non-reference node in turn, starting with node 1, we get

$$2 = 2V_1 + 3V_{12} = 2V_1 + 3(V_1 - V_2) = 5V_1 - 3V_2$$

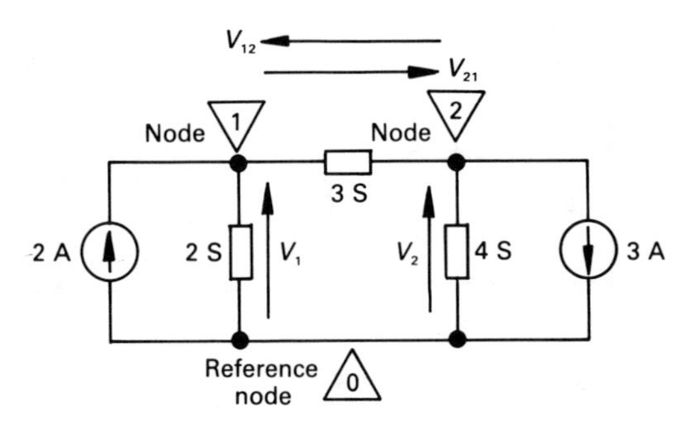

Figure 2.10 *Simple example of nodal analysis.*

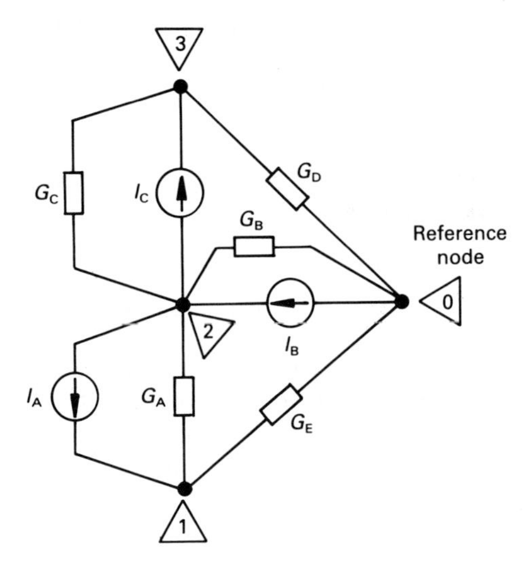

Figure 2.11 *Nodal analysis.*

In the above equation, the current entering the node appears on the left-hand side of the equals sign, and the current leaving the node appears on the right-hand side. For node 2 we have

$$-3 = 3V_{21} + 4V_2 = 3(V_2 - V_1) + 4V_2 = -3V_1 + 7V_2$$

Solving the resulting pair of simultaneous equations gives V_1 and V_2, enabling other data to be calculated. We merely quote the solutions $V_1 =$ 0.192 V and $V_2 = -0.346$ V here, and will proceed to a more useful circuit which enables us to write down the circuit equations by observation.

Consider the circuit in figure 2.11. When applying KCL to node 1, we see that I_A flows towards it and a current of $(G_E\,(V_1 - V_0) + G_A\,(V_1 - V_2))$ flows away from it. Since $V_0 = 0$, the node equation is

$$I_A = (G_A + G_E)V_1 - G_AV_2$$

At node 2, a current of $(I_B - I_A - I_C)$ enters the node and $(G_A(V_2 - V_1) + G_B(V_2 - V_0) + G_C(V_2 - V_3))$ leaves it. As before, $V_0 = 0$, hence the equation at this node is

$$I_B - I_A - I_C = -G_AV_1 + (G_A + G_B + G_C)V_2 - G_CV_3$$

The current entering node 3 is I_C, and the current leaving it is $(G_C(V_3 - V_2) + G_D(V_3 - V_0))$. The equation for node 3 is, therefore

$$I_C = -G_CV_2 + (G_C + G_D)V_3$$

The three node voltage equations describing the circuit are therefore

$$\begin{aligned} I_A &= (G_A + G_E)V_1 - G_AV_2 \\ I_B - I_A - I_C &= -G_AV_1 + (G_A + G_B + G_C)V_2 - G_CV_3 \\ I_C &= - G_CV_2 + (G_C + G_D)V_3 \end{aligned}$$

These can, conveniently, be written in the following generalised form

$$\begin{aligned} I_1 &= G_{11}V_1 + G_{12}V_2 + G_{13}V_3 \\ I_2 &= G_{21}V_1 + G_{22}V_2 + G_{23}V_3 \\ I_3 &= G_{31}V_1 + G_{32}V_2 + G_{33}V_3 \end{aligned}$$

where I_1 $(= I_A)$ is the current *entering* node 1, I_2 $(= I_B - I_A - I_C)$ is the current entering node 2, and I_3 $(= I_C)$ is the current entering node 3.

We may write the conductance elements in matrix form (see chapter 15 for details) as follows.

$$\begin{bmatrix} G_{11} & G_{12} & G_{13} \\ G_{21} & G_{22} & G_{23} \\ G_{31} & G_{32} & G_{33} \end{bmatrix}$$

Each of the terms on the major diagonal of the *conductance matrix* (which is a square matrix) is the sum of the conductances terminating on node 1, 2 and 3, respectively. That is G_{11} $(= G_A + G_E)$ is the total conductance terminating on node 1, G_{22} $(= G_A + G_B + G_C)$ is the total conductance terminating on node 2, etc.

The voltages V_1, V_2 and V_3 are, respectively, the unknown voltage at node 1, 2 and 3.

If G_{ij} is the conductance in row i and column j of the matrix containing the conductance elements, we see that G_{ij} (for $i \neq j$) is

$(-1) \times$ the conductance linking node i to node j

That is, $G_{12} = -G_A$, $G_{32} = -G_C$, etc.

The conductance matrix is a square matrix, and is symmetrical about the major diagonal, that is for all ij $(i \neq j)$, $G_{ij} = G_{ji}$, that is, $G_{12} = G_{21}$, $G_{23} = G_{32}$, etc.

In the case of a bilateral network, all the elements on the major diagonal of the conductance matrix are positive: elements not on the major diagonal are negative or zero.

2.8 General rules for writing nodal equations

Depending on the circuit, there are either three or four steps to be carried out:

1. Draw a carefully labelled circuit diagram.
2. Mark the principal nodes on the circuit, and select a reference node. If there are n principal nodes, $(n - 1)$ simultaneous equations are needed to solve the circuit.
3. If the circuit contains only independent current sources, apply KCL to each non-reference node. If the circuit contains dependent current sources, relate the source current to the unknown node voltages.
4. If the circuit contains voltage sources, we cannot deal with it in the normal way because its internal resistance is zero. The following rule explains how to handle it, and is fairly technical; the reader should study worked example 2.9.2 in association with the following.

Replace each voltage source by a short-circuit; the voltages assigned in step 2 should not be changed. Each source voltage should then be related to the unknown node voltages. If the circuit contains a practical voltage source, it can be converted to its equivalent practical current source (see chapter 3) and dealt with as a normal current source

2.9 Worked examples using nodal analysis

In the following we analyse a circuit and illustrate its solution firstly when it contains only independent current sources, secondly when it contains independent current and voltage sources and, thirdly, when it contains independent and dependent current sources. Finally we will analyse an operational amplifier circuit.

Worked example 2.9.1

Determine the node voltages in figure 2.12.

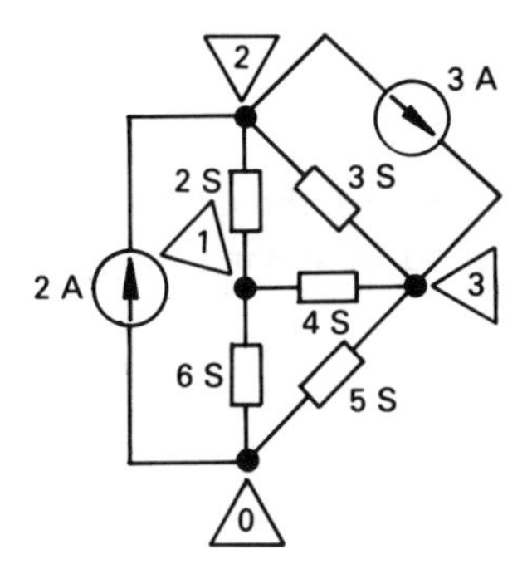

Figure 2.12 *Circuit for worked example 2.9.1.*

Solution

This circuit is relatively straightforward, as it contains only independent current sources. We nominate node 0 as the reference node (as we do for the other circuits of this kind). The first two steps outlined in section 2.8 have already been carried out, and are shown in figure 2.12. Next we will write down by inspection (see section 2.7) the node voltage equations. For node 1 the equation is

$$0 = 12V_1 - 2V_2 - 4V_3$$

and for node 2 the equation is

$$-1 = -2V_1 + 5V_2 - 3V_3$$

For node 3 the equation is

$$3 = -4V_1 - 3V_2 + 12V_3$$

The equations are

$$\begin{aligned} 0 &= 12V_1 - 2V_2 - 4V_3 \\ -1 &= -2V_1 + 5V_2 - 3V_3 \\ 3 &= -4V_1 - 3V_2 + 12V_3 \end{aligned}$$

Solving by any of the methods described earlier yields

$$\begin{aligned} V_1 &= 0.0963 \text{ V} \\ V_2 &= 0.0092 \text{ V} \\ V_3 &= 0.2844 \text{ V} \end{aligned}$$

Note: all voltages are relative to node 0.

Worked example 2.9.2

In the second worked example in this section, an independent 4 V voltage source is connected in parallel with the 4 S conductance in figure 2.12 (see also figure 2.13). One method of solving this type of circuit is described below.

Solution

Since the nodal analysis deals with current sources, it is not possible to handle voltage sources directly. In order to deal with this type of circuit element, the notion of a *supernode* is introduced. What we do in this case is to regard nodes which are connected by the voltage source as though they were connected by the internal resistance of the voltage source, namely zero ohms. That is, nodes 1 and 3 in figure 2.13 become a supernode (which is enclosed by a broken line). The total current flowing towards the supernode is 3 A, and the total current leaving it is

$$2(V_1 - V_2) + 6(V_1 - V_0) + 3(V_3 - V_2) + 5(V_3 - V_0)$$

The reader should note that although nodes 1 and 3 are combined in the supernode, the voltage of both nodes is maintained in the above equation. The nodal equation (remember, node 0 is the reference nodc, so that $V_0 = 0$) for the supernode is therefore

$$\begin{aligned} 3 &= 2(V_1 - V_2) + 6V_1 + 3(V_3 - V_2) + 5V_3 \\ &= 8V_1 - 5V_2 + 8V_3 \end{aligned}$$

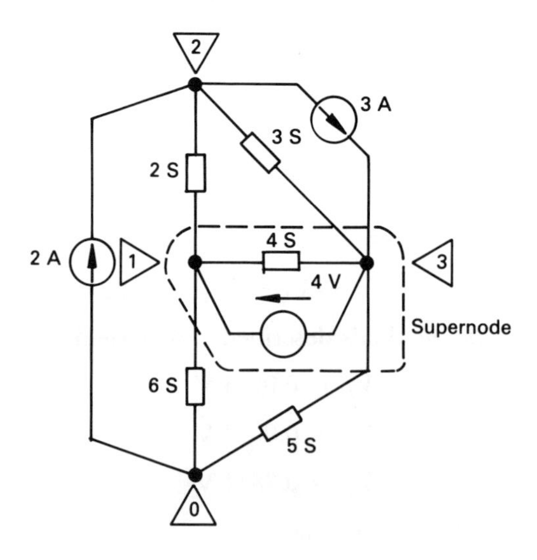

Figure 2.13 *The concept of a supernode.*

Applying KCL to node 2 gives (see also worked example 2.9.1)

$$-1 = -2V_1 + 5V_2 - 3V_3$$

The process of combining two nodes into one supernode means that we have lost one of the nodes to which we could apply KCL! However all is not lost because, at the supernode, nodes 1 and 3 are actually separated by a potential of 4 V, that is

$$4 = V_1 - V_3$$

which leaves us with three equations which are written below

$$\begin{aligned} 3 &= 8V_1 - 5V_2 \quad 8V_3 \\ -1 &= -2V_1 + 5V_2 - 3V_3 \\ 4 &= V_1 \qquad\quad - V_3 \end{aligned}$$

Solving by one of the methods outlined earlier gives

$$V_1 = 2\text{ V} \quad V_2 = -0.6\text{ V} \quad V_3 = -2\text{ V}$$

The 'supernode' is a convenient fiction which allows us to get out of some difficult situations. However, study of the circuit shows that if node 1 is chosen as the reference node, the analysis follows the normal pattern. Not all lecturers are convinced of the usefulness of the supernode concept.

Worked example 2.9.3

In this case, the 4 S conductance in figure 2.12 is shunted by a voltage-dependent current source, as shown in figure 2.14. The analysis is as follows.

Solution

Since all sources in the circuit are current sources, we can apply KLC directly to each node. At node 1 we have

$$-1.5V_{32} = 12V_1 - 2V_2 - 4V_3$$

However $V_{32} = V_3 - V_2$. Inserting this into the above equation gives, for node 1

$$0 = 12V_1 - 3.5V_2 - 2.5V_3$$

Applying KCL to node 2 gives

$$-1 = -2V_1 + 5V_2 - 3V_3$$

Finally, at node 3

$$3 + 1.5V_{32} = -4V_1 - 3V_2 + 12V_3$$

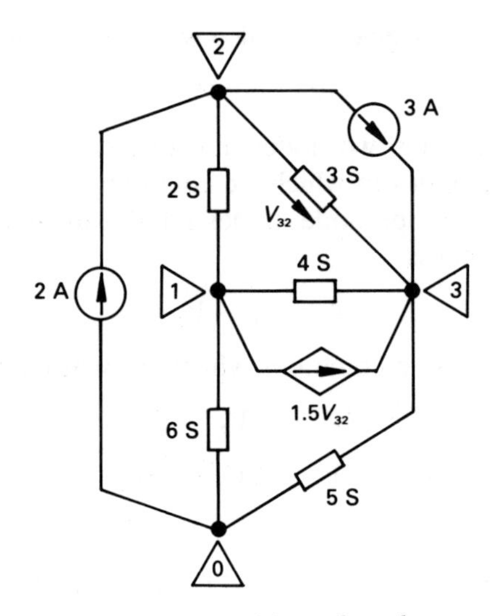

Figure 2.14 *Nodal analysis of a circuit containing a dependent current source.*

or

$$3 = -4V_1 - 1.5V_2 + 10.5V_3$$

and the circuit equations are

$$\begin{aligned} 0 &= 12V_1 - 3.5V_2 - 2.5V_3 \\ -1 &= -2V_1 + 5V_2 - 3V_3 \\ 3 &= -4V_1 - 1.5V_2 + 10.5V_3 \end{aligned}$$

and the solution is

$$\begin{aligned} V_1 &= 0.0707 \text{ V} \\ V_2 &= 0.0174 \text{ V} \\ V_3 &= 0.3151 \text{ V} \end{aligned}$$

Worked example 2.9.4

The operational amplifier in the circuit in figure 2.15(a) has the following parameters

input resistance = 50 kilohms

voltage gain = −1000

output resistance = 1.5 kilohms

Calculate the overall gain (V_2/V_1) of the circuit.

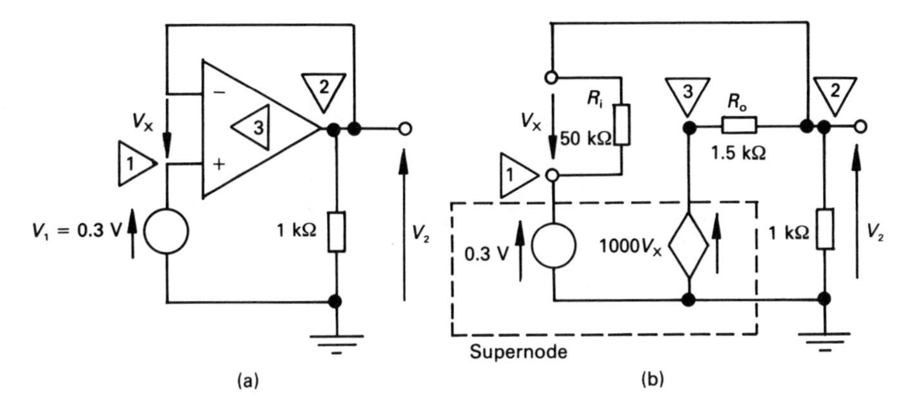

Figure 2.15 *Figure for worked example 2.9.4.*

Solution

The equivalent circuit of the amplifier is shown in figure 2.15(b); since the values differ from those of an ideal operational amplifier, we will analyse the circuit using nodal analysis.

In this case, we place the 0.3 V independent source and the $1000V_X$ dependent source (corresponding to the amplifier gain) in the supernode in diagram (b), thereby reducing the number of nodes by two. Applying KCL to node 2 yields

$$\frac{V_2 - V_1}{50\,000} + \frac{V_2 - V_3}{1500} + \frac{V_2}{1000} = 0$$

Now at the supernode $V_1 = 0.3$ V, and

$$V_3 = 1000V_x = 1000(V_1 - V_2) = 1000(0.3 - V_2)$$

That is

$$\frac{V_2 - 0.3}{50\,000} + \frac{V_2 - 1000(0.3 - V_2)}{1500} + \frac{V_2}{1000} = 0$$

giving $V_2 = 0.299$ V. The voltage gain therefore is

$$V_2/V_1 = 0.299/0.3 = 0.997$$

2.10 Network topology

It was stated earlier that mesh current analysis is applicable only to planar networks. However there is a similar approach – known as *loop analysis*

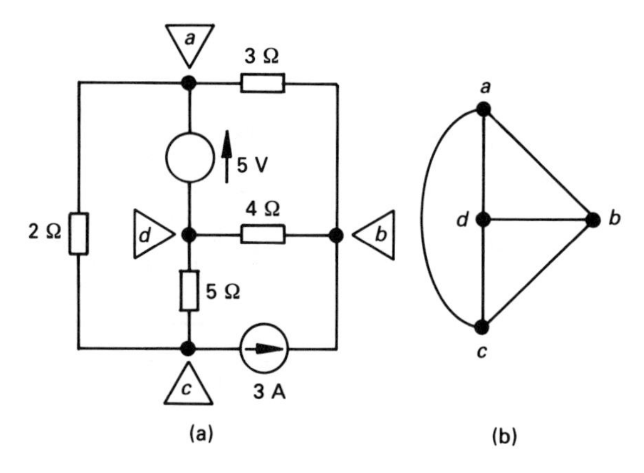

Figure 2.16 *The circuit in (a) has the graph in (b).*

(see also section 2.11) – which allows us to solve non-planar networks. In order to understand some of the techniques involved we need to have a brief introduction to the subject of network topology.

Network topology is concerned with a mathematical discipline known as *graph theory* with special reference to electrical circuits. The 'graph' referred to here is not a conventional graph, but is a collection of points (nodes) and connecting lines (branches). When drawing the 'graph' of a network, the nature of the element in the branch between a pair of nodes is suppressed, and is replaced by a line or 'edge'.

The circuit in figure 2.16(a) has four nodes and six branches; the corresponding *graph* (or *connected graph*) is shown in diagram (b).

Given a graph, we define a *tree* (or *spanning tree*) as any set of branches in the graph which connect every node to all other nodes in the graph, but not necessarily directly. Moreover, *the tree does not contain a loop of any kind*.

If the graph has N nodes, each tree has $(N-1)$ branches in it. The four-node graph in figure 2.16(b) contains sixteen trees, four of which are shown in full line figure 2.17. In loop current analysis, we select a *normal tree*, that is one *containing all the voltage sources* in the network, together with the maximum number of voltage-controlled dependent sources.

A *cotree* is a set of branches which do not belong to a tree; the cotrees corresponding to each of the four trees in figure 2.17 are shown in broken lines. A branch in a cotree is known as a *link*. A cotree is the complement of a tree, and a tree and its cotree form the complete graph of a network. An N-node network contains a number of trees, each with $(N-1)$ branches; if B is the number of branches *in the network*, and L is the number of links in the cotree, the relationship between them is

$$B = L + (N - 1)$$

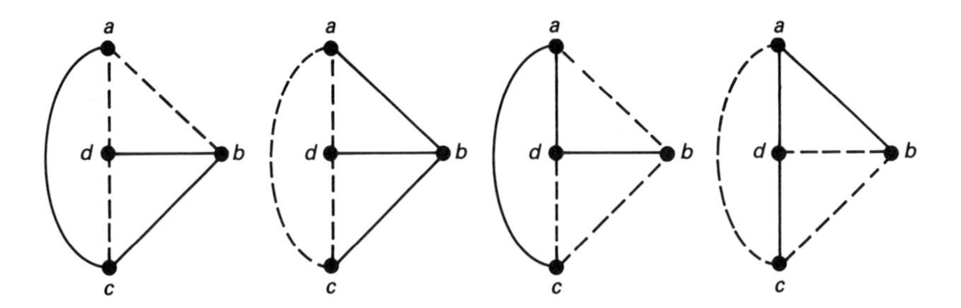

Figure 2.17 *Examples of trees and cotrees within the graph of figure 2.16(b).*

For example, in a four-node ($N = 4$), six-branch ($B = 6$) network, there are a number of trees each containing $(N - 1) = 3$ branches, and the number of links in each cotree is

$$L = B - (N - 1) = 6 - (4 - 1) = 3$$

If we re-position any one of the links from the cotree in the tree, we will form a loop in the tree. Consider for the moment the left-hand tree in figure 2.17. If we add the link connecting *d* to *c* we get the *independent loop bcdb*. This process is repeated throughout the tree as follows. If we add link *ad* we get the independent loop *adbca*; adding link *ab* gives the independent loop *abca*.

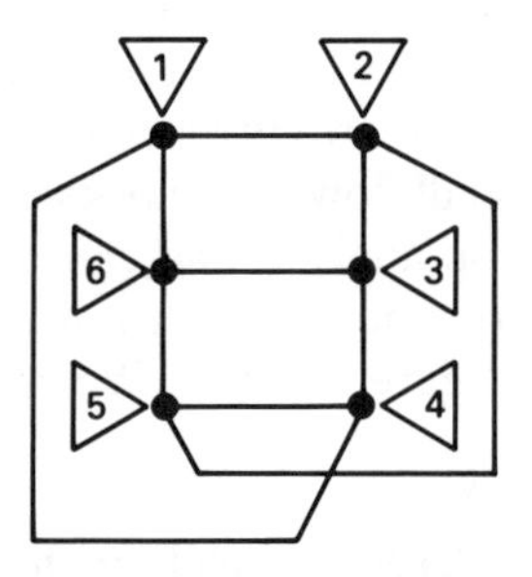

Figure 2.18 *The graph of a six-node non-planar network.*

Since adding a link to the tree forms a loop, the number of links is equal to the number of independent voltage equations we need to form the loop equations of the network. For example, the six-node nonplanar network in figure 2.18 contains a number of trees, each having five branches, and we need $L = 9 - (6 - 1) = 4$ independent equations to solve the network. In some cases, however, the situation is not always that difficult because (as we shall see from an example in section 2.11) the number of equations needed to solve the circuit is reduced when there are current sources in the links.

2.11 Loop analysis

The following steps allow us to write a set of loop current equations for a circuit:

1. Draw a graph of the network and identify a normal tree.
2. Ensure that all voltage sources and, if possible, all control-voltage branches for voltage-controlled dependent sources are in the tree.
3. Ensure that all current sources and, if possible, all control-current branches for current-controlled dependent sources are in the cotree.
4. Reposition in the tree, one at a time, each link in the cotree. Using KVL, write down for each loop the associated loop current equation; solve the equations.

To illustrate the general principles involved we will solve, using loop analysis, the non-planar circuit in figure 2.19. The circuit has six nodes and nine branches, and requires a set of four loop current equations for its solution. Since the circuit is non-planar, it cannot be solved by mesh analysis. The selected normal tree is shown in full line in diagram (b).

Since the voltage source must be included in the tree, the branch *be* must be included in the tree. Thereafter we can select any four other connected nodes (provided that they do not include the current source) to complete the tree.

Next we insert the links (shown broken in figure 2.19(b)), one at a time, from the cotree into the tree in order to produce four *fundamental loops* in the graph; these will provide us with the required equations. These loops are shown in figure 2.20 for loop currents I_A, I_B, I_C, and I_D. We have decided that each current shall flow in a clockwise direction; we need not choose this direction for loop analysis and can, alternatively, be counter-clockwise, or in either direction. The currents are also shown on the branches of the circuit in figure 2.19(a). We now write down the loop current equations.

For loop I_B (the loop *abcda*)

It will be seen that, in figure 2.19(a), each branch of the normal tree carries more than one loop current, and that each link of the cotree carries only one loop current. Since the 3 A current source is in link *da*, we can simply say that

$$I_A = 3 \text{ A}$$

This has the effect of reducing by unity the number of equations required to solve the circuit.

For loop I_B (the loop *abcfa*)

The equation is

$$0 = 7I_B + 5(I_A + I_B - I_D) + 4(I_A + I_B + I_C) + 3(I_B + I_C)$$

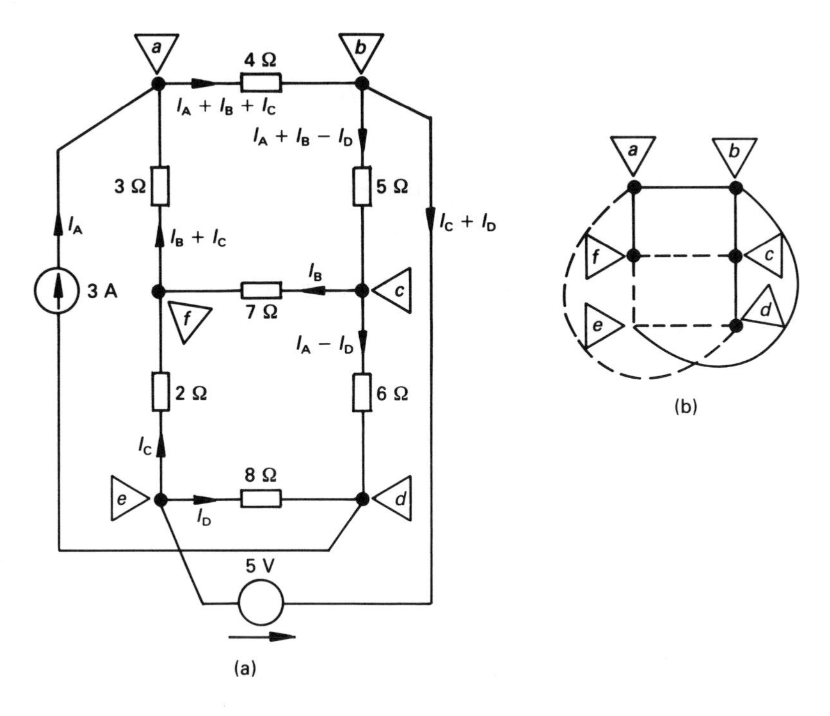

Figure 2.19 *The solution of a non-planar circuit using loop analysis.*

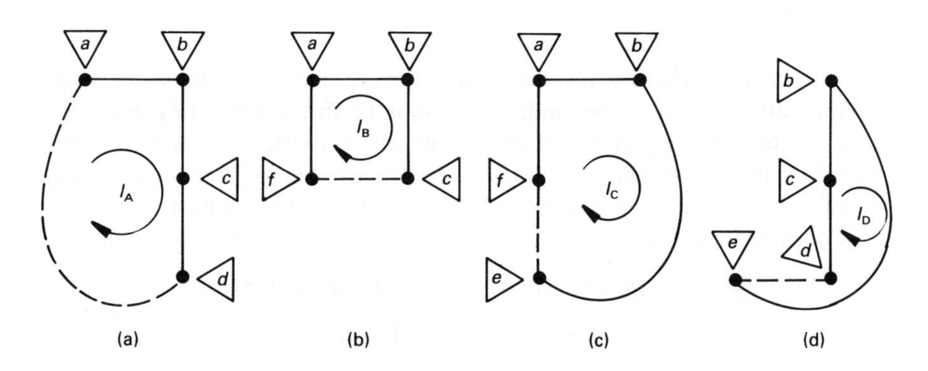

Figure 2.20 *The fundamental loops for figure 2.19.*

or

$$9I_A = -27 = 19I_B + 7I_C - 5I_D$$

For loop I_C (the loop *abefa*)

$$0 = 5 + 4(I_A + I_B + I_C) + 3(I_B + I_C) + 2I_C$$

or

$$-5 - 4I_A = -17 = 7I_B + 9I_C$$

For loop I_D (the loop *bedca*)

$$0 = 5 - 5(I_A + I_B - I_D) - 6(I_A - I_D) + 8I_D$$

or

$$-5 + 11I_A = 28 = -5I_B + 19I_D$$

The equations are regrouped below

$$-27 = 19I_B + 7I_C - 5I_D$$

$$-17 = 7I_B + 9I_C$$

$$28 = -5I_B \qquad + 19I_D$$

Solving gives

$$I_B = -0.5236 \text{ A}$$

$$I_C = -1.4816 \text{ A}$$

$$I_D = 1.336 \text{ A}$$

2.12 Duality

Two circuits are *duals* if the mesh equations of one circuit have the same mathematical form as the nodal equations of the other. They are *exact duals* if the mesh equations of one circuit are numerically identical to the nodal equations of the other. A limitation of duality is that it is only possible to produce the dual of a network if it is a planar network.

Consider the circuit in figure 2.21; the equations for the two circuits are

Circuit 2.21(a)	*Circuit 2.21(b)*
$v = v_1 + v_2$	$i = i_1 + i_2$
$= Ri + L\dfrac{di}{dt}$	$= Gv + C\dfrac{dv}{dt}$

where R is a resistance, G is a conductance, L is an inductance*, and C is a capacitance.

The circuits are duals of one another because the mesh current equation of circuit (a) has the same form as the node voltage equation of circuit (b).

* A full description of inductance and capacitance is given in chapter 4.

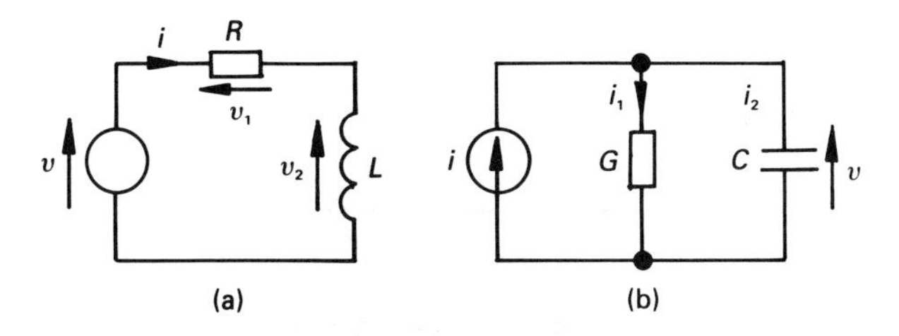

Figure 2.21 *Introduction to dual circuits.*

The two would be exact duals if the voltage of v in volts in circuit (a) was equal to the current i in circuit (b), and if the value of R in ohms in (a) was equal to the value of G in siemens in (b), and if the value of L in henrys in (a) was the same as the capacitance of C in farads.

Clearly, once the (say) mesh equations of one circuit have been solved, then the nodal equations of the exact dual have also been solved.

The relationship between a circuit and its dual are summarised in table 2.1. Multiples and submultiples are also exchanged between the circuit and its dual; for example, mV in a circuit become mA in the dual, μF in the circuit become μH in the dual, etc.

Table 2.1

Circuit element	*Dual circuit element*
Series connection	Parallel connection
Parallel connection	Series connection
Voltage source	Current source
Current source	Voltage source
Voltage of n volts	Current of n amperes
Current of n amperes	Voltage of n volts
Resistance of n ohms	Conductance of n siemens
Conductance of n siemens	Resistance of n ohms
Inductance of n henrys	Capacitance of n farads
Capacitance of n farads	Inductance of n henrys

The reader should note that *the dual is not the equivalent of the original circuit*. That is, if a current of 5 A in an element in the original circuit produces a voltage of 2 V across that element then, in the dual circuit, a voltage of 5 V across the dual of the element produces a current of 2 A in it.

We will now study how a planar circuit in figure 2.22(a) is converted into its dual. The steps followed are:

1. Place a node (we use node 0) in the space outside the circuit.
2. Place a node (nodes 1–4) inside each mesh of the circuit.

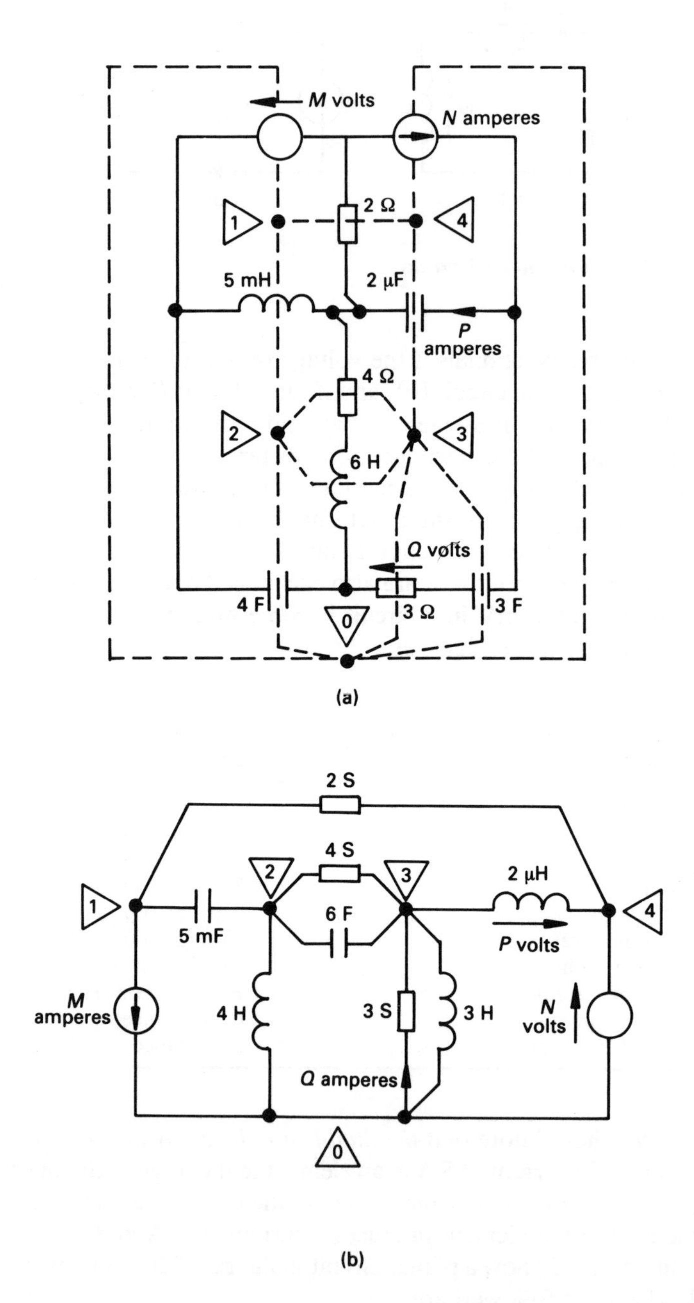

Figure 2.22 *(a) An electrical network and (b) its dual.*

3. Draw a broken line joining adjacent pairs of nodes, each line passing through *one circuit element* on the perimeter of each mesh.
4. Draw the dual circuit by inserting the dual of each element in the original circuit on the broken line linking a pair of nodes.
5. Assign directions to current and voltage sources in the dual using the method outlined below. This method can also be used to assign initial values of current and voltage in the dual.

Taking a look at the broken links between nodes 2 and 3 in the original circuit we see that, in the dual, the circuit comprises a conductance of 4 S in parallel with a capacitor of capacitance 6 F. This process is repeated until all the elements have been replaced by their dual. Finally, it only remains to assign voltage and current direction in the dual, as follows.

Rotate the voltage (or current) arrow in the original circuit *in a clockwise direction* until it lies in the broken line linking the nodes in the meshes on the original circuit. The new direction of the arrow indicates the direction of the corresponding current (or voltage) arrow on the dual.

Unworked problems

2.1. Calculate the currents I_1 and I_2 in figure 2.4, and determine the total power consumed.
[I_1 = 0.636 A; I_2 = 0.182 A; 10.9 W]

2.2. If, in figure 2.5, V_1 = 10 V, V_2 = 20 V, R_1 = 1 ohm, R_2 = 2 ohm, R_3 = 3 ohm, R_4 = 4 ohm and R_5 = 5 ohm, calculate the mesh currents.
[I_1 = 7.7 A; I_2 = −0.875 A; I_3 = 2.35 A]

2.3. Calculate V_1 and V_2 in figure 2.10, and compute the total power consumed.
[V_1 = 0.192 V; V_2 = − 0.346 V; 1.42 W]

2.4. Using mesh analysis, calculate I_1 and I_2 in figure 2.23.
[I_1 = 0.555 A; I_2 = − 0.803 A]

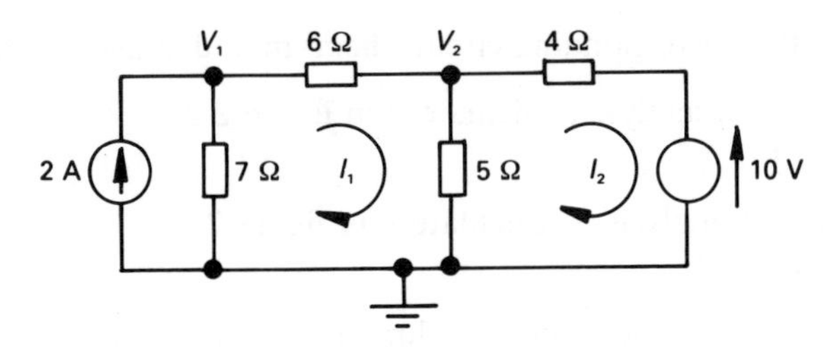

Figure 2.23

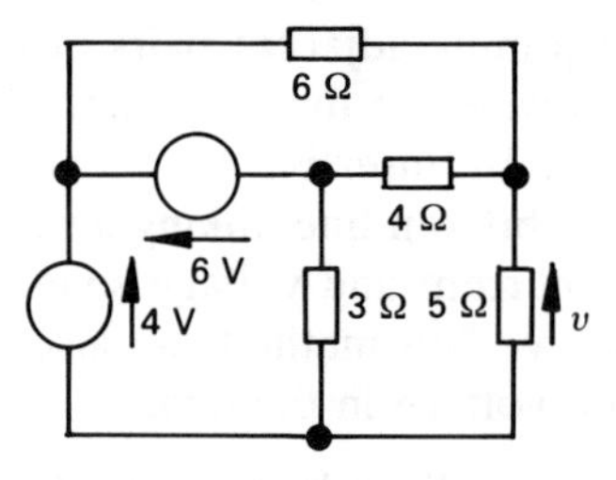

Figure 2.24

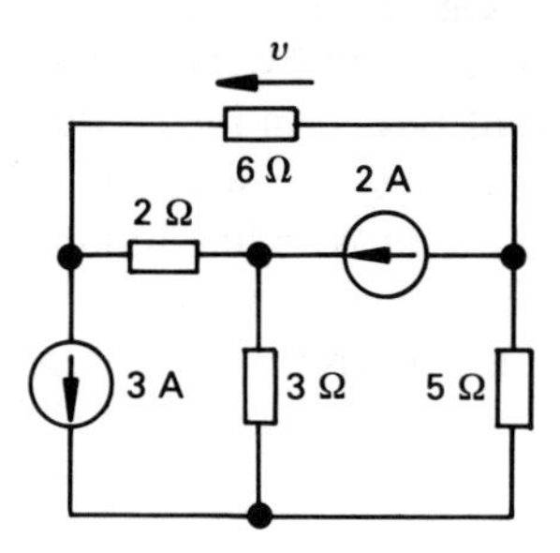

Figure 2.25

2.5. Using nodal analysis, calculate V_1 and V_2 in figure 2.23.
[V_1 = 10.12 V; V_2 = 6.79 V]

2.6. Construct the dual of the circuit in figure 2.23.

2.7. The mesh equations of a network are as follows:

$$\begin{aligned} -10.7 &= 11I_1 - 4I_2 - 2I_3 \\ 14.9 &= -4I_1 + 14I_2 - 3I_4 \\ -5.3 &= -2I_1 + 17I_3 - 8I_4 \\ -1.1 &= -3I_2 - 8I_3 - 17I_4 \end{aligned}$$

Draw the corresponding circuit diagram and construct its dual.

2.8. Using mesh analysis, calculate v in figure 2.24.
[0.27 V]

2.9. Use nodal analysis to calculate v in figure 2.25.
[0.36 V]

2.10. In figure 2.24, the 3 ohm, 5 ohm and 6 ohm resistors form a tree. Use loop analysis with respect to this tree to calculate v.
[0.27 V]

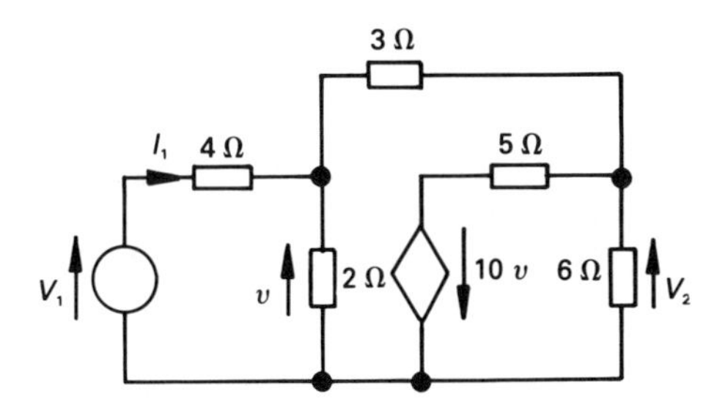

Figure 2.26

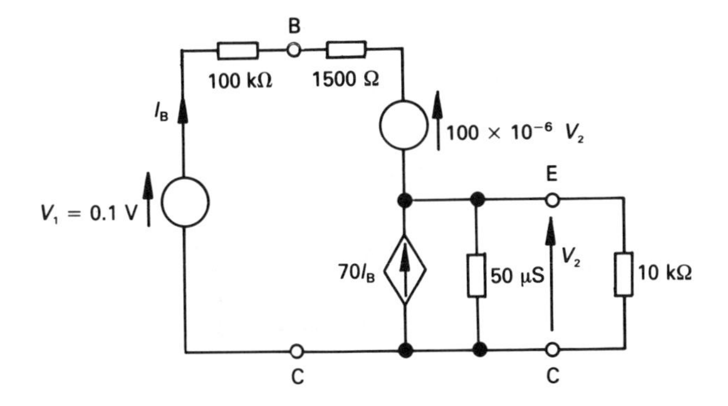

Figure 2.27

2.11. Use mesh analysis to calculate v in figure 2.25.
[0.36 V]

2.12. For the circuit in figure 2.26, use mesh analysis to calculate (a) the voltage gain V_2/V_1 and (b) the input resistance ($= V_1/I_1$) of the circuit.
[Gain = –0.32; input resistance = 4.61 ohms]

2.13. For the simplified emitter follower equivalent circuit in figure 2.27, use mesh analysis to calculate (a) the voltage gain of the circuit ($= V_2/V_1$) and (b) the input resistance ($= V_1/I_1$).
[(a) 0.823; (b) 10.9 kilohm]

2.14. Use nodal analysis to solve problem 2.12.
[gain = –0.32; input resistance = 4.61 ohm]

2.15. Solve problem 2.13 using nodal analysis.
[(a) 0.823; (b) 10.9 kilohm]

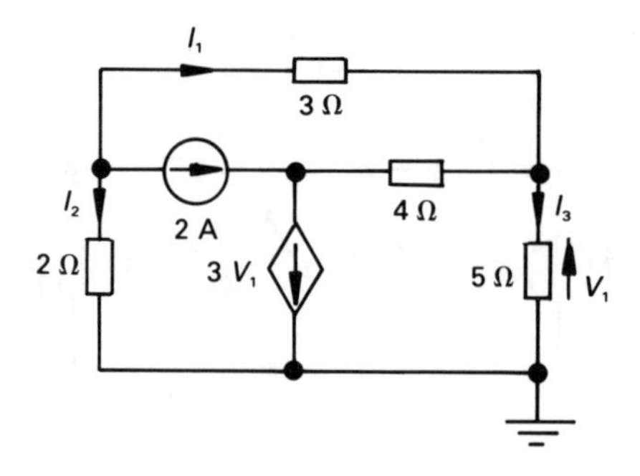

Figure 2.28

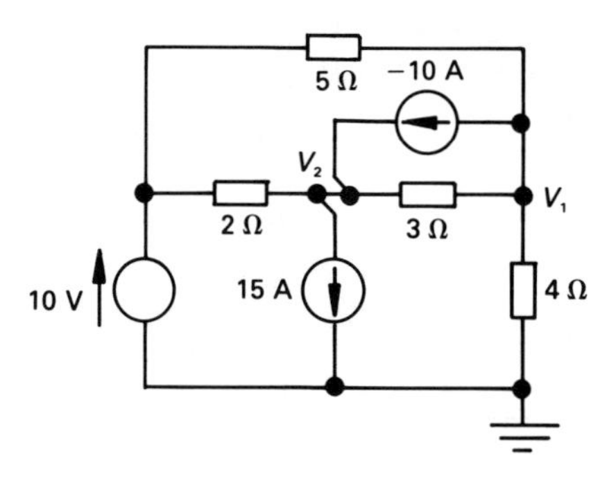

Figure 2.29

2.16. Use nodal analysis to calculate I_1, I_2 and I_3 in the circuit in figure 2.28.
[$I_1 = -0.87$ A; $I_2 = -1.128$ A; $I_3 = 0.071$ A]

2.17. Use nodal analysis to determine the node voltages V_1 and V_2 in figure 2.29.
[$V_1 = 6.154$ V; $V_2 = -21.54$ V]

3

Circuit Theorems

3.1 Introduction

All electrical and electronic circuits can be solved by the application of basic circuit laws such as Ohm's law, Kirchhoff's laws, etc. However, it is useful to have a collection of theorems which, for a particular application, encapsulate appropriate laws; this allows us to obtain a speedy solution to these problems. We look at the more important theorems in this chapter.

3.2 Linearity

Many network theorems are based on the concept of *linearity*; an element, or system is said to be linear if the effect (such as the output voltage from a circuit) is directly proportional to the stimulus (such as the input voltage).

3.3 Principle of superposition

The *principle of superposition* states that in a linear system, *having more than one independent source*, the response (either a voltage or a current) can be obtained from the sum of the responses produced by each source acting alone.

While it is not possible here to provide a formal proof of the principle of superposition, a simple demonstration of the principle is given.

Consider the linear resistor R connected to the independent voltage source V_1 in figure 3.1(a). The resulting current is $I_1 = V_1/R$. When V_1 is removed and replaced by its internal resistance (a short-circuit), and a second independent source V_2 is connected in the circuit (see diagram (b)), the resulting current in the circuit is $I_2 = V_2/R$. Finally, when V_1 is

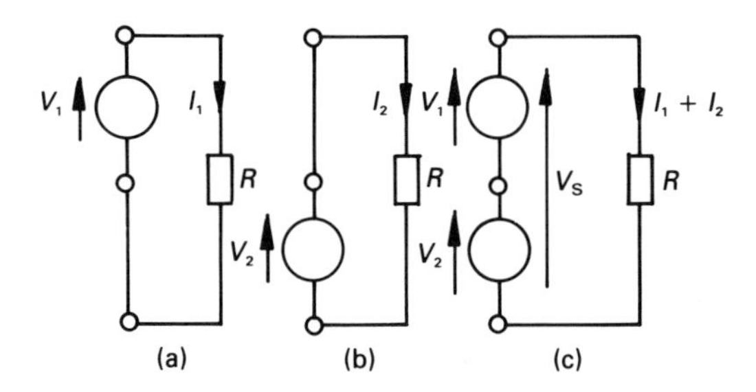

Figure 3.1 *The principle of superposition.*

re-connected in the circuit (diagram (c)), the current in the circuit is $(V_1 + V_2)/R = I_1 + I_2$, and

$$V_S = R(I_1 + I_2) = V_1 + V_2$$

That is, the response is the sum of the stimulating signals.

Moreover, if the stimulus is scaled by a factor K, then the response is also scaled by the same factor. That is, if the applied voltage is KV, then the response is KI, where $KV = R(KI)$. The principle of superposition can be stated as follows:

In any linear bilateral network containing several independent sources, the voltage across (or the current in) any element or source is the sum of the individual voltages (or currents) produced by each individual source acting alone (other sources in the network meanwhile being replaced by their internal resistance)

While the principle of superposition can be applied to many d.c. circuits, it is particularly useful in its application to a.c. circuits where, for example, sources of differing frequencies are involved.

Worked example 3.3.1

Calculate I in figure 3.2 using the superposition theorem.

Solution

We need to calculate the current in the 10 Ω resistor produced by the individual resources.

Current I_1 produced by the 10 V source: Initially the 5 A current source is removed and replaced by its internal resistance, namely an open-circuit; this is illustrated in figure 3.3. The current I_1 (flowing *downwards*) is

$$I_1 = 10/(5 + 10) = 0.6667 \text{ A}$$

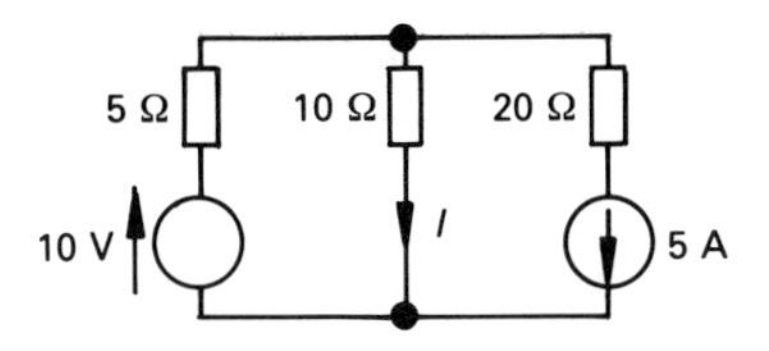

Figure 3.2 *Figure for worked example 3.3.1.*

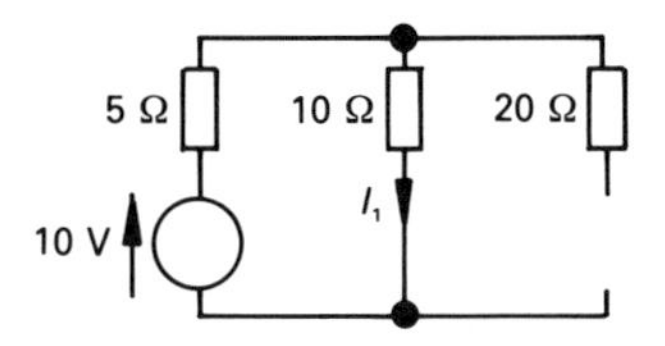

Figure 3.3 *Current* I_1 *produced in the 10 Ω resistor by the 10 V source acting alone.*

Current I_2 *produced by the 5 A source*: In this case the 10 V independent voltage source is replaced by its internal resistance, namely a short-circuit (see figure 3.4). Using the work developed in chapter 1 on the current division in parallel circuits we get

$$I_2 = 5 \times 5/(10 + 5) = 1.6667 \text{ A}$$

which flows *upwards* through the 10 Ω resistor.
Complete solution: The superposition theorem states that the current i flowing *downwards* through the 10 Ω resistor in figure 3.2 is

$$I = I_1 + (-I_2) = 0.6667 - 1.6667 = -1 \text{ A}$$

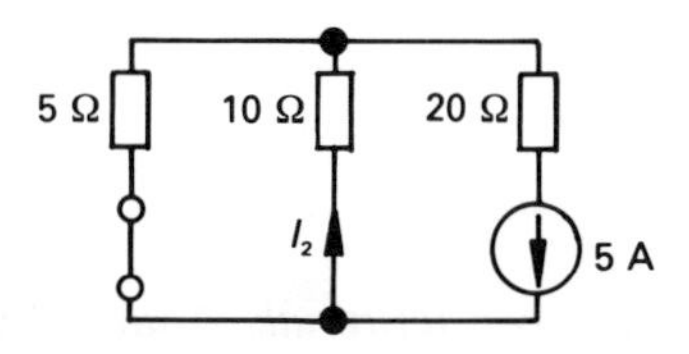

Figure 3.4 *Current* I_2 *produced by the 5 A source acting alone.*

3.4 Thévenin's theorem

This theorem states that any two-terminal active network, no matter how complex, can be replaced by a practical voltage source of the type described in section 1.12 (see figure 3.5). *Thévenin's theorem* may be summarised as follows:

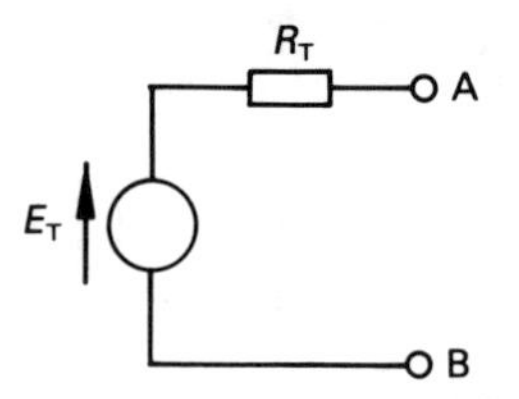

Figure 3.5 *Thévenin's equivalent circuit.*

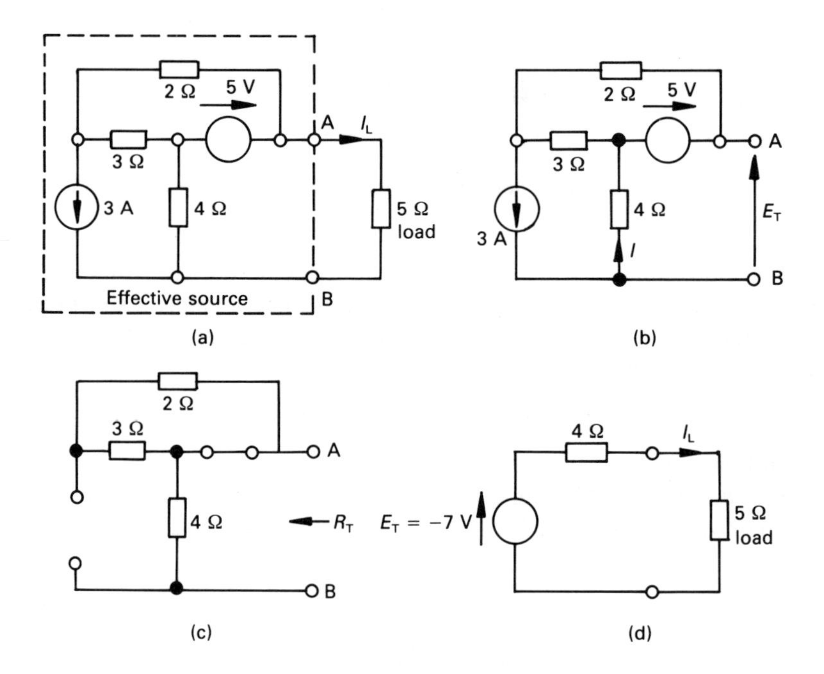

Figure 3.6 *Diagram for worked example 3.4.1.*

An active network, having two terminals A and B to which a load may be connected, behaves as though it contains a single source of e.m.f. E_T, *of internal resistance* R_T. *The e.m.f.* E_T *is the voltage between A and B when the load is disconnected, and* R_T *is the resistance measured between A and B when the load is disconnected and each internal source within the original network is replaced by its internal resistance*

Worked example 3.4.1.

Using Thévenin's theorem, calculate the current I_L in the 5 Ω load resistor in figure 3.6(a).

Solution

Initially we will calculate E_T when the load is disconnected from the circuit (see figure 3.6(b)). Applying KCL to the bottom node in figure 3.6(b), by observation we note that $I = 3$ A, hence

$$E_T = -4I + 5 = -(3 \times 4) + 5 = -7 \text{ V}$$

Next, with the load disconnected, we replace each source in the circuit by its internal resistance, leaving the circuit in diagram (c). Hence

$$R_T = 4\ \Omega$$

That is, Thévenin's equivalent circuit for this problem consists of a $E_T = V_{AB} = -7$ V source in series with a 4 Ω resistor, so that the circuit in diagram (d) is equivalent to the original circuit in diagram (a). Hence

$$\begin{aligned} I_L &= E_T/(R_T + 5) = -7/(4 + 5) \\ &= -0.7778 \text{ A} \end{aligned}$$

3.5 Norton's theorem

This theorem states that any two-terminal active network, no matter how complex, can be replaced by a practical current source of the type described in section 1.12 and illustrated in figure 3.7. *Norton's theorem* may be summarised as follows:

> *Any active network, having terminals A and B to which a load may be connected, behaves as though it contains a current source,* I_N*, of internal conductance* G_N*. The current* I_N *is the current which would flow from terminal A to B when they are short-circuited, and* G_N *is the conductance measured between A and B with the load disconnected and each source within the network replaced by its internal conductance*

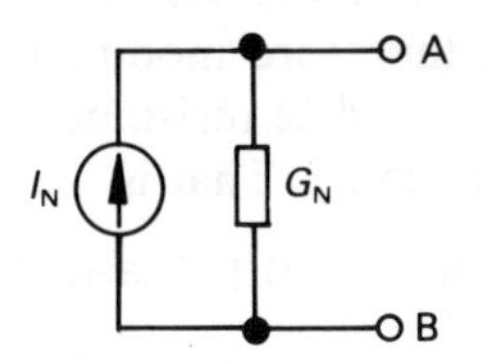

Figure 3.7 *Norton's theorem equivalent circuit.*

Worked example 3.5.1

Using Norton's theorem, calculate the current I_L in the 2 Ω load in figure 3.8.

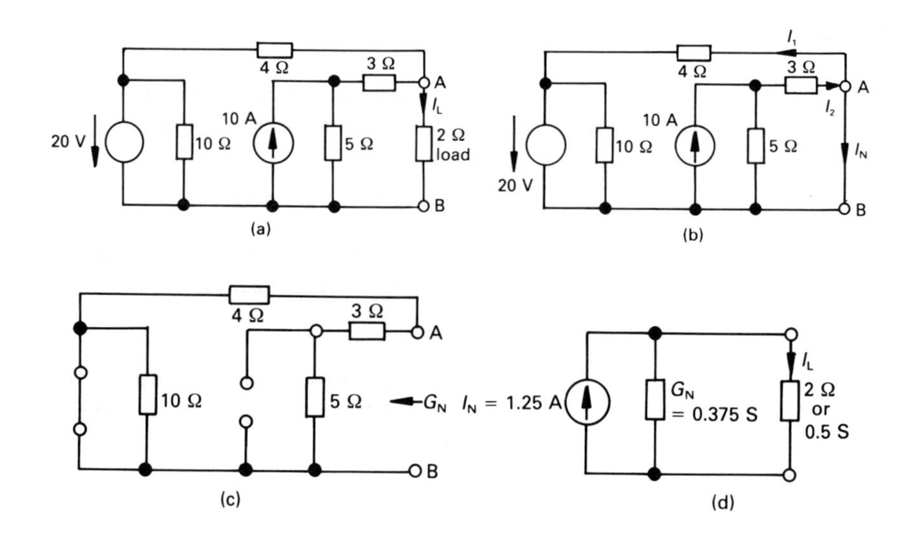

Figure 3.8 *Circuit diagram for worked example 3.5.1.*

Solution

We will first evaluate I_N by short-circuiting the load, as shown in diagram (b). Now $I_N = I_2 - I_1$; we will calculate I_1 and I_2 separately. Since terminals A and B are short-circuited together, then

$$I_1 = 20/4 = 5 \text{ A}$$

and

$$I_2 = 10 \times 5/(5 + 3) = 6.25 \text{ A}$$

hence

$$I_N = I_2 - I_1 = 6.25 - 5 = 1.25 \text{ A}$$

which leaves terminal A.

The internal conductance is evaluated when the external load is disconnected, and each source within the circuit is replaced by its internal conductance. The circuit is therefore modified as shown in diagram (c). The internal circuit comprises a 4 Ω resistance (0.25 S) in parallel with a (3 + 5) = 8 Ω resistance (0.125 S). That is

$$G_N = 0.25 + 0.125 = 0.375 \text{ S}$$

When the original network is replaced by the Norton equivalent circuit, we arrive at the circuit in diagram (d). Using the work covered in chapter 1 we calculate

$$I_L = I_N \times G_L/(G_N + G_L) = 1.25 \times 0.5/(0.375 \times 0.5)$$

$$= 0.714 \text{ A}$$

3.6 Relationship between Thévenin's and Norton's circuits

We showed in section 1.13 that a practical voltage source and a practical current source were interrelated. Using the relationships deduced in that section, we may say that

$$R_T = 1/G_N$$

and

$$E_T = I_N/G_N = I_N R_T$$

3.7 Reciprocity theorem

Up to this point we have discussed two-terminal or *one-port networks*. There is a range of four-terminal or *two-port networks* (see chapter 8 for details), which includes filters, transformers, semiconductor devices, etc.; these require more than a single relationship between the terminal voltage and current to specify their operation. The reciprocity theorem can be applied to these networks or, more specifically, to linear bilateral single-source networks, and may be stated in two ways as follows:

1. *If the single voltage source* V_X *in branch* X *produces the current* I_Y *in branch* Y *then, when the voltage source is removed from branch* X *and inserted in branch* Y, *it produces current* I_Y *in branch* X.
2. *If the single current source* I_X *connected between nodes* X *and* X′ *produces the voltage* V_Y *between nodes* Y *and* Y′ *then, when the current source is removed from between nodes* X *and* X′ *and inserted between nodes* Y *and* Y′, *it produces voltage* V_Y *between nodes* X *and* X′.

The reader should note that the current and voltage at other points in the network change when the single voltage (or current) source changes position.

Worked example 3.7.1.

Calculate I_Y in the single-source linear bilateral network in figure 3.9(a). Remove the source V_X and replace it in the branch in which I_Y flows, and verify the prediction of the reciprocity theorem.

Solution

The mesh currents I_1, I_2 and I_Y are shown in diagram (a), and the corresponding mesh equations are

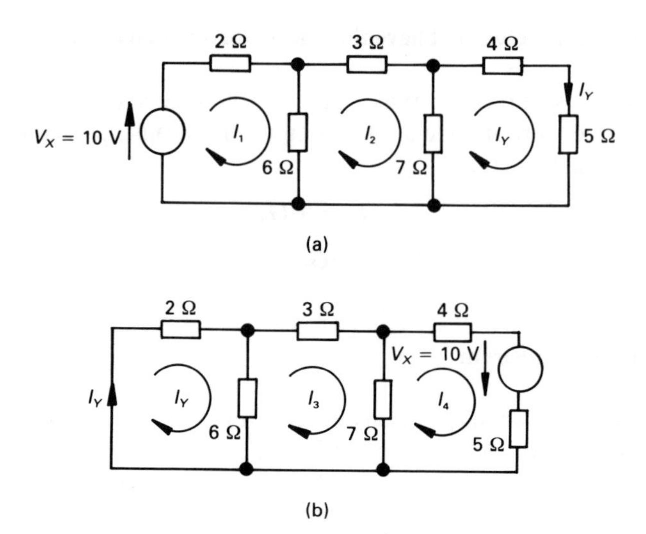

Figure 3.9 *Diagram for worked example 3.7.1.*

$$10 = 8I_1 - 6I_2$$
$$0 = -6I_1 + 16I_2 - 7I_Y$$
$$0 = - 7I_2 + 16I_Y$$

Solving for I_Y by one of the methods outlined earlier gives

$$I_Y = 0.389 \text{ A}$$

We now demonstrate the reciprocity theorem by interchanging the excitation voltage and the response current, as shown in figure 3.9(b). The reader should note that V_X is connected so that it acts in the same direction as I_Y in figure 3.9(a). The mesh equations for the latter circuit are

$$0 = 8I_Y - 6I_3$$
$$0 = -6I_Y + 16I_3 - 7I_4$$
$$10 = - 7I_3 + 16I_4$$

and solving for I_Y (see section 2.5) yields

$$I_Y = 0.389 \text{ A}$$

Since the value of I_Y is the same in both cases, the reciprocity theorem is demonstrated to be correct.

3.8 The maximum power transfer theorem

A practical source of electricity has internal resistance, and when a load is connected to its terminals, the p.d. in its internal resistance causes the terminal voltage to fall. Clearly, if the load resistance is zero, no power is dissipated in the load (even though there may be a large current flowing in it). If the load resistance is very high, very little current flows in the load, and very little power is consumed by the load. Between these two extreme values of load resistance, there will be a particular value of load resistance which consumed maximum power from the supply source. It is this we look at here.

The *maximum power transfer theorem* states that, if the supply source can be described in terms of a Thévenin or of a Norton equivalent circuit, *maximum power is absorbed by a resistive load,* R_L, *when the resistance of the load is equal to the internal resistance of the source.*

Consider the circuit in figure 3.10. The current in the load is $I_L = V_S/(R_S + R_L)$, and the power absorbed by the load is

$$P = I^2R_L = V_S^2R_L/(R_S + R_L)^2$$

When the conditions for maximum power are investigated, that is, when $dP/dR_L = 0$, we find that maximum power transfer occurs when $R_L = R_S$.

The reader will note that when $R_L = R_S$, *the same amount of power is absorbed in the load as in the source*. That is, the efficiency of power transfer (when maximum power transfer occurs) is only 50 per cent.

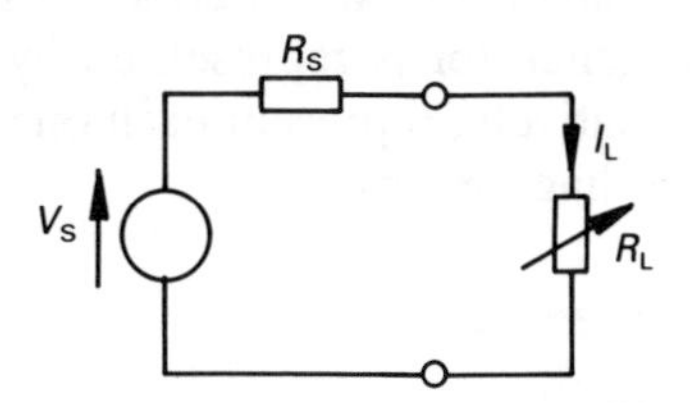

Figure 3.10 *Maximum power transfer theorem (d.c. circuits).*

The conditions for maximum power in an a.c. circuit are somewhat more complex, and are described in chapter 6.

Worked example 3.8.1.

If, in figure 3.8(a), the 2 Ω load resistor (see also worked example 3.5.1) is replaced by a variable resistor R_L, what value of R_L absorbs maximum power, and what is the value of this power?

Solution

In worked example 3.5.1 we showed that the equivalent circuit of the source was a 1.25 A current source shunted by a 0.375 S conductance (or 2.667 Ω resistance).

Maximum power is delivered to the load when the resistance of the load is

$$R_L = 1/G_N = 1/0.375 = 2.667\ \Omega$$

and the current in the load at this time is

$$1.25/2 = 0.625\ \text{A}$$

The maximum power absorbed by the load therefore is

$$(0.625)^2 \times 2.667 = 1.042\ \text{W}$$

3.9 The parallel-generator (Millman's) theorem

This theorem is a special case of nodal analysis and is particularly useful not only in the case of parallel generators, but also in electronic amplifier circuits, and in the solution of unbalanced three-phase three-wire a.c. circuits (see chapter 7 for details). We will concentrate here on the general principles involved so far as d.c. circuits are concerned.

Consider the case of the three parallel-connected generators E_1, E_2 and E_3 in figure 3.11(a), each having its own internal resistance R_1, R_2 and R_3, respectively. Since each generator is represented by its Thévenin equivalent circuit we can, alternatively, represent each one by its Norton equivalent circuit, as shown in diagram (b).

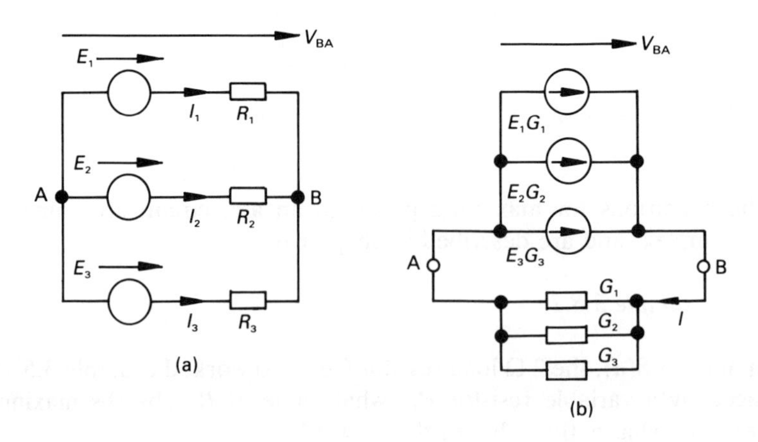

Figure 3.11 *The parallel-generator theorem or Millman's theorem.*

The current, I, leaving the idealised Norton generators at node B is

$$I = E_1G_1 + E_2G_2 + E_3G_3$$

where $G_N = 1/R_N$. By KCL, this is equal to the current entering the three parallel-connected conductances, that is

$$V_{BA}(G_1 + G_2 + G_3)$$

hence

$$V_{BA} = \frac{E_1G_1 + E_2G_2 + E_3G_3}{G_1 + G_2 + G_3}$$

If there are n practical voltage sources in parallel with one another, we may say

$$V_{BA} = \frac{\sum_{k=1}^{n} E_kG_k}{\sum_{k=1}^{n} G_k} = \frac{\sum_{k=1}^{n} \frac{E_k}{R_k}}{\sum_{k=1}^{n} 1/R_k}$$

or, more simply, though not quite as comprehensively

$$V_{BA} = \frac{\Sigma EG}{\Sigma G}$$

Worked example 3.9.1

If, in figure 3.11(a), $E_1 = 10$ V, $E_2 = 20$ V, $E_3 = -25$ V, $R_1 = 20\ \Omega$, $R_2 = 15\ \Omega$ and $R_3 = 10\ \Omega$, calculate V_{BA} and the current in each generator.

Solution

From the above theory

$$V_{BA} = \frac{\frac{E_1}{R_1} + \frac{E_2}{R_2} + \frac{E_3}{R_3}}{\frac{1}{R_1} + \frac{1}{R_2} + \frac{1}{R_3}} = \frac{\frac{10}{20} + \frac{20}{15} - \frac{25}{10}}{\frac{1}{20} + \frac{1}{15} + \frac{1}{10}}$$

$$= \frac{-0.6667}{0.2167} = -3.077 \text{ V}$$

If I_1, I_2 and I_3 flow towards node B (see figure 3.10(a)), then

$$I_1 = (-E_1 + V_{BA})/R_1 = (-10 - 3.077)/20$$
$$= -0.654 \text{ A}$$

$$I_2 = (-E_2 + V_{BA})/R_2 = (-20 - 3.077)/15$$
$$= -1.538 \text{ A}$$

$$I_3 = (-E_3 + V_{BA})/R_3 = (25 - 3.077)/10$$
$$= 2.192 \text{ A}$$

3.10 Rosen's theorem or the general star–mesh transformation

It is possible to transform a network of N conductances which are connected to a common star point, S, as shown in figure 3.12(a), to a mesh of conductances which are connected between the N terminals as shown in diagram (b).

The relationship between the two sets of conductances can be determined as follows. Consider the case where node 1 is connected to ground (zero potential), and other terminals are energised. Applying Millman's theorem to figure 3.12(a) gives

$$V_{S1} = \frac{\sum_{k=1}^{n} E_{1k}G_k}{\sum_{k=1}^{n} G_k} = \frac{E_{12}G_2 + E_{13}G_3 + \ldots + E_{1n}G_n}{G_1 + G_2 + \ldots + G_n}$$

and the current entering node 1 is

$$I_1 = V_{S1}G_1 = \frac{E_{12}G_1G_2 + E_{13}G_1G_3 + \ldots + E_{1n}G_1G_n}{\sum_{k=1}^{n} G_k}$$

$$= E_{12}\frac{G_1G_2}{\Sigma G} + E_{13}\frac{G_1G_3}{\Sigma G} + \ldots + E_{1n}\frac{G_1G_n}{\Sigma G}$$

This is the same value of current that would flow into node 1 if it were connected to node 2 by a conductance $G_{12} = G_1G_2/\Sigma G$, and to node 3 by a conductance $G_{13} = G_1G_3/\Sigma G$, etc. That is, the star and mesh circuits in

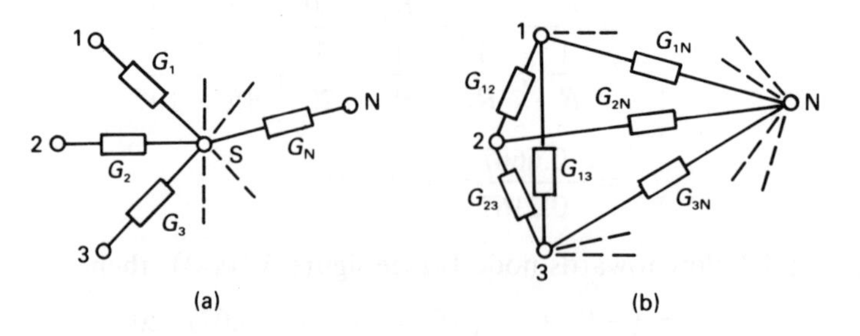

Figure 3.12 *Rosen's theorem or the general star–mesh transformation.*

diagrams (a) and (b), respectively, of figure 3.12 are equivalent if the conductance connected between, say, nodes f and g is

$$G_{fg} = \frac{G_f G_g}{\sum_{k=1}^{n} G_k}$$

3.11 The star–delta, tee–wye or tee–pi transformation

This is a version of the general star–mesh transformation, and is restricted to three elements – see figure 3.13.

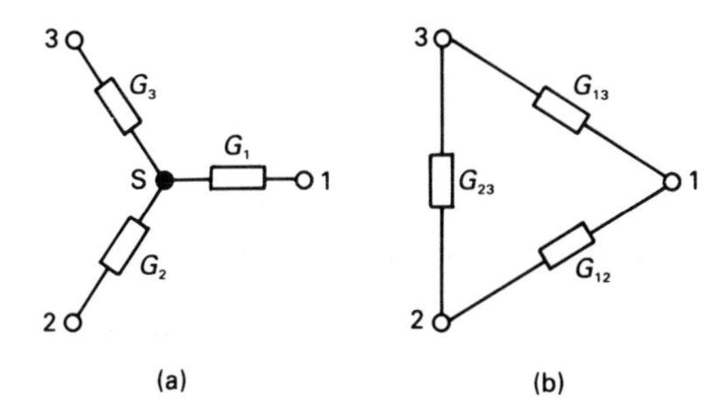

Figure 3.13 *The star–delta and delta–star transformations.*

Using the results of the general star–mesh transformation, we see that

$$G_{ij} = \frac{G_i G_j}{\Sigma G}$$

that is

$$G_{12} = \frac{G_1 G_2}{G_1 + G_2 + G_3} \quad \text{or} \quad R_{12} = R_1 + R_2 + \frac{R_1 R_2}{R_3} \tag{3.1}$$

where $R_n = 1/G_n$. Also if $\Sigma G = G_1 + G_2 + G_3$ then

$$G_{23} = \frac{G_2 G_3}{\Sigma G} \quad \text{or} \quad R_{23} = R_2 + R_3 + \frac{R_2 R_3}{R_1} \tag{3.2}$$

$$G_{13} = \frac{G_1 G_3}{\Sigma G} \quad \text{or} \quad R_{13} = R_1 + R_3 + \frac{R_1 R_3}{R_2} \tag{3.3}$$

3.12 The delta–star, wye–delta or pi–tee transformation

While it is not possible to obtain a general mesh–star transformation, a delta–star transformation for three elements of the type in figure 3.13 can be deduced as follows.
From equation (3.1)

$$G_{12}(G_1 + G_2 + G_3) = G_1G_2 \tag{3.4}$$

From equation (3.2)

$$G_{23}(G_1 + G_2 + G_3) = G_2G_3 \tag{3.5}$$

and from equation (3.3)

$$G_{13}(G_1 + G_2 + G_3) = G_1G_3 \tag{3.6}$$

Dividing (3.4) by (3.5) yields

$$G_1 = G_3G_{12}/G_{23}$$

and dividing (3.4) and (3.6) gives

$$G_2 = G_3G_{12}/G_{13}$$

Substituting for G_1 and G_2 in equation (3.3) shows that

$$G_3 = G_{13} + G_{23} + \frac{G_{13}G_{23}}{G_{12}} \quad \text{or} \quad R_3 = \frac{R_{13}R_{23}}{\Sigma R}$$

where $\Sigma R = R_{12} + R_{23} + R_{13}$. Similarly it may be shown that

$$G_1 = G_{12} + G_{13} + \frac{G_{12}G_{13}}{G_{23}} \quad \text{or} \quad R_1 = \frac{R_{12}R_{13}}{\Sigma R}$$

and

$$G_2 = G_{12} + G_{23} + \frac{G_{12}G_{23}}{G_{13}} \quad \text{or} \quad R_2 = \frac{R_{12}R_{23}}{\Sigma R}$$

3.13 Summary of star–delta and delta–star transformations

For the star–delta transformation

$$G_{\mathrm{ij}} = \frac{G_iG_j}{\Sigma G}$$

and for the delta–star transformation

$$R_i = \frac{R_{ij}R_{ik}}{\Sigma R}$$

where $i \neq j$ and $i \neq k$.

Unworked problems

3.1. Using the principle of superposition, calculate I_1 in figure 3.14.
[−2.667 A]

3.2. Using the principle of superposition, calculate the value of I_1, I_2 and I_3 in figure 3.15. Calculate the power consumed in the 4 Ω resistor, and show that the principle of superposition does not hold for power.
[−2.4 A; 0.553 A; −2.953 A; 1.22 W]

3.3. Determine Thévenin's equivalent circuit with respect to terminals A and B for the circuit in figure 3.16. What current would flow in a resistor of 10 Ω resistance connected between A and B?
[$E_T = -13$ V, $R_T = 10$ Ω; 0.65 A (B to A)]

3.4. Determine Thévenin's equivalent circuit with respect to terminals A and B of figure 3.17. Hence calculate the power which would be developed in an 8 Ω resistor connected between terminals A and B.
[$E_T = 1.17$ V, $R_T = 4.44$ Ω; 70.7 mW]

3.5. Determine Norton's equivalent circuit with respect to terminals A and B of figure 3.15. A resistance of 8 Ω is connected between these terminals; calculate the current in the resistor.
[$I_N = 2.833$ A, $G_N = 1.283$ S; 0.2515 A]

3.6. Deduce Norton's equivalent circuit with respect to terminals A and B of figure 3.18 if element X is (a) a 4 A current source with the current flowing upwards, (b) a 10 V voltage source with the positive pole connected to terminal B.

[(a) 4 A, 0.5833 S; (b) 5 A, 1.083 S]

3.7. In problems 3.3, 3.4 and 3.5, determine the resistance of the load resistor connected to terminals A and B which dissipates maximum power, together with the power in each case.
[10 Ω, 4.225 W; 4.44 Ω, 0.077 W; 0.779 Ω, 1.57 W]

3.8. Calculate the maximum power which may be delivered to a load connected between A and B in problem 3.6.
[(a) 6.86 W; (b) 5.77 W]

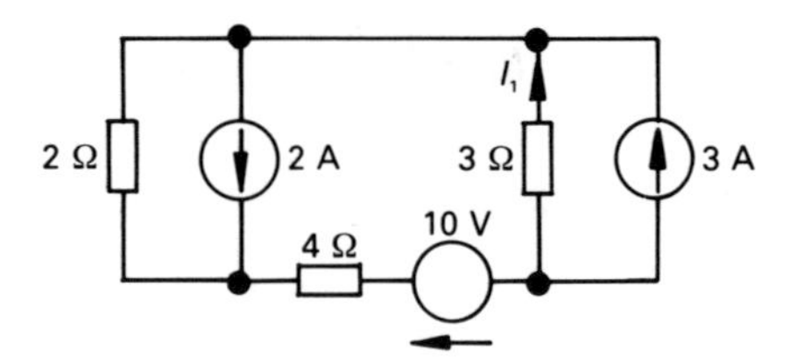

Figure 3.14

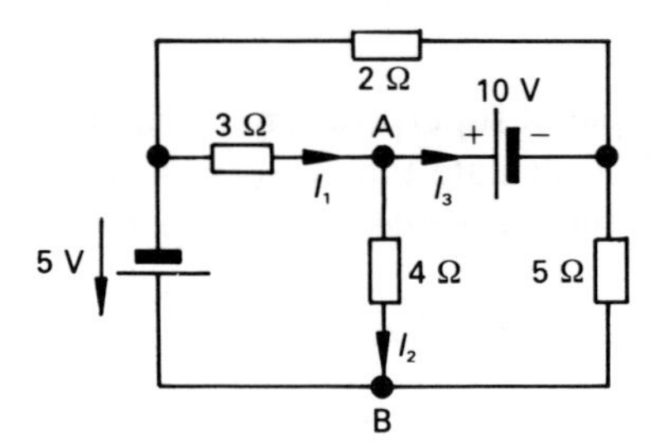

Figure 3.15

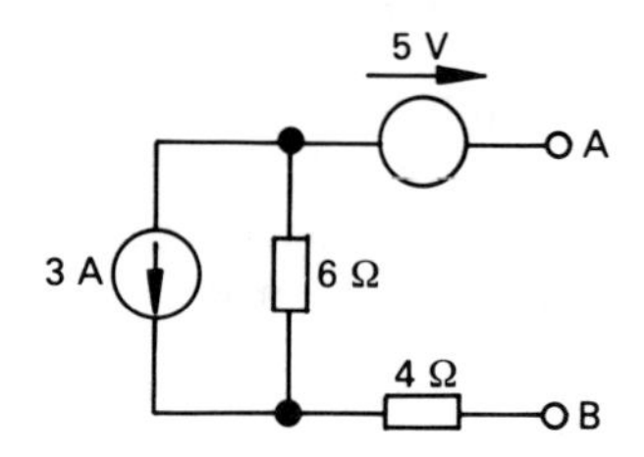

Figure 3.16

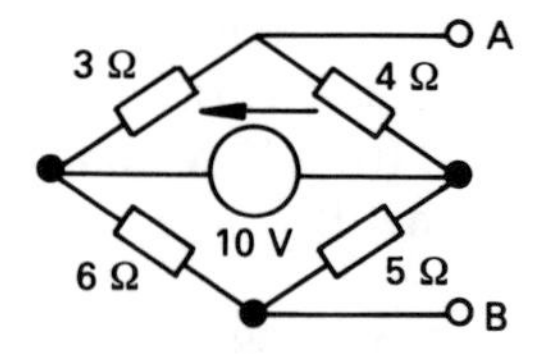

Figure 3.17

3.9. For the circuit in figure 3.19, calculate I_1, I_2, I_3 and V_{BA}.
[1.091 A; −1.455 A; 0.364 A; 4.54 V]

3.10. Convert the star network in figure 3.20 into its equivalent generalised mesh network.

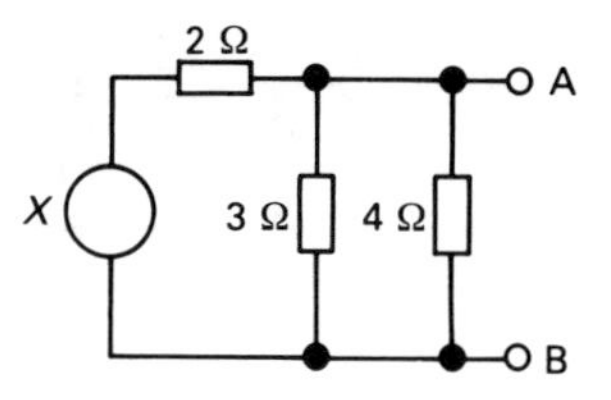

Figure 3.18

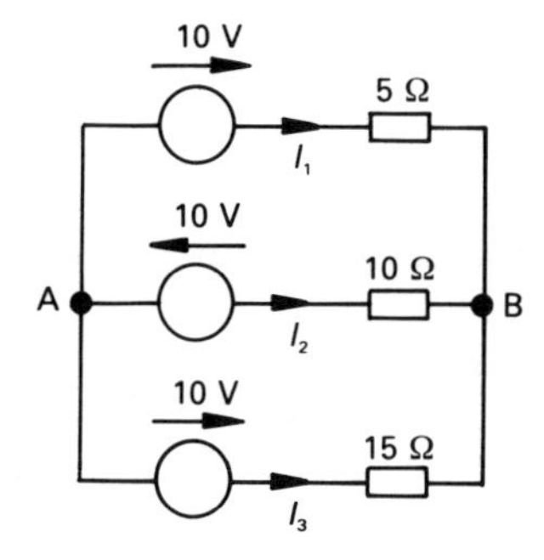

Figure 3.19

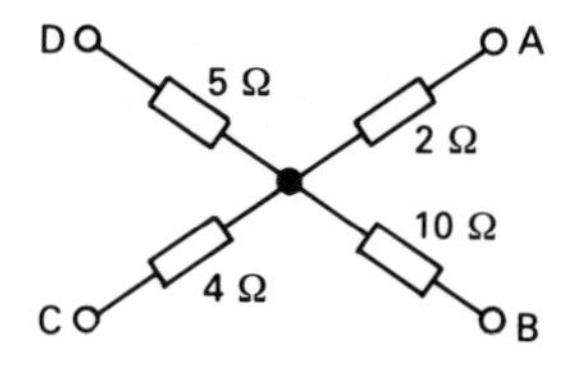

Figure 3.20

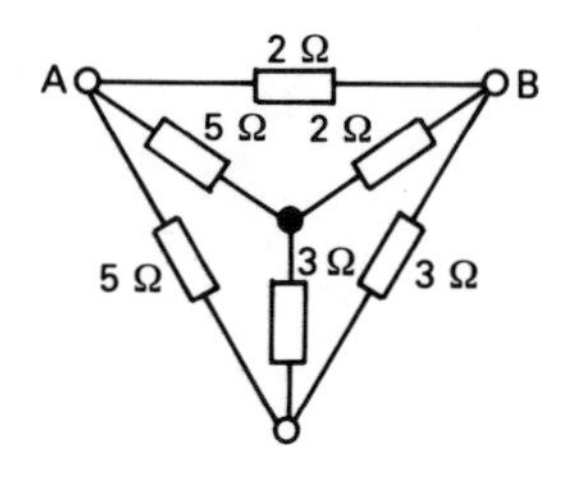

Figure 3.21

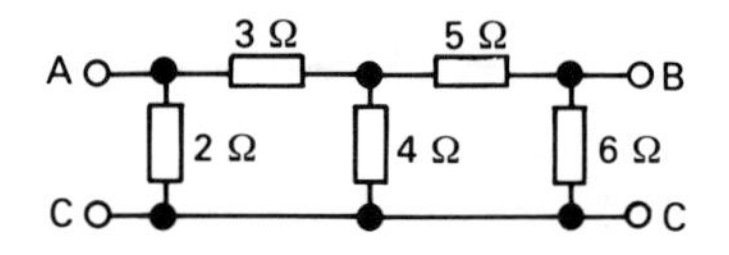

Figure 3.22

[G_{AD} = 0.0952 S; G_{AB} = 0.0476 S; G_{AC} = 0.119 S;
G_{BD} = 0.019 S; G_{BC} = 0.024 S; G_{CD} = 0.048 S]

3.11. Convert the star-connected set of resistors in figure 3.21 into their equivalent delta network, and hence calculate the resistance between A and B.
[1.3 Ω]

3.12. Evaluate the equivalent (a) delta network, (b) star network for the circuit in figure 3.22.
[(a) R_{AC} = 1.65 Ω; R_{BC} = 4.34 Ω; R_{AB} = 11.75 Ω;
(b) R_{AS} = 1.09 Ω; R_{BS} = 2.87 Ω; R_{CS} = 0.403 Ω]

4

Energy Storage Elements

4.1 Introduction

So far, our discussions have covered elements which are either energy sources or energy dissipators. However, elements such as capacitors and inductors have the property of being able to *store energy*, whose *V–I* relationships contain either time integrals or derivatives of voltage or current. As one would suspect, this means that the response of these elements is not instantaneous.

4.2 Capacitors

A simple *capacitor* comprises parallel conducting plates separated by a *dielectric*. In an *ideal capacitor*, the charge q stored in the dielectric is

$$q = Cv$$

where v is the voltage across the capacitor, and C is the *capacitance* of the capacitor in farads (F). It is of interest to point out that the abbreviation for the unit of charge (the coulomb) is also C; the reader should be careful not to confuse the symbol for capacitance with that for the unit of charge. The current, i, which charges the capacitor is

$$i = \frac{\mathrm{d}q}{\mathrm{d}t} = C\frac{\mathrm{d}v}{\mathrm{d}t}$$

Consider the circuit in figure 4.1, in which the capacitor has been connected to a d.c. supply long enough to be fully charge. The differential relationship for the capacitor at this time is

$$i = C\frac{\mathrm{d}v}{\mathrm{d}t} = 3\frac{\mathrm{d}[10]}{\mathrm{d}t} = 0 \text{ A}$$

that is, an ideal capacitor is an open-circuit to a d.c. source. Using this relationship in the circuit in figure 4.2(a), and assuming that the capacitors

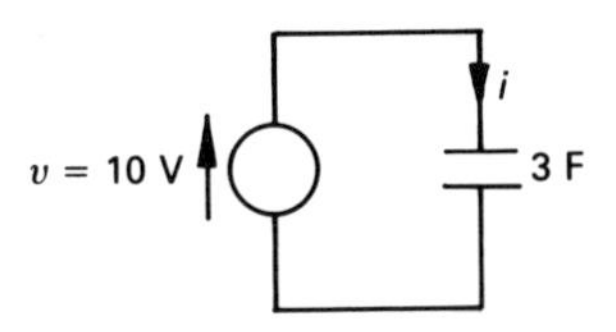

Figure 4.1 *Current in a capacitor in a d.c. circuit.*

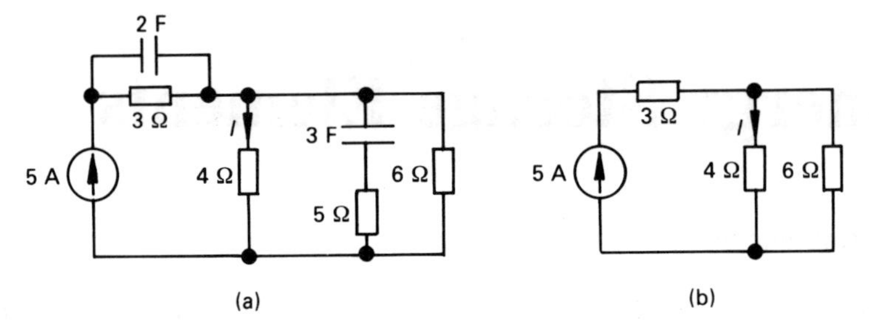

Figure 4.2 *Capacitors in a d.c. network.*

are fully charged, the circuit can be reduced to that in figure 4.2(b) for the purpose of the calculation of the steady-state current, I, in the 4 Ω resistor. That is

$$I = 5 \times 6/(4 + 6) = 3 \text{ A}$$

Worked example 4.2.1

The voltage waveform, v, applied to the circuit in figure 4.3(b) is described by

$$v = \begin{cases} 0 \text{ for } t < 0 \\ 1.5t \text{ for } 0 \leqslant t < 2 \\ (6 - 1.5t) \text{ for } 2 \leqslant t < 4 \\ 0 \text{ for } 4 \leqslant t < \infty \end{cases}$$

and is illustrated in figure 4.3(b). Sketch the waveform of the current through the capacitor.

Solution

The current in the circuit is

$$i = C\frac{\mathrm{d}v}{\mathrm{d}t} = 3\frac{\mathrm{d}v}{\mathrm{d}t} = \begin{cases} 0 \text{ for } t < 0 \\ 4.5 \text{ for } 0 \leqslant t < 2 \\ -4.5 \text{ for } 2 \leqslant t < 4 \\ 0 \text{ for } 4 \leqslant t < \infty \end{cases}$$

and is illustrated in figure 4.3(c).

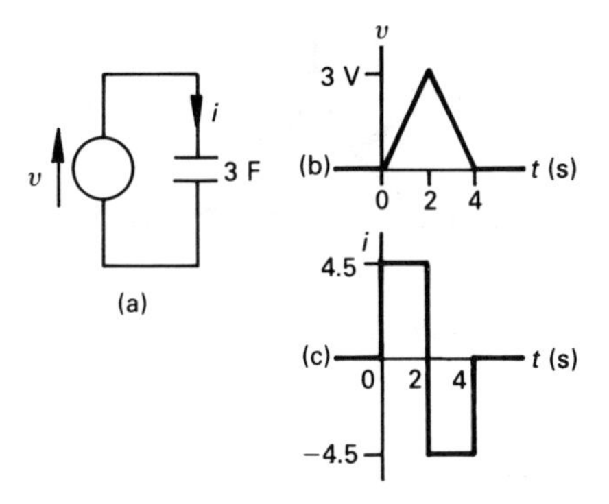

Figure 4.3 *Figure for worked example 4.2.1.*

4.3 Energy stored in capacitor

Energy is stored in the electric field of the capacitor, and the instantaneous energy supplied to a capacitor of capacitance C in time dt is

$$\mathrm{d}W = p\ \mathrm{d}t = vi\ \mathrm{d}t = vC\frac{\mathrm{d}v}{\mathrm{d}t}\ \mathrm{d}t = Cv\ \mathrm{d}v$$

The total energy supplied to the capacitor is the time integral of this expression, as follows

$$W = \int_0^V Cv\ \mathrm{d}v = \tfrac{1}{2}CV^2$$

Worked example 4.3.1

For worked example 4.2.1, sketch to a base of time the graph of energy stored in the capacitor.

Solution

The energy stored is

$$W = \frac{1}{2}Cv^2 = 1.5v^2 = \begin{cases} 0 \text{ for } t < 0 \\ 3.375t^2 \text{ for } 0 \leqslant t < 2 \\ 1.5(6 - 1.5t)^2 \text{ for } 2 \leqslant t < 4 \\ 0 \text{ for } 4 \leqslant t < \infty \end{cases}$$

The resulting graph is shown in figure 4.4.

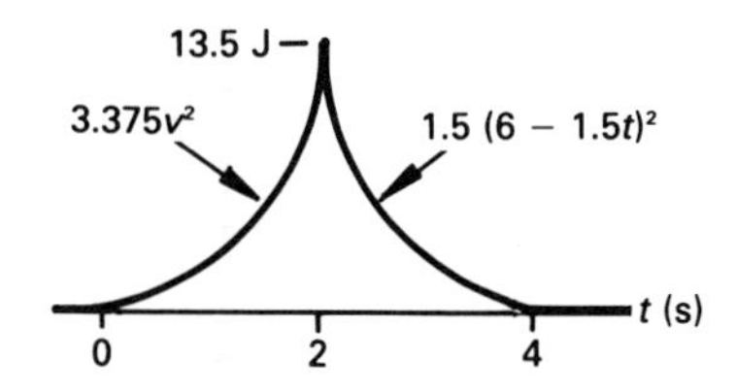

Figure 4.4 *Figure for worked example 4.3.1.*

4.4 Capacitors in parallel

When n ideal capacitors are connected in parallel with one another, each supports the supply voltage, V_S, between its terminals – see figure 4.5. That is, the charge stored by each capacitor is

$$Q_1 = C_1 V_S, \quad Q_2 = C_2 V_S, \quad Q_n = C_n V_S$$

and the total charge, Q, stored by the parallel-connected capacitors is

$$Q = Q_1 + Q_2 + \ldots + Q_n = V_S(C_1 + C_2 + \ldots + C_n)$$

If the parallel-connected capacitors are replaced by an *equivalent capacitor*, C_E, then

$$Q = C_E V_S = (C_1 + C_2 + \ldots + C_n) V_S$$

That is, the effective capacitance of the parallel circuit is

$$C_E = C_1 + C_2 + \ldots + C_n$$

For a parallel circuit, the effective capacitance is always greater than the largest value of capacitance in the circuit.

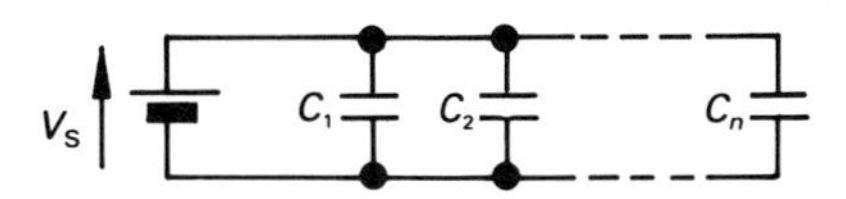

Figure 4.5 *Capacitors in parallel.*

4.5 Capacitors in series

When capacitors are connected in series with one another (known as a *string* of capacitors), the same value of charging current flows through each capacitor for the same length of time (see figure 4.6). That is, *each capacitor supports the same value of charge*, Q, hence

$$Q = Q_1 = Q_2 = \ldots = Q_n$$

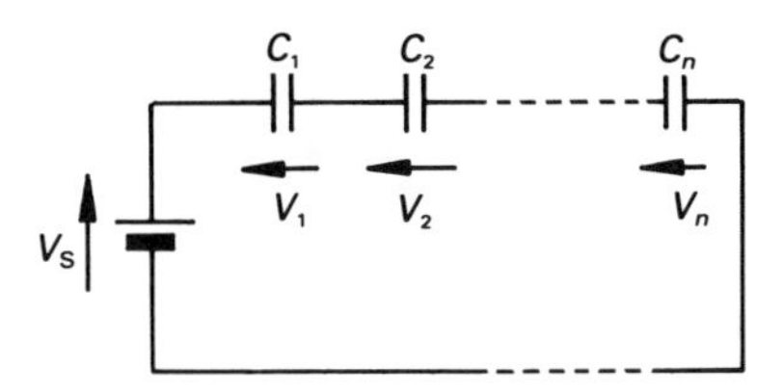

Figure 4.6 *Capacitors in series.*

where $Q = C_1V_1 = C_2V_2 = \ldots = C_nV_n$. Applying KVL to the circuit in figure 4.6 yields

$$V_S = V_1 + V_2 + \ldots + V_n = \frac{Q}{C_1} + \frac{Q}{C_2} + \ldots + \frac{Q}{C_n}$$

$$= Q\left[\frac{1}{C_1} + \frac{1}{C_2} + \ldots + \frac{1}{C_n}\right]$$

If C_E is the effective capacitance of the series circuit, then

$$V_S = \frac{Q}{C_E}$$

or

$$\frac{1}{C_E} = \frac{1}{C_1} + \frac{1}{C_2} + \ldots + \frac{1}{C_n}$$

For the special case of two capacitors in parallel

$$C_E = \frac{C_1C_2}{C_1 + C_2}$$

In the case of series-connected capacitors, the effective capacitance is always less than the lowest value of capacitance in the circuit.

4.6 Potential division in series-connected capacitors

As stated earlier for series-connected capacitors, the total charge stored by the string is equal to the charge stored by each capacitor in the string, that is

$$C_nV_n = C_EV_S$$

or

$$V_n = V_S\frac{C_E}{C_n}$$

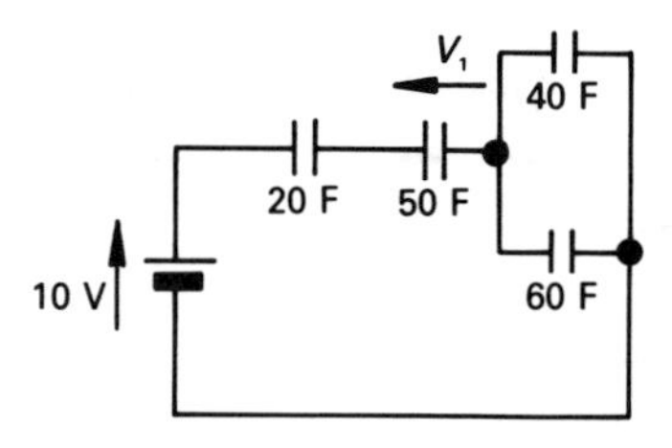

Figure 4.7 *Diagram for worked example 4.6.1.*

Consequent upon this piece of theory, *the capacitor with the smallest value of capacitance supports the largest voltage*!

Worked example 4.6.1

Calculate the effective capacitance of the circuit in figure 4.7 and the value of V_1.

Solution

The capacitance, C_p, of the parallel section of the circuit is

$$C_p = 40 + 60 = 100 \text{ F}$$

The complete circuit effectively consists of a string of capacitors of 20 F, 50 F and C_p F. The reciprocal of the effective capacitance of the circuit therefore is

$$\frac{1}{C_E} = \frac{1}{20} + \frac{1}{50} + \frac{1}{100} = 0.05 + 0.02 + 0.01$$

$$= 0.08 \text{ F}^{-1}$$

or

$$C_E = 1/0.08 = 12.5 \text{ F}$$

The voltage across the 50 F capacitor can be calculated as follows

$$V_n = V_S \frac{C_E}{C_n} = 10 \frac{12.5}{50} = 2.5 \text{ V}$$

4.7 Inductance

When current flows in a wire, it produces a magnetic field around the wire; when the wire is wound into a *coil* or *inductor*, the resulting magnetic field is strengthened. The *self-inductance*, L henrys (unit symbol H), of the coil is defined as the ratio of the magnetic flux, Φ weber (unit symbol Wb), to the current I amperes (also known as the *excitation current*) which produces the flux. That is

$$L = \frac{\Phi}{I} \text{ H}$$

When the current in the inductor changes, the resulting change in magnetic flux associated with the circuit produces a *self-induced e.m.f.*, e, in the coil, that is

$$e = -L\frac{\mathrm{d}i}{\mathrm{d}t} \text{ V}$$

where $\mathrm{d}i/\mathrm{d}t$ is the rate of change of current in the circuit. Since electrical engineers regard an inductor as a passive element rather than a source of e.m.f., we write the p.d. across the inductor as

$$v_L = L\frac{\mathrm{d}i}{\mathrm{d}t} \text{ V}$$

That is to say, e and v_L act in opposite directions, as shown in figure 4.8.

Let us look for the moment at the circuit in figure 4.9 in which a constant current of 2 A has been flowing long enough to allow steady-state operating conditions to be reached. The volt–ampere relationship for the circuit at this time is

$$e = L\frac{\mathrm{d}i}{\mathrm{d}t} = 3\frac{\mathrm{d}[2]}{\mathrm{d}t} = 0 \text{ V}$$

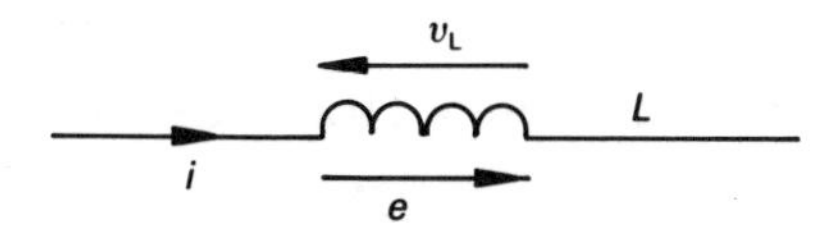

Figure 4.8 *e.m.f. and p.d. associated with an inductor.*

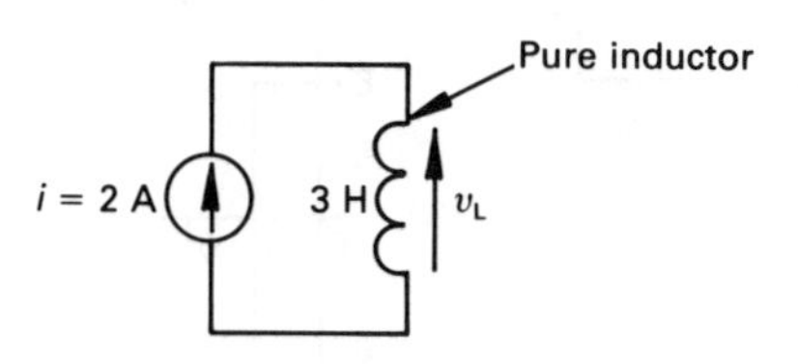

Figure 4.9 *The steady-state self-induced e.m.f. in an ideal inductor in a d.c. circuit.*

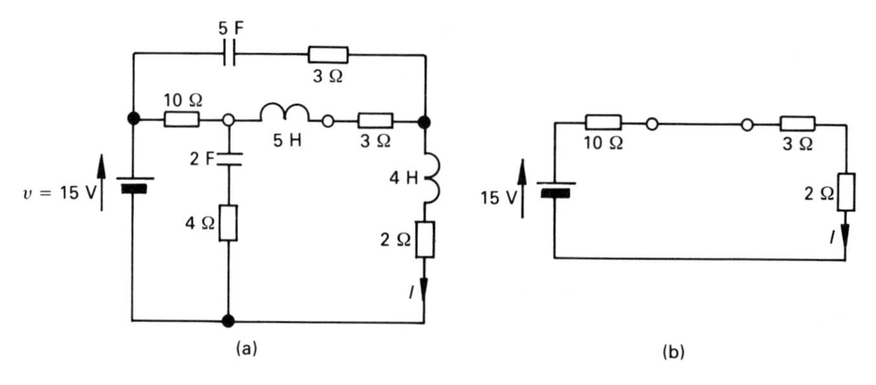

Figure 4.10 *A more complex d.c. circuit.*

That is, under steady-state conditions in a d.c. circuit, an *ideal inductor* acts as though it were a short-circuit.

Looking now at the somewhat more complex d.c. circuit in figure 4.10(a) involving both capacitors and inductors, we will calculate the steady-state value of the current I. For this calculation we replace inductors by short-circuits and capacitors by open-circuits, leaving the 'steady-state' d.c. circuit in figure 4.9(b). Clearly, the steady-state value of I is

$$I = 15/(10 + 3 + 2) = 1 \text{ A}$$

While the above discussion is in order for steady-state d.c. conditions, there may be other factors operating in the circuit because we have two types of energy storage elements in the circuit. We will discuss these factors in chapter 10.

Worked example 4.7.1

The current in the circuit in figure 4.11(a) is described as follows

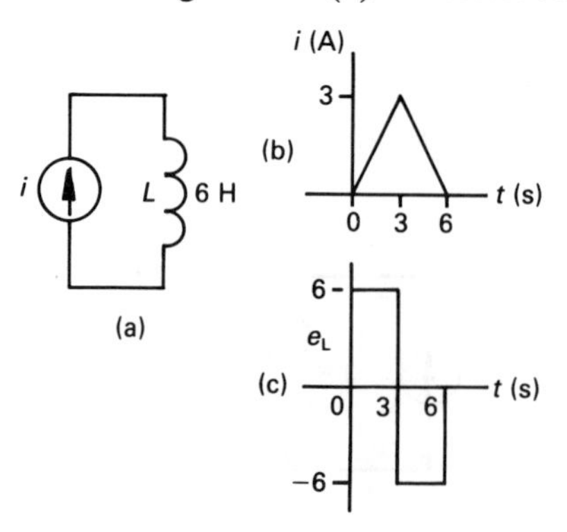

Figure 4.11 *Diagram for worked example 4.7.1.*

$$i = \begin{cases} 0 \text{ for } t < 0 \\ t \text{ for } 0 \leqslant t < 3 \\ (6 - t) \text{ for } 3 \leqslant t < 6 \\ 0 \text{ for } 6 \leqslant t < \infty \end{cases}$$

Determine the waveshape of the voltage across the inductor.

Solution

The current waveform is shown in figure 4.11(b), and the self-induced e.m.f. is defined by

$$v_L = L\frac{di}{dt} = 6\frac{di}{dt} = \begin{cases} 0 \text{ for } t < 0 \\ 6 \text{ for } 0 \leqslant t < 3 \\ -6 \text{ for } 3 \leqslant t < 6 \\ 0 \text{ for } 6 \leqslant t < \infty \end{cases}$$

and is shown in figure 4.11(c).

4.8 Energy stored in an inductor

The instantaneous energy supplied to an inductor is

$$dW = p\ dt = v_L i\ dt = L\frac{di}{dt} \times i\ dt = Li\ di$$

and the total energy supplied is the time integral of this expression as follows

$$W = \int_0^I Li\ di = \frac{1}{2}Li^2$$

4.9 Inductors in series

For the series circuit in figure 4.12

$$v_1 = L_1\ di/dt, \quad v_2 = L_2\ di/dt, \quad v_n = L_n\ di/dt$$

Applying KVL to the circuit shows that

$$v_S = v_1 + v_2 + \ldots + v_n = (L_1 + L_2 + \ldots + L_n)\ di/dt$$

If L_E is the effective inductance of the circuit, then

$$v_S = L_E\ di/dt$$

hence the effective inductance of the circuit is

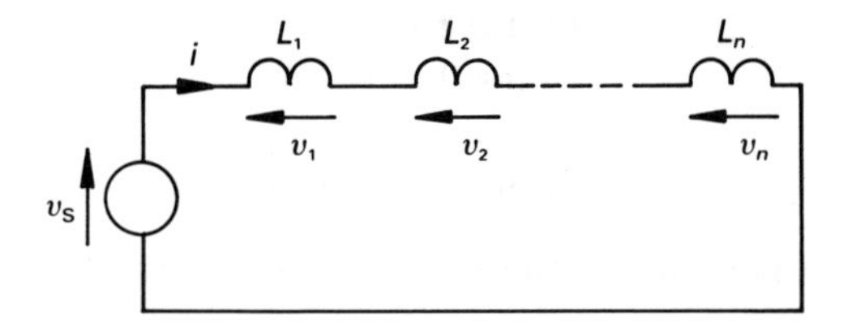

Figure 4.12 *Inductors in series.*

$$L_E = L_1 + L_2 + \ldots + L_n$$

That is, *the effective inductance of a series circuit is greater than the largest individual inductance in the circuit.*

4.10 Inductors in parallel

Applying KCL to the parallel-connected inductors in figure 4.13 yields

$$i_S = i_1 + i_2 + \ldots + i_n$$

hence

$$\frac{di_S}{dt} = \frac{di_1}{dt} + \frac{di_2}{dt} + \ldots + \frac{di_n}{dt}$$

and, since the voltage across the circuit is

$$v = L_1\frac{di_1}{dt} = L_2\frac{di_2}{dt} = \ldots = L_n\frac{di_n}{dt}$$

that is

$$\frac{di_n}{dt} = \frac{v}{L_n}$$

or

$$\frac{di_S}{dt} = \frac{v}{L_1} + \frac{v}{L_2} + \ldots + \frac{v}{L_n}$$

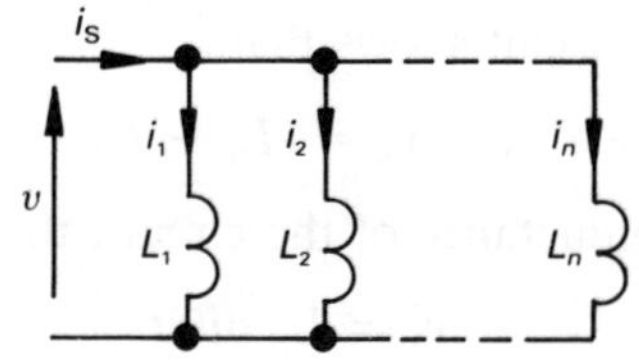

Figure 4.13 *Inductors in parallel.*

If the effective inductance of the circuit is L_E, then

$$v = L_E \frac{di_S}{dt} \quad \text{or} \quad \frac{di_S}{dt} = \frac{v}{L_E}$$

hence

$$\frac{v}{L_E} = \frac{v}{L_1} + \frac{v}{L_2} + \ldots + \frac{v}{L_n}$$

or

$$\frac{1}{L_E} = \frac{1}{L_1} + \frac{1}{L_2} + \ldots + \frac{1}{L_n}$$

In the case of parallel-connected inductors, *the effective inductance is always less than the lowest value of inductance in the circuit.*

For the special case of two inductors in parallel

$$L_E = \frac{L_1 L_2}{L_1 + L_2}$$

Worked example 4.10.1

Calculate the effective inductance between terminals A and B in figure 4.14.

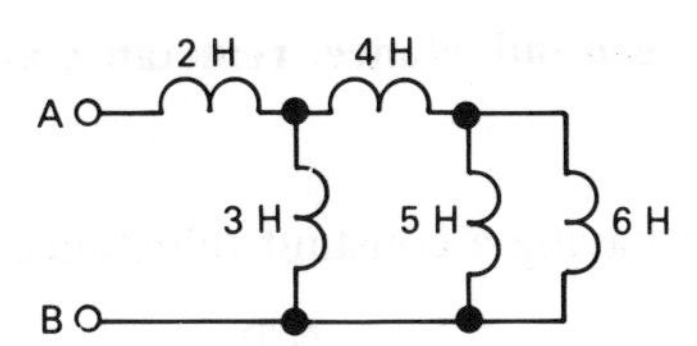

Figure 4.14 *Figure for worked example 4.10.1.*

Solution

We must begin to analyse this circuit at the point which is most remote from the input terminals, and then work towards the input. For the parallel combination of 5 H and 6 H inductors

$$L_{p1} = 5 \times 6/(5 + 6) = 2.727 \text{ H}$$

The effective inductance of 4 H and L_{p1} in series is

$$L_{s1} = 4 + 2.727 = 6.727 \text{ H}$$

The parallel combination of 3 H and L_{s1} results in an inductance of

$$L_{p2} = \frac{3 \times 6.727}{3 + 6.727} = 2.075 \text{ H}$$

and the effective inductance between A and B is

$$L_E = 2 + L_{p2} = 4.075 \text{ H}$$

4.11 Duality between inductors and capacitors

Two circuits are said to be the *dual* of one another if the mesh current equations of one circuit have the same mathematical form as the node voltage equations of the other.

The differential equations for inductors and capacitors are

Inductors	*Capacitors*
$v = L\frac{di}{dt}$	$i = C\frac{dv}{dt}$
$i = \frac{1}{L}\int_0^t v \, dt$	$v = \frac{1}{C}\int_0^t i \, dt$

That is, inductors and capacitors are dual quantities. An inductor of 3 H is the exact dual of a 3 F capacitor (see also section 2.12).

4.12 Relationship between inductance, reluctance and the number of turns on a coil

For a magnetic circuit having a constant reluctance, that is, an air-core

$$L = \frac{N\Phi}{I}$$

Now, the magnetic flux produced is

$$\Phi = BA = \mu HA = \mu\frac{NI}{l}A = \frac{NI}{S}$$

where B is the magnetic flux density, A is the cross-sectional area of the magnetic circuit, H is the magnetic field intensity, N is the number of turns on the coil, I is the current in the coil, μ is the permeability of the magnetic path, l is the length of the magnetic circuit and S is the reluctance of the magnetic path. Hence

$$L = N^2/S$$

That is, if the reluctance of the magnetic path is constant then doubling the number of turns on the coil quadruples the inductance.

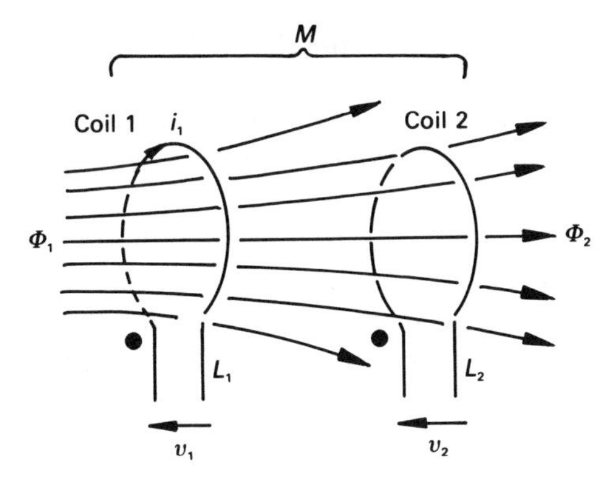

Figure 4.15 *Magnetically coupled circuits; Φ_1 is the flux leaving coil 1 and Φ_2 is the flux reaching coil 2.*

4.13 Mutual inductance

When current flows in a coil, a magnetic flux is established and energy is stored in the magnetic field. If two coils are wound on the same magnetic circuit (see figure 4.15) and current flows, say, in coil 1, then some of the magnetic flux produced by that coil links with coil 2. If the current in coil 1 is altered, the flux entering coil 2 also changes, resulting in a *mutually-induced e.m.f.*, v_2, in coil 2. If M_{12} is the *mutual inductance* existing between the coils, then

$$M_{12} = \Phi_2/i_1$$

where Φ_2 is the flux reaching coil 2. Similarly, if coil 2 is excited and coil 1 is the magnetically coupled coil, then the mutual inductance M_{21} between coil 2 and coil 1 is

$$M_{21} = \Phi_1/i_2$$

where i_2 is the current in coil 2 and Φ_1 is the magnetic flux entering coil 1. Since the two coils are wound on the same magnetic circuit

$$M_{12} = M_{21} = M$$

where M is the mutual inductance between the two coils.

When only one of the coils is excited, that coil is known as the *primary winding*, and the coil which has the mutually induced e.m.f. in it is known as the *secondary winding*. The two circuits are described as *magnetically coupled circuits*.

4.14 Direction of the mutually induced e.m.f. – the dot notation

In figure 4.15, the current i_1 is produced by the applied voltage v_1, which has the polarity shown. The direction of the current induced in coil 2 is deduced as follows.

Lenz's law states that the direction of the induced e.m.f. (whether self or mutually induced) opposes the change producing it. Consequently, the current induced in coil 2 must produce a magnetic flux which opposes the flux developed by coil 1. Since, in figure 4.15, the flux enters coil 2 from the left, the current induced in coil 2 must produce a magnetic flux which leaves the left-hand side of coil 2. The result is that the mutually induced e.m.f., v_2, has the polarity shown in the figure.

Engineers have developed the concept of the *dot notation* which allows us to specify ends of coils having similar polarity as follows. One end (any end) of the primary coil is marked with a dot, and the end of each secondary coil (there may be several of these) having the same instantaneous polarity as the 'dotted' end of the primary coil is also marked with a dot.

Thus, in figure 4.15, if we mark the left-hand end of the left-hand coil with a dot, then we must mark the left-hand end of the right-hand coil with a dot since they both have the same instantaneous polarity (one polarity arising from the forcing voltage, and the other being the induced polarity).

If we apply this reasoning to the coupled circuit in figure 4.16(a), we will quickly be able to deduce a set of mesh equations to solve the circuit. Lenz's law enables us to say that, for each mutually coupled coil, there is a mutually induced e.m.f. in that coil. Since each winding carries a current, there is a mutually induced e.m.f. (a current-controlled voltage source) in each coupled coil. Thus we can separate the two magnetically coupled circuits in figure 4.16(a) into two separate circuits, as shown in diagram (b).

We now investigate a method of deducing the direction of the mutually induced e.m.f.s. When drawing diagram (b), we transfer a copy of the 'dots'

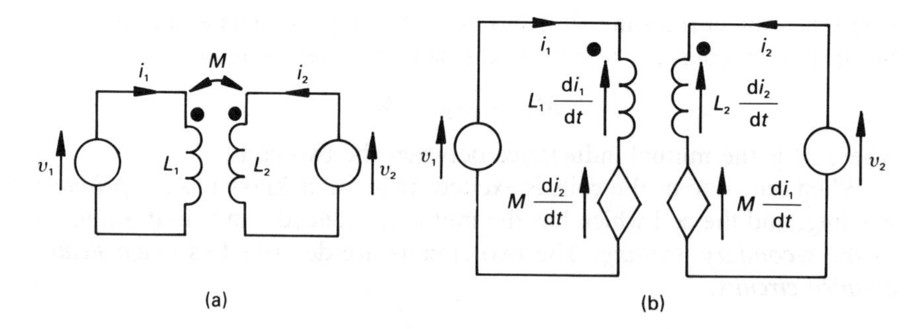

Figure 4.16 *(a) Coupled circuits energised by independent sources, (b) an equivalent circuit.*

from diagram (a), and insert a dependent voltage source in each winding for each mutually coupled coil (*there is one such voltage source for each mutually coupled coil*). The possibility always exists for the introduction of errors at this point; from experience, the author suggests that the dependent voltage source(s) should be drawn at the *opposite end* to the dot on the coil – see figure 4.15(b). The value of the mutually induced e.m.f. is

$$M\,\mathrm{d}i_n/\mathrm{d}t$$

where i_n is, in this case, either i_1 or i_2. At this point the reader should note that *a mutually induced e.m.f. only exists so long as the current is changing in the exciting coil.*

Since we have assumed that i_1 enters the dotted end of L_1, that is, the dotted end of L_1 is assumed to be connected to the positive pole of the forcing voltage, the magnitude of the mutually induced e.m.f. in coil 2 is $M\,\mathrm{d}i_1/\mathrm{d}t$, and this acts to make the dotted end of coil 2 positive.

By applying a similar reasoning to the mutually induced e.m.f. in coil 1, the reader will confirm that the direction of the e.m.f. is as shown in figure 4.16(b), and that its value is $M\,\mathrm{d}i_2/\mathrm{d}t$. In addition to the mutually induced e.m.f. in each coil, there is also a self induced e.m.f. of value $L_n\,\mathrm{d}i_n/\mathrm{d}t$ in each coil. The mesh equations for the two circuit are therefore

$$v_1 = L_1\frac{\mathrm{d}i_1}{\mathrm{d}t} + M\frac{\mathrm{d}i_2}{\mathrm{d}t}$$

$$v_2 = M\frac{\mathrm{d}i_1}{\mathrm{d}t} + L_2\frac{\mathrm{d}i_2}{\mathrm{d}t}$$

4.15 Coefficient of coupling

Suppose that coil 1 (of inductance L_1) in figure 4.15 produces flux Φ_1, and that a proportion $k\Phi_1$ links with coil 2 (of inductance L_2). The parameter k is known as the *magnetic coupling coefficient*, where $0 \leqslant k \leqslant 1$. Also, if a current flows in L_2 and produced flux Φ_2, then a flux $k\Phi_2$ links with L_1. Now

$$M = \frac{N_2 k\Phi_1}{i_1} \quad \text{and} \quad M = \frac{N_1 k\Phi_2}{i_2}$$

Multipyling the two equations gives

$$M^2 = \frac{N_1 k\Phi_2}{i_2}\,\frac{N_2 k\Phi_1}{i_1} = k^2\frac{N_1\Phi_1}{i_1}\,\frac{N_2\Phi_2}{i_2} = k^2 L_1 L_2$$

or $$k = M/\surd(L_1 L_2)$$

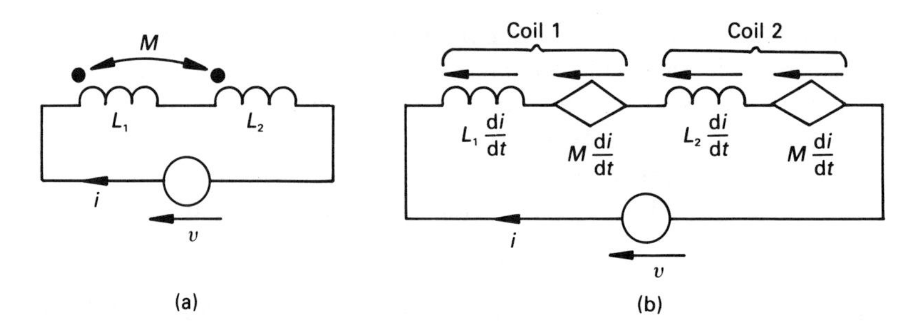

Figure 4.17 *Series-connected mutually coupled coils.*

4.16 Mutually coupled coils in series and in parallel

Series-connected coils

Using the work on the dot notation, the circuit in figure 4.17(a) can be re-drawn as shown in figure 4.17(b), the equation for which is

$$v = L_1\frac{di}{dt} + M\frac{di}{dt} + L_2\frac{di}{dt} + M\frac{di}{dt}$$

$$= (L_1 + L_2 + 2M)\frac{di}{dt}$$

That is, the effective inductance of the circuit is $L_1 + L_2 + 2M$ H. In the connection shown, the coils are said to be *series-aiding* since the flux from one coil assists or aids the flux produced by the second coil.

If the coils are re-connected so that the flux produced by the coils oppose one another, they are said to be *series-opposing*, and the effective inductance of the circuit is $L_1 + L_2 - 2M$ H.

Parallel-connected coils

Consider the two *parallel-aiding* magnetically coupled coils in figure 4.18(a). The corresponding equivalent circuit is drawn in diagram (b) (the reader will find it an interesting exercise to verify the circuit), and the corresponding equations for the circuits are

$$v = L_1\frac{di_1}{dt} + M\frac{di_2}{dt}$$

$$v = M\frac{di_1}{dt} + L_2\frac{di_2}{dt}$$

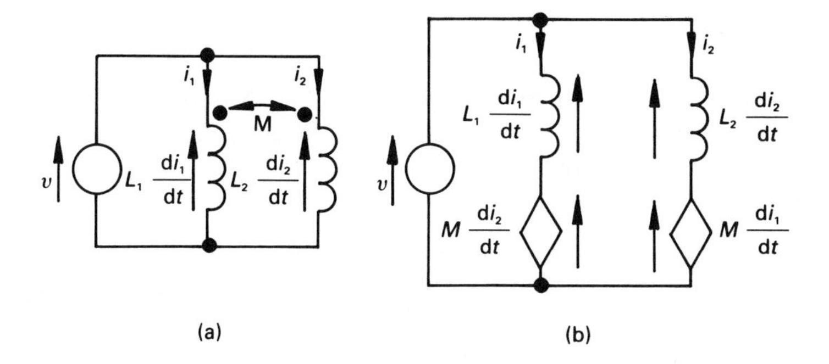

Figure 4.18 *Parallel-connected mutually coupled coils.*

Eliminating di_2/dt between the two equations gives

$$v\left(1 - \frac{M}{L_2}\right) = \left(L_1 - \frac{M^2}{L_2}\right)\frac{di_1}{dt}$$

or

$$v = \frac{L_1L_2 - M^2}{L_2 - M}\frac{di_1}{dt} = L_1'\frac{di_1}{dt}$$

where L_1' is the equivalent inductance of the branch containing L_1. Similarly, eliminating di_2/dt between the equations gives

$$v = \frac{L_1L_2 - M^2}{L_1 - M}\frac{di_2}{dt} = L_2'\frac{di_2}{dt}$$

where L_2' is the effective inductance in the branch containing L_2. Hence

$$L_1' = \frac{L_1L_2 - M^2}{L_2 - M} \quad \text{and} \quad L_2' = \frac{L_1L_2 - M^2}{L_1 - M}$$

For example, if $L_1 = 1$ H, $L_2 = 4$ H and $M = 0.2$ H, then

$$L_1' = (1 \times 4 - 0.2^2)/(4 - 0.2) = 1.042 \text{ H}$$

and

$$L_2' = (1 \times 4 - 0.2^2)/(1 - 0.2) = 4.95 \text{ H}$$

The effective inductance, L_E, of the complete circuit is

$$L_E = \frac{L_1'L_2'}{L_1' + L_2'} = \frac{1.042 \times 4.95}{1.042 + 4.95} = 0.861 \text{ H}$$

The reader will observe that when $L_2 = M$ then $L_1' \to \infty$, and when $L_1 = M$ then $L_2' \to \infty$. If, for example, in the above calculation $M = 1$ H, then $L_1' = 1$ H, $L_2' = \infty$ and $L_E = 1$ H.

Unworked problems

4.1. A pulse of direct current of amplitude 30 mA is applied for 10 ms to an ideal capacitor of 5 μF capacitance. What is the change of voltage between its terminals at the end of the pulse?
[60 V]

4.2. A practical capacitor can be represented by an ideal capacitor shunted by a leakage resistor. If a 5 μF capacitor has a leakage resistance of 1 MΩ, determine an expression for (a) the current through the leakage resistor, (b) the current through the 'ideal' capacitor and (c) the energy stored in the capacitor when a voltage of 200 sin $120\pi t$ V is applied.
[(a) 2×10^{-4} sin $120\pi t$ A; (b) 0.12π cos $120\pi t$ A;
(c) 0.1 $\sin^2 120\pi t$ J]

4.3. The operational amplifier circuit in figure 4.19(a) has the equivalent circuit in diagram (b). If V_A is a voltage which rises steadily from zero volts at the rate of 1 V/s for 1 s, and then falls at the rate of 1 V/s until it reaches zero, thereafter remaining at zero, deduce the waveform for v_B. The voltage gain, A of the op-amp is very large, and it may be assumed that the op-amp does not saturate.
[$v_B = -1$ V for 1 s, when it becomes 1 V for 1 s, remaining at zero thereafter.]

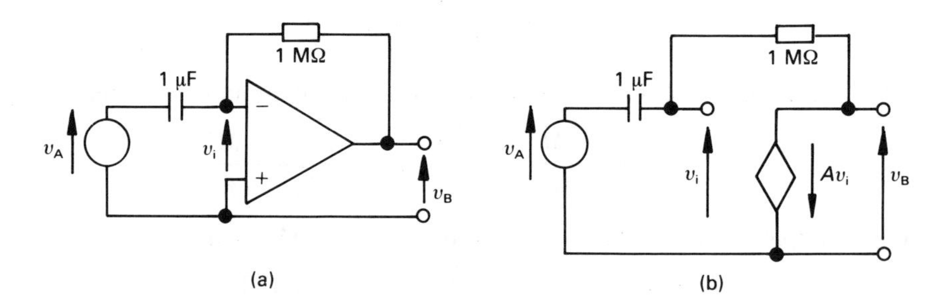

Figure 4.19

4.4. The voltage across a 20 μF capacitor during the time interval $0 \leqslant t \leqslant 3.333$ s is given by $v(t) = 50t^2(5 - t)$ V. Deduce an expression for $i(t)$, and calculate the maximum current and the instant at which it occurs.
[$t(10 - 3t)$ mA; 8.333 mA; 1.6667 s]

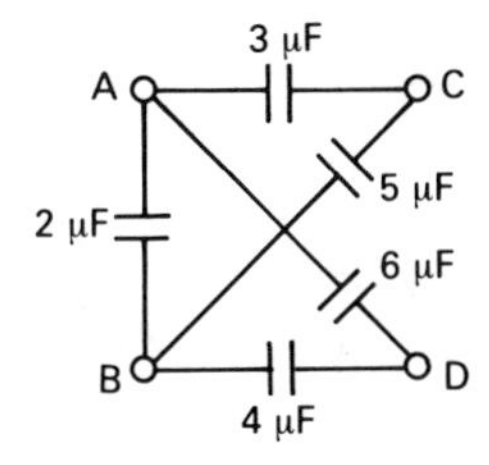

Figure 4.20

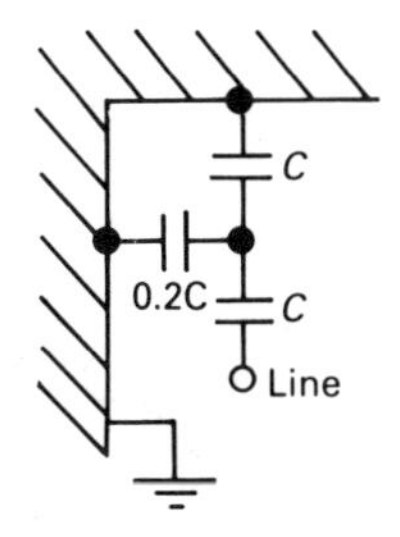

Figure 4.21

4.5. Calculate the effective capacitance between terminals A and B in figure 4.20 when C and D are (a) open-circuited, (b) short-circuited.
[(a) 6.275 μF; (b) 6.5 μF]

4.6. Figure 4.21 is a simplified diagram of a string of suspension insulators. If the maximum voltage per unit (unit capacitance C) is 18 kV, calculate the maximum line voltage.
[33 kV]

4.7. A current $i(t) = 100(2 + 10 \sin 100\,\pi t)$ A flows in a pure inductor of 0.1 H inductance during the time interval $0 \leqslant t \leqslant 0.02$ s. Calculate the maximum voltage across the inductor.
[$10^4\pi$ V]

4.8. If, in figure 4.22(a), (i) element A is a pure inductor and element B is a pure resistor, (ii) element A is a pure resistor and element B is a pure inductor, deduce an expression for the voltage $v_B(t)$ for the case where the voltage gain, A, of the op-amp (see diagram (b)) is infinity.

$$\left[\text{(i)} - \frac{R}{L}\int v_A\,\mathrm{d}t;\ \text{(ii)} - \frac{L}{R}\frac{\mathrm{d}v_A}{\mathrm{d}t}\right]$$

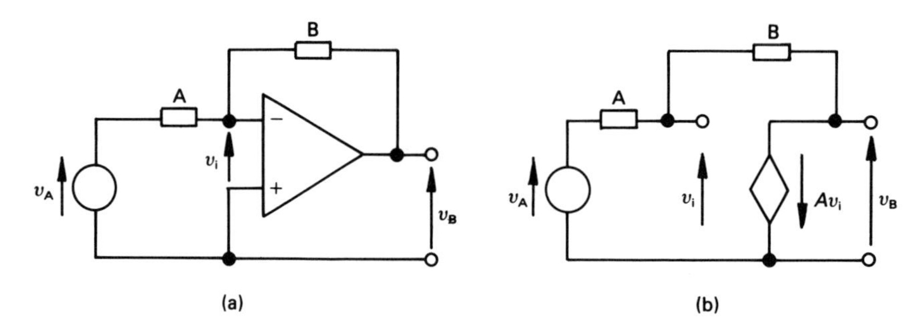

Figure 4.22

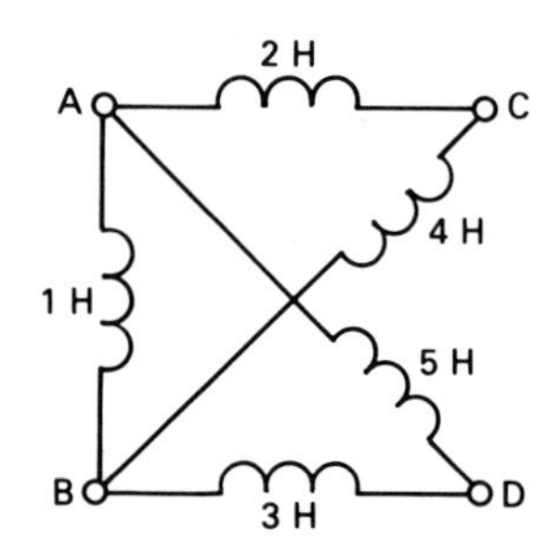

Figure 4.23

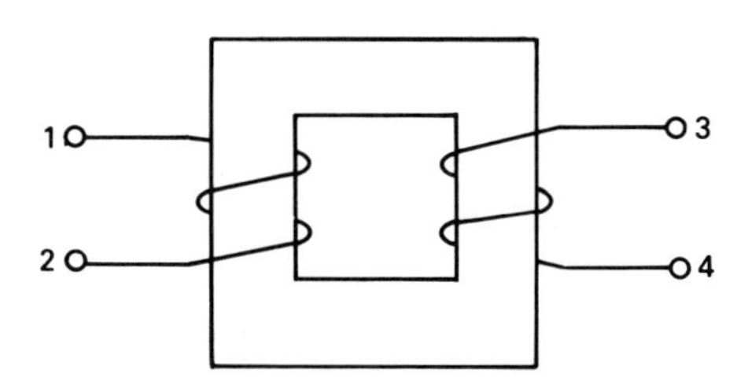

Figure 4.24

4.9. A solenoid 1 m in length and 15 cm diameter has 4000 turns of wire on it; the coil has an air core ($\mu_0 = 4\pi \times 10^{-7}$ H/m). Calculate (a) its approximate inductance and (b) the energy stored when a current of 3 A flows in the coil.
[(a) 0.355 H; (b) 1.6 J]

4.10. Calculate the equivalent inductance between terminals A and B of figure 4.23 when C and D are (a) open-circuited, (b) short-circuited.
[(a) 0.774 H; (b) 0.758 H]

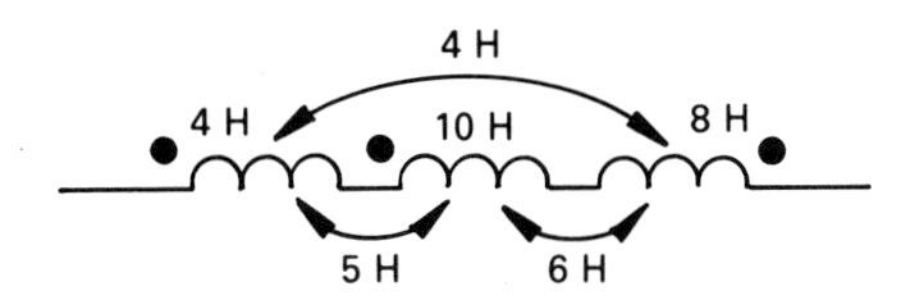

Figure 4.25

4.11. What are the possible alternative locations for the dot notation for the coupled circuit in figure 4.24?
[1 and 4 or 2 and 3]

4.12. Calculate the effective inductance of the circuit in figure 4.25.
[12 H]

5

The Sinewave, Phasors and Power

5.1 Introduction

Advantages of an alternating supply over a direct current supply include not only the simplicity by which it can be 'transformed' or altered in magnitude, but also the way in which it can be transmitted and controlled.

When a given amount of power is transmitted over a great distance from a generating source to a load, the voltage is transformed to a high value so that the current is correspondingly low. The net result is that the power loss (I^2R) in the transmission system is reduced when compared with the case when the power is transmitted at a lower voltage (and a higher current). At the receiving end, the voltage is transformed to a lower value before being connected to the consumer.

While there is an infinite variety of alternating waveshapes, such as, rectangular, triangular, etc., the sine and cosine waves are the best for electrical power transmission.

When describing alternating quantities, the letters 'a.c.' should simply be interpreted as meaning 'alternating'. For example 'a.c. current' means 'alternating current', 'a.c. voltage' means 'alternating voltage', etc.

5.2 Mean or average value of an alternating quantity

When the alternating voltage waveform in figure 5.1(a) is applied to a resistance, R ohms, by Ohm's law the current is proportional to the voltage. The current waveform therefore has the sinusoidal waveform in diagram (b). The *instantaneous voltage* at time t is

$$v = V_m \sin \omega t$$

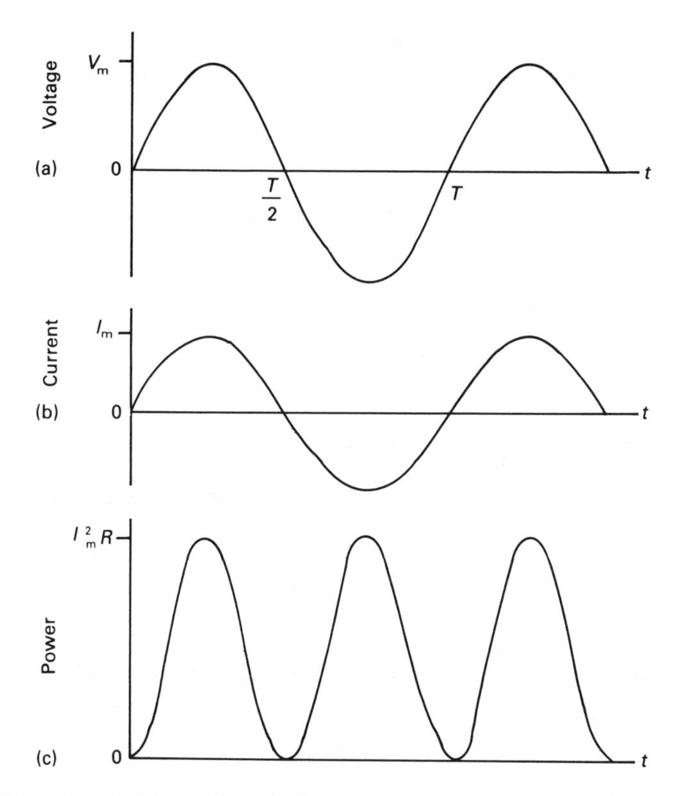

Figure 5.1 *Sinusoidal (a) voltage and (b) current waveforms associated with a resistive element. (c) Power waveform in a resistive element.*

where ω is the *angular frequency* in rad/s. The current, from Ohm's law is

$$i = \frac{v}{R} = \frac{V_m}{R} \sin \omega t$$

Since both the voltage and the current waves are sinusoidal, the area under the positive half-cycle is equal to the area under the negative half-cycle. That is, the mathematical average value of both of them over one complete cycle is zero!

In cases such as the sinewave, electrical engineers take a special viewpoint about the average or mean value of the wave. It is, in fact, taken to be the average value of the *rectified wave*; that is, the negative part of the wave is inverted so that it appears in the positive half. The average value of the sinewave is then taken to be the average value of the rectified sinewave over the complete cycle; since both positive and 'negative' half-cycles in this case have the same shape, we may either calculate the average value over the complete cycle, or over the first half-cycle! We will do the latter. If $a = A \sin \omega t$, then the average value (A_{av}) or mean value (A_{mean}) is

$$A_{av} \text{ or } A_{mean} = \frac{1}{T/2}\int_0^{T/2} A \sin \omega t \, dt$$

where T $(= 2\pi/\omega)$ is the *periodic time* of the wave, hence

$$A_{av} \text{ or } A_{mean} = \frac{A}{T/2}\left[-\frac{1}{\omega}\cos \omega t\right]_0^{T/2}$$

$$= -\frac{A}{T/2}\left[\frac{1}{2\pi/T}\cos\frac{2\pi}{T}t\right]_0^{T/2} = -\frac{A}{\pi}\left[\cos\frac{2\pi}{T}t\right]_0^{T/2}$$

$$= 2A/\pi = 0.637A$$

The corresponding value for a sinewave of voltage and current, respectively, are

$$V_{av} = 2V_m/\pi = 0.637\ V_m$$

$$I_{av} = 2I_m/\pi = 0.637\ I_m$$

The *frequency* of the wave in hertz (Hz) is given by the number of cycles per second, and is

$$f = 1/T \text{ Hz}$$

The corresponding *angular frequency* or *radian frequency* is

$$\omega = 2\pi f \text{ rad/s}$$

5.3 The effective value or r.m.s. value of a periodic wave

The *effective value* of an alternating current is equal to the value of direct current which, when flowing in a resistor, delivers the same power to the resistor. The waveform of the instantaneous power in a resistor when a sinusoidal voltage is applied is shown in figure 5.1(c). The *instantaneous power* consumed by a resistor R is

$$p = i^2R \quad \text{W}$$

and the *average power* consumed during one cycle is

$$P = \frac{1}{T}\int_0^T i^2R \, dt = \frac{R}{T}\int_0^T i^2 \, dt$$

The power delivered by the corresponding direct current, I_{eff}, is

$$P = RI_{eff}^2$$

hence

$$I_{\text{eff}} = \sqrt{\left[\frac{1}{T}\int_0^T i^2\,\mathrm{d}t\right]}$$

and similarly

$$V_{\text{eff}} = \sqrt{\left[\frac{1}{T}\int_0^T v^2\,\mathrm{d}t\right]}$$

The values I_{eff} and V_{eff}, respectively, are the *root-mean-square value* (or *r.m.s. value*, meaning the square *r*oot of the *m*ean of the sum of the instantaneous *s*quare values) of the current and the voltage waves. When dealing with periodic waves, the subscript 'eff' is usually dropped, and the symbols I and V are taken to mean the r.m.s. or effective value of the current and voltage wave, respectively.

r.m.s. value of a sinewave

The instantaneous current in a sinewave is

$$i = I_{\text{m}} \sin \omega t$$

which has a periodic time of $T = 2\pi/\omega$. The r.m.s. value of the wave is

$$I = \sqrt{\left[\frac{1}{T}\int_0^T I_{\text{m}}^2 \sin^2 \omega t\,\mathrm{d}t\right]}$$

$$= I_{\text{m}}\sqrt{\left[\frac{\omega}{2\pi}\int_0^T \frac{1}{2}(1 - \cos 2\omega t)\,\mathrm{d}t\right]}$$

$$= I_{\text{m}}/\sqrt{2} = 0.707\, I_{\text{m}}$$

Similarly, for a voltage sinusoid

$$V = V_{\text{m}}/\sqrt{2} = 0.707\, V_{\text{m}}$$

5.4 Phase angle

The general equation for a sinusoidal wave is

$$a = A \sin (\omega t + \theta)$$

which includes a *phase angle*, θ, in its argument $(\omega t + \theta)$. A sinewave can be thought of as the plot of the vertical displacement of a line of length A which rotates in a *counterclockwise direction* at a constant speed ω rad/s (see figure 5.2). If the radial line has moved through angle θ at $t = 0$, then

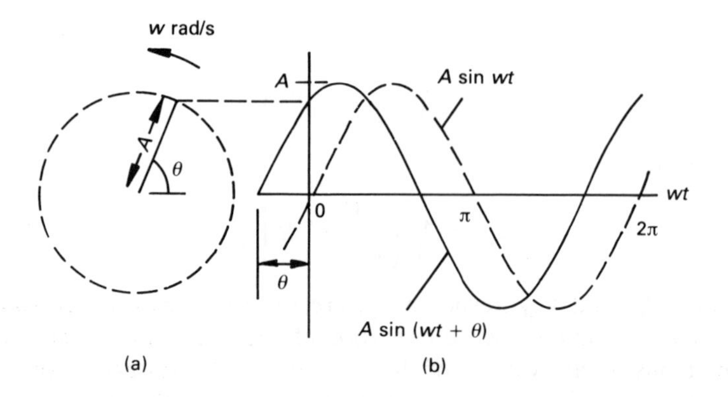

Figure 5.2 *The sinewaves A sin* (ωt + θ) *and* A *sin* ωt.

we get the waveform shown in full line in diagram (b); this corresponds to the general sinusoid $a = A \sin (\omega t + \theta)$. If the radial line is horizontal at $t = 0$, that is, $\theta = 0$, we get the sinewave shown in the broken line in diagram (b); this corresponds to the curve $a = A \sin \omega t$. The two sinewaves in figure 5.2 are said to be *out of phase* with one another.

The sinewave $A \sin \omega t$ can be though of as a *reference sinewave*, since its value is zero when $t = 0$, and increases when t is greater than zero. That is, a sinewave represents the displacement of the tip of a line of length A, which rotates in a counterclockwise direction. When the line is horizontal, engineers say that it is in the *reference direction*.

Since sinewaves are drawn out by the tip of a line rotating in a counterclockwise direction, then the line which traces the curve $A \sin (\omega t + \theta)$ passes a given position (say the horizontal) before the line which traces the curve $A \sin \omega t$. That is, the waveform $A \sin (\omega t + \theta)$ *leads* the waveform $A \sin \omega t$ by θ, the phase angle being measured relative to the reference direction.

In electrical engineering it is customary (but not mandatory) to give the phase angle in degrees, and no confusion should arise if we express a current in any of the following ways

$$\begin{aligned} i &= 10 \sin (\omega t - \pi/3) \text{ A} \\ &= 10 \sin (\omega t - 60°) \text{ A} \\ &= 10 \sin (2\pi \times 100t - \pi/3)\text{A} \\ &= 10 \sin (628.3t - 60°) \text{ A} \end{aligned}$$

In the case of the latter two expressions, the supply frequency is 100 Hz (or 628.3 rad/s).

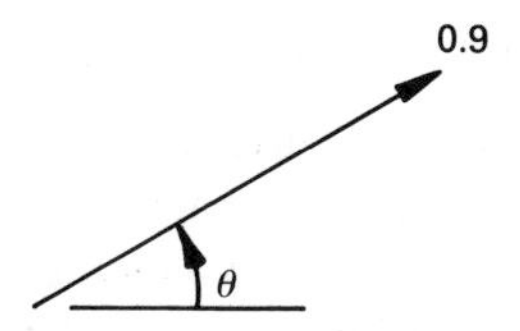

Figure 5.3 *Phasor representation of 1.273 sin (ωt + θ).*

5.5 Phasors and phasor diagrams

At a given frequency, a sinusoidal quantity is characterised by two parameters, namely its *magnitude* or *amplitude* and its *phase angle*. In the UK we usually mean the r.m.s. value of a wave when we refer to the magnitude (in some US texts it may mean the maximum value). We can therefore describe the waveform 1.273 sin $(\omega t + \theta)$ as the *phasor quantity*

$$0.9\angle\theta$$

where the value 0.9 represents the r.m.s. value of the sinewave of maximum value $0.9\sqrt{2} = 1.273$, and we can represent it as shown in figure 5.3. Thus the voltage $v = 141.4 \sin (\omega t + \pi/4)$ V can be represented as

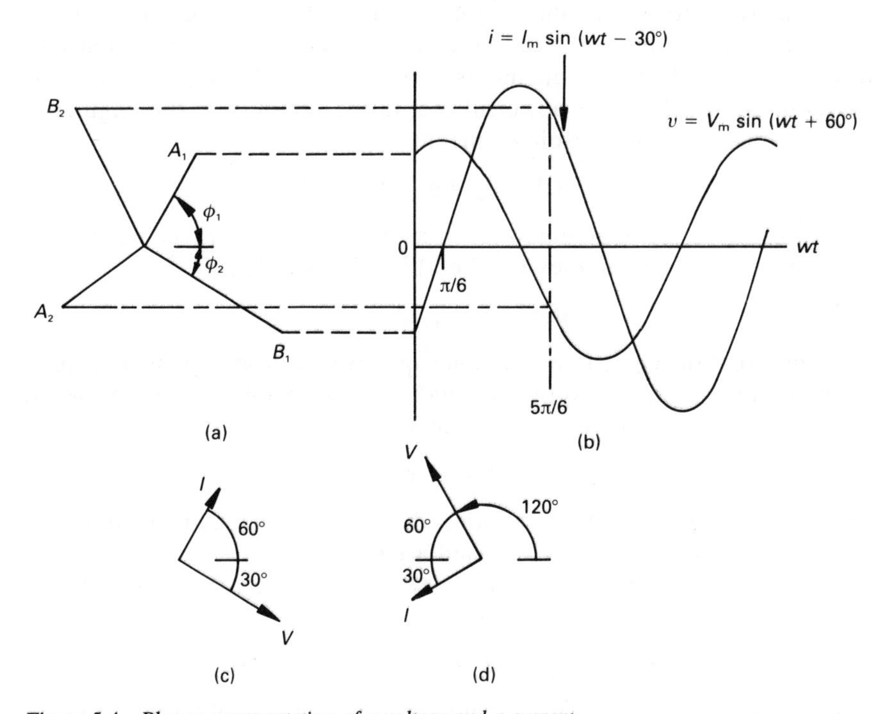

Figure 5.4 *Phasor representation of a voltage and a current.*

$$100\angle\pi/4 \text{ V} \quad \text{or as} \quad 100\angle 45° \text{ V}$$

Suppose that we represent the current and voltage waveforms in figure 5.4(b) by the pair of rotating lines in diagram (a). If we choose to 'freeze' or 'fix' the rotating lines at $\omega t = 0$, the voltage is represented by line A_1 and the current by B_1. Since these lines represent the maximum value of the respective quantities, the *phasor diagram* at $t = 0$ is obtained by scaling the length of each line by a factor of $1/\sqrt{2}$ to give the corresponding r.m.s. phasors, shown in diagram (c).

Alternatively, we can select some other angle at which to 'freeze' or look at the phasors. If we choose $\omega t = 5\pi/6$ rad or 150°, we get the phasor diagram in figure 5.4(d). The reader will note that in both cases I leads V by 90° (or V lags behind I by 90°).

5.6 Representation of a sinusoidal quantity as a complex value

Sinusoidal quantities are time-varying and, strictly speaking, we cannot state the value of, say, a current or voltage that will be correct for all values of time. Moreover, we need a technique which will allow us to apply laws and theorems to alternating current circuits. We therefore introduce a technique here, namely the *complex value*, which is of vital importance to electrical and electronic engineers. The concept of complex values is, without doubt, the most powerful tool in the armoury of the engineer.

The *time-domain* sinusoidal current

$$i(t) = I_m \sin(\omega t + \phi)$$

is expressed as the imaginary part of a complex quantity (see chapter 15 for information on complex numbers) by Euler's identity

$$i(t) = \text{Im}(I_m \, e^{j(\omega t + \phi)})$$

We represent the current as a complex quantity by dropping the imaginary (Im) wording from the expression, and a further simplification is achieved by suppressing the factor $e^{j\omega t}$ as follows.

$$\mathbf{I} = I_m \, e^{j\phi}$$

Since phasors are complex quantities, that is, they embody magnitude and phase angle, they are printed in **boldface type**.

In addition, the maximum value is converted to its r.m.s. value, I, and the whole is written in its polar form as follows

$$\mathbf{I} = I\angle\phi$$

The quantity $i(t)$ varies sinusoidally, and contains amplitude and phase information as a function of time and, accordingly, it is known as the *time-domain representation* of the current. The phasor representation, $\boldsymbol{I}$, is the *frequency-domain representation* of the current, and does not explicitly include the frequency. (The 'frequency' aspect of the frequency-domain representation is emphasised by the absence of frequency in the expression!)

Similarly for voltages, we may represent the time-domain sinusoidal voltage

$$v(t) = V_{\mathrm{m}} \sin (\omega t + \theta)$$

as the complex phasor quantity

$$\boldsymbol{V} = V \angle \theta$$

where V is the r.m.s. value of the voltage.

Worked example 5.6.1

Transform the current $i(t) = 150 \sin (300t - \pi/6)$ A into its frequency-domain complex polar value.

Solution

The general steps to be followed are:

1. Convert the maximum amplitude to its r.m.s. value.
2. If necessary, convert the phase angle to degrees.
3. Write down the frequency-domain representation.

Since $\pi/6$ rad $= 30°$, then

$$\boldsymbol{I} = \frac{150}{\sqrt{2}} \angle -30° = 106.07 \angle -30° \text{ A}$$

To convert from the frequency-domain representation to the time-domain representation, it is merely necessary to reverse the steps. If the frequency is unknown, then it is simply necessary to write ωt where $300t$ appears in the equation in the worked example.

5.7 Impedance of elements

The *impedance*, $\boldsymbol{Z}$, of an electrical element is its total opposition to flow of (sinusoidal) current. If a complex voltage $\boldsymbol{V}$ gives rise to the complex current $\boldsymbol{I}$ in an element then, by Ohm's law, the impedance of the element is

$$\boldsymbol{Z} = \frac{\boldsymbol{V}}{\boldsymbol{I}} \ \Omega$$

5.7.1 Resistance

A resistance does not store energy, and a change in the voltage across the resistor gives rise to an instantaneous change in current in the resistor. That is *the sinusoidal waveform of the voltage across the resistor is in phase with the current in the resistor*. The impedance of the resistor is, therefore

$$\boldsymbol{Z}_R = \frac{\boldsymbol{V}}{\boldsymbol{I}} = R\angle 0° = R + \mathrm{j}0\ \Omega$$

Worked example 5.7.1

Calculate the complex polar expression for the current in a 5 Ω resistor when a voltage of $v(t) = 10 \sin (50t - 50°)$ V is applied to it.

Solution

The frequency-domain representation of the voltage is

$$\boldsymbol{V} = \frac{10}{\sqrt{2}}\angle -50° = 7.071\angle -50°\ \mathrm{V}$$

hence

$$\boldsymbol{I} = \frac{7.071}{5}\angle -50° = 1.414\angle -50°\ \mathrm{A}$$

5.7.2 Pure inductance

Since an inductor stores energy in its magnetic field, a change in current through the inductor causes the stored energy to change and, when this has occurred, the voltage across the inductor changes. That is, the current through the inductor and the voltage across it are not in phase with one another. The phase relationship between the voltage and current associated with the inductor can be deduced as follows.

The v–i relationship for an inductor is $v_L = L\,\mathrm{d}i/\mathrm{d}t$, and if the current is of the form $i = I_m \sin \omega t$, then

$$v_L = L\frac{\mathrm{d}i}{\mathrm{d}t} = L\frac{\mathrm{d}(I_m \sin \omega t)}{\mathrm{d}t} = \omega L I_m \cos \omega t$$

$$= \omega L I_m \cos \omega t = \omega L I_m \sin(\omega t + 90°)$$

$$= V_m \sin(\omega t + 90°)$$

where V_m is the maximum voltage across the inductor.
Hence

$$V_m = \omega L I_m = X_L I_m$$

where $X_L = \omega L$ is known as the *inductive reactance* at frequency ω. Moreover, since v_L has a cosine waveform, it *leads the current waveform by 90°*. Engineers tend to think of the voltage across the inductor as being in the reference direction, and we prefer to say that *the current in the inductor lags behind the voltage across it by 90°*.

The complex expression for the impedance of the inductor is, therefore

$$\boldsymbol{Z}_L = \frac{\boldsymbol{V}_L}{\boldsymbol{I}} = \frac{V_L}{I}\angle 90° = \mathrm{j}\omega L = \mathrm{j}X_L$$

Worked example 5.7.2

A current of $i(t) = 14.14 \sin(100t + 15°)$ A flows in a 0.5 H inductor. Determine the phasor voltage across the inductor.

Solution

The polar complex current in the inductor is

$$\boldsymbol{I} = \frac{14.14}{\sqrt{2}}\angle 15° = 10\angle 15° \text{ A}$$

and $\quad X_L = \omega L = 100 \times 0.5 = 50\ \Omega$

hence $\quad \boldsymbol{Z}_L = 50\angle 90°\ \Omega$

From Ohm's law

$$\boldsymbol{V} = \boldsymbol{I}\boldsymbol{Z}_L = 10\angle 15° \times 50\angle 90° = 500\angle 105° \text{ V}$$

5.7.3 Pure capacitance

Since a capacitor stores energy in its electric field, a change in the voltage between the capacitor terminals causes the stored energy to change; when the energy changes, the current flowing through the capacitor changes. That is, the voltage between the capacitor terminals and the current flowing through it are not in phase with one another. The phase relationship between the voltage and the current can be deduced as follows.

The v–i relationship for a capacitor is $i = C\,\mathrm{d}v_C/\mathrm{d}t$, and if the capacitor voltage is $v_C = V_m \sin \omega t$, then

$$i = C\,\mathrm{d}(V_m \sin \omega t)/\mathrm{d}t = \omega C V_m \cos \omega t$$

$$= \omega C V_m \sin (\omega t + 90°) = I_m \sin (\omega t + 90°)$$

where $I_m = \omega C V_m$, and is the maximum current in the capacitor. From Ohm's law

$$I_m = \frac{V_m}{X_C} = \frac{V_m}{1/\omega C}$$

where X_C is the *reactance* of the capacitor at frequency ω. That is

$$X_C = \frac{1}{\omega C}$$

Morever, since the current has a cosine form, *it leads the voltage across the capacitor by 90°* (alternatively, we may say that the voltage lags behind the current by 90°).

The *impedance* of the capacitor is given by Ohm's law as follows

$$\mathbf{Z}_C = \frac{\mathbf{V}}{\mathbf{I}} = \frac{V}{I}\angle -90° = -\,\mathrm{j}X_C = X_C\angle -90° = \frac{1}{\mathrm{j}\omega C}$$

The manipulation of complex numbers is fully explained in chapter 15.

Worked example 5.7.3

If a voltage $3\angle -60°$ V is applied to an ideal 2 μF capacitor at a frequency of 60 rad/s, calculate the current flowing through the capacitor.

Solution

The capacitive reactance is

$$X_C = 1/\omega C = 1/(60 \times 2 \times 10^{-6}) = 8.333\ \mathrm{k\Omega}$$

and the current in the capacitor is

$$\mathbf{I} = \frac{\mathbf{V}}{\mathbf{Z}_C} = \frac{3\angle -60°}{8.333 \times 10^3\angle -90°} = 0.36\angle 30°\ \mathrm{mA}$$

5.7.4 CIVIL – an a.c. mnemonic

In the above context, the mnemonic CIVIL is very useful, as follows

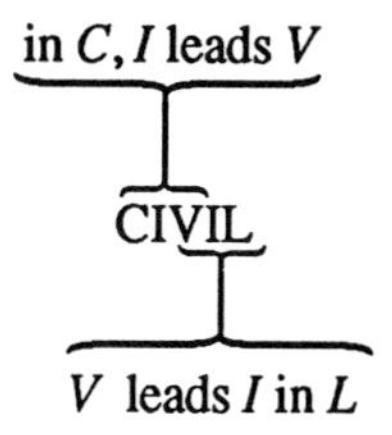

5.8 The susceptance of elements

The *susceptance* of an element is the reciprocal of its reactance. That is

$$\boldsymbol{B} = 1/\boldsymbol{X}$$

For an *inductance*

$$B_{\mathrm{L}} = \frac{1}{X_{\mathrm{L}}} = \frac{1}{\omega L}$$

and for a *capacitance*

$$B_{\mathrm{C}} = \frac{1}{X_{\mathrm{C}}} = \frac{1}{1/\omega C} = \omega C$$

5.9 The admittance of elements

The *admittance*, $\boldsymbol{Y}$, of an element is the reciprocal of its complex impedance; since it is a complex value, it is written in ***bold typeface***. That is

$$\boldsymbol{Y} = \frac{1}{\boldsymbol{Z}} = G + \mathrm{j}B$$

where G is the conductance and B is admittance.
For a *resistance*

$$\boldsymbol{Y}_{\mathrm{R}} = \frac{1}{R + \mathrm{j}0} = \frac{1}{R} + \mathrm{j}0 = G_{\mathrm{R}} + \mathrm{j}B_{\mathrm{R}}$$

That is, $G_{\mathrm{R}} = 1/R$ (as it is in d.c. circuits) and $B_{\mathrm{R}} = 0$. The process of division of complex numbers is fully described in chapter 15.
For an *inductance*

$$\boldsymbol{Y}_{\mathrm{L}} = \frac{1}{0 + \mathrm{j}X_{\mathrm{L}}} = 0 + \frac{1}{\mathrm{j}X_{\mathrm{L}}} = 0 - \mathrm{j}/X_{\mathrm{L}} = G_{\mathrm{L}} - \mathrm{j}B_{\mathrm{L}}$$

That is $G_L = 0$ and $B_L = 1/X_L = 1/\omega L$.
For a *capacitance*

$$Y_C = \frac{1}{0 - j/X_C} = 0 + j/X_C = G_C + jB_C$$

or $G_C = 0$ and $B_C = 1/X_C = \omega C$.

The corresponding representation of Ohm's law is

$$\boldsymbol{I} = \boldsymbol{V}\boldsymbol{Y} = \boldsymbol{V}(G \pm jB)$$

5.10 The impedance of elements in series

If n elements $\mathbf{Z_1}, \mathbf{Z_2}, \ldots, \mathbf{Z_n}$ are connected in series, and current $\boldsymbol{I}$ flows through them, by KVL

$$\boldsymbol{V}_S = \mathbf{Z}_1\boldsymbol{I} + \mathbf{Z}_2\boldsymbol{I} + \ldots + \mathbf{Z}_n\boldsymbol{I} = \boldsymbol{I}(\mathbf{Z}_1 + \mathbf{Z}_2 + \ldots + \mathbf{Z}_n)$$

where $\boldsymbol{V}_S$ is the supply voltage. The effective complex impedance of the circuit is

$$\mathbf{Z}_E = \mathbf{Z}_1 + \mathbf{Z}_2 + \ldots + \mathbf{Z}_n = \sum_{k=1}^{n} \mathbf{Z}_k \quad \Omega$$

and the complex voltage $\boldsymbol{V}_n$ across the nth element is

$$\boldsymbol{V}_n = \boldsymbol{V}_s \frac{\mathbf{Z}_n}{\mathbf{Z}_E}$$

Worked example 5.10.1

A voltage of $10\angle 20°$ V is applied to a series circuit containing the following elements

$$\mathbf{Z}_1 = 10\angle 45°\ \Omega, \quad \mathbf{Z}_2 = 10\angle 0°\ \Omega, \quad \mathbf{Z}_3 = 15\angle -80°\ \Omega$$

Calculate the current in the circuit and the voltage across $\mathbf{Z}_2$.

Solution

This example illustrates the advantage of having complex numbers in rectangular form when adding (or subtracting) complex values, and of having them in polar form when multiplying (or dividing) complex values.

In order to add complex impedance together, it is necessary to convert them to their rectangular complex values; this is described fully in chapter 15.

$$Z_1 = 10\angle 45° = 7.071 + j7.071\ \Omega$$

$$Z_2 = 10\angle 0° = 10 + j0\ \Omega$$

$$Z_3 = 15\angle -80° = 2.605 - j14.772\ \Omega$$

and $$Z_E = \Sigma Z = 19.676 - j7.071 = 21.13\angle -21.37°\ \Omega$$

From Ohm's law

$$I = \frac{V}{Z_E} = \frac{10\angle 20°}{21.13\angle -21.37°} = 0.473\angle 41.37°\text{A}$$

and the voltage across Z_2 is

$$V_2 = V_s\frac{Z_2}{Z_E} = 10\angle 20°\ \frac{10\angle 0°}{21.13\angle -21.37°}$$

$$= 4.73\angle 41.37°\ \text{V}$$

The reader should note that, since Z_2 is a resistor, the voltage, V_2, across it is in phase with the current.

5.11 The admittance and impedance of elements in parallel

If n admittances $Y_1, Y_2, \ldots, Y_n$ are connected to the voltage V_s, then by KCL

$$I_s = V_sY_1 + V_sY_2 + \ldots + V_sY_n = V_s(Y_1 + Y_2 + \ldots + Y_n)$$

where I_s is the supply current. The effective admittance of the parallel circuit is

$$Y_E = Y_1 + Y_2 + \ldots + Y_n \sum_{k=1}^{n} Y_k$$

and the corresponding effective impedance, Z_E, is calculated from

$$Y_E = \frac{1}{Z_E} = \frac{1}{Z_1} + \frac{1}{Z_2} + \ldots + \frac{1}{Z_n} = \sum_{k=1}^{n}\frac{1}{Z_k}$$

The current, I_n, in the nth branch is

$$I_n = I_s\frac{Y_n}{Y_E} = I_s\frac{Z_E}{Z_n}$$

In the special case of a *two-branch parallel circuit*

$$Y_E = Y_1 + Y_2$$

and

$$Z_E = \frac{Z_1Z_2}{Z_1 + Z_2}$$

The current in each of the two branches is

$$I_1 = I_s \frac{Y_1}{Y_1 + Y_2} = I_s \frac{Z_2}{Z_1 + Z_2}$$

$$I_2 = I_s \frac{Y_2}{Y_1 + Y_2} = I_s \frac{Z_1}{Z_1 + Z_2}$$

Worked example 5.11.1

Impedances of $10\angle 20°\ \Omega$, $15\angle 0°\ \Omega$ and $(17.32 - j10)\ \Omega$ are connected in parallel to a $100\angle 30°$ V a.c. supply.

Calculate (a) the effective admittance and impedance of the circuit, (b) the current drawn by the circuit and (c) the current in the branch containing the capacitor.

Solution

(a) We can solve the problem using either admittances or impedances. Selecting the former gives

$$Y_1 = \frac{1}{10\angle 20°} = 0.1\angle -20° = 0.094 - j0.0342 \text{ S}$$

$$Y_2 = \frac{1}{15\angle 0°} = 0.0667 - j0 \text{ S}$$

$$Y_3 = \frac{1}{17.32 - j10} = \frac{1}{20\angle -30°} = 0.05\angle 30°$$

$$= 0.0433 + j0.025 \text{ S}$$

and $$Y_E = \Sigma Y = 0.204 - j0.0092 = 0.2042\angle -2.58° \text{ S}$$

hence

$$Z_E = 1/Y_E = 4.9\angle 2.58° = 4.89 + j0.22\ \Omega$$

(b) From Ohm's law

$$I_s = V_sY_E = 100\angle 30° + 0.2042\angle -2.58° = 20.42\angle 27.42° \text{ V}$$

(c) Since the third branch has a negative value of reactance, it must contain the capacitance. From the theory outlined in section 5.10

$$I_3 = I_s \frac{Y_3}{Y_E} = 20.42\angle 27.42° \frac{0.05\angle 30°}{0.2042\angle -2.58°}$$

$$= 5\angle 60° \text{ A}$$

Alternatively, I_3 could have been worked out as follows

$$I_3 = V_S/Z_3 = 100\angle 30°/20\angle -30° = 5\angle 60° \text{ A}$$

5.12 Impedance and admittance of series–parallel circuits

The impedance (or the admittance) of an a.c. series–parallel circuit is worked out in a similar manner to that for an otherwise similar circuit containing pure resistances (or conductances). The only exception is that it is necessary to not only add the *complex impedances* of series elements, but also add the *complex admittances* of parallel elements.

5.13 Power and power factor

If the voltage across an element is

$$v(t) = V_m \sin(\omega t + \theta_1)$$

and the current through the element is

$$i(t) = I_m \sin(\omega t + \theta_2)$$

then the instantaneous power consumed by the element is

$$p(t) = [V_m \sin(\omega t + \theta_1)][I_m \sin(\omega t + \theta_2)]$$

$$= V_m I_m \sin(\omega t + \theta_1) \cdot \sin(\omega t + \theta_2)$$

The trigonometric identity

$$\sin A \cdot \sin B = \tfrac{1}{2}[\cos(A - B) - \cos(A + B)]$$

leads to the following expression for the instantaneous power consumed by the element

$$P(t) = \tfrac{1}{2} V_m I_m [\cos(\theta_1 - \theta_2) - \cos(2\omega t + \theta_1 + \theta_2)]$$

The *average power* consumed is obtained by integrating this expression over one cycle. Since the second term in the expression is a sinusoid of frequency 2ω, its average value is zero, and it may be ignored. The average power, P, is therefore

$$P = \tfrac{1}{2} V_m I_m \cos(\theta_1 - \theta_2) = VI \cos\phi \quad \text{W}$$

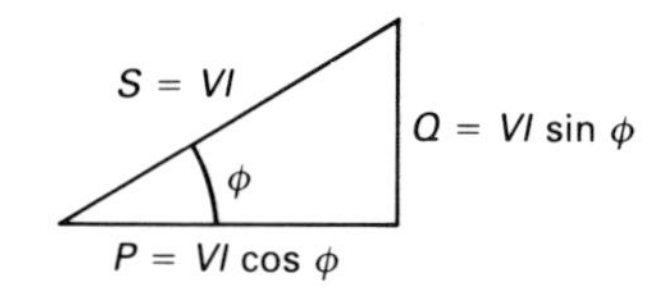

Figure 5.5 *The power triangle.*

where V and I are the r.m.s. value of the voltage across and the current in the element, respectively. The angle ϕ is the phase angle between the voltage and the current.

The situation in an a.c. circuit is illustrated by the *power triangle* in figure 5.5. The *average power* (or actual power), P, absorbed by the circuit or element is

$$P = VI \cos \phi \quad \text{W}$$

As measured on an ammeter and a voltmeter, the *apparent power* absorbed by the circuit is

$$S = VI \text{ volt-amperes (VA)}$$

In the above equation for power, $\cos \phi$ is known as the *power factor* of the circuit, where

$$\text{power factor} = \cos \phi = \frac{P}{VI} = \frac{\text{power}}{\text{volt amperes}}$$

$$= \frac{\text{power}}{\text{apparent power}}$$

Since $\cos \phi$ has a positive mathematical sign for either $+\phi$ or $-\phi$, it is customary to define the power factor as follows. In engineering it is usual to think of the voltage as being the reference phasor, and to state whether I lags or leads V. If I lags behind V, we say that the circuit has a *lagging power factor* and, if I leads V we say that the circuit has a *leading power factor*.

The third side of the power triangle is the *reactive VA* (VAr) or *reactive power*, Q, as follows

$$Q = VI \sin \phi \text{ volt-amperes reactive (VAr)}$$

and, finally

$$S = \sqrt{(P^2 + Q^2)} \text{ VA}$$

5.14 Power, VA and VAr absorbed by ideal elements

In the following we will assume that the r.m.s. voltage across an element is V, the current in it is I, and the phase angle quoted assume that V lies in the *reference direction*.

Resistors

In this case $\phi = 0°$, hence

$$P_R = VI \cos 0° = VI$$

$$S_R = VI \text{ VA}$$

$$Q_R = VI \sin 0° = 0$$

or $P_R = S_R$, and *an ideal resistor does not consume reactive VA*. Also $\mathbf{V} = \mathbf{I}R$ so that

$$P_R = VI = V \times V/R = V^2/R$$

and

$$P_R = VI = IR \times R = I^2R$$

Inductors

In this case $\phi = 90°$ (lagging) so that

$$P_L = VI \cos 90° = 0$$

$$S_L = VI$$

$$Q_L = VI \sin 90° = VI \text{(lagging)}$$

Also $\mathbf{V} = \mathbf{I}X_L$ so that

$$S_L = VI = V \times V/X_L = V^2/X_L$$

and

$$S_L = VI = IX_L \times I = I^2X_L$$

hence $S_L = Q_L$, and *an ideal inductor does not absorb power*.

Capacitors

In a capacitor $\phi = 90°$ (leading), and

$$P_C = VI \cos 90° = 0$$

$$S_C = VI$$

$$Q_C = VI \sin 90° = VI(\text{leading})$$

hence $S_C = Q_C$, and *an ideal capacitor does not absorb power*. Also $\mathbf{V} = \mathbf{I}X_C$ so that

$$S_C = VI = V \times V/X_C = V^2/X_C$$

and

$$S_C = VI = IX_C \times I = I^2X_C$$

5.15 *v–i* waveforms

From work in earlier sections we know that, in a resistor, the voltage and current are in phase with one another. These waveforms, together with that of the *vi* product are shown in diagrams (a) and (b), respectively, of figure 5.6. The instantaneous value of the *vi* product wave is either positive or zero at all times, and the average power consumed is finite. The reader will note that the *v–i* wave is a sinusoid of twice the supply frequency (as it is for the other cases considered here).

In the case of a pure inductor, the current lags behind the voltage by 90° as shown in figure 5.6(c). The corresponding *v–i* product wave is drawn in diagram (d), and its average value over the complete cycle is zero! This confirms the work in section 5.14, where we showed that the average power consumed by an inductor is zero. The physical reason is that the energy absorbed by the inductor during the period when the magnetic field is being built up is returned to the supply when the magnetic field collapses.

Similarly for a pure capacitor, the average power consumed is zero (see diagrams (e) and (f) of figure 5.6).

When there is a phase difference of ϕ between the *v* and *i* waveforms, as shown in figure 5.7(a), the corresponding *v–i* product wave is as shown in diagram (b). In this case, more energy is consumed by the circuit than is returned to the supply, resulting in a net average power consumption by the circuit, which is

$$P = VI \cos \phi \quad \text{W}$$

– see also section 5.13.

5.16 Power consumed in an a.c. circuit

We will use the information in the previous sections to evaluate the apparent power, the power and the reactive power absorbed in each element in figure 5.8.

The impedance of the parallel section of the circuit is

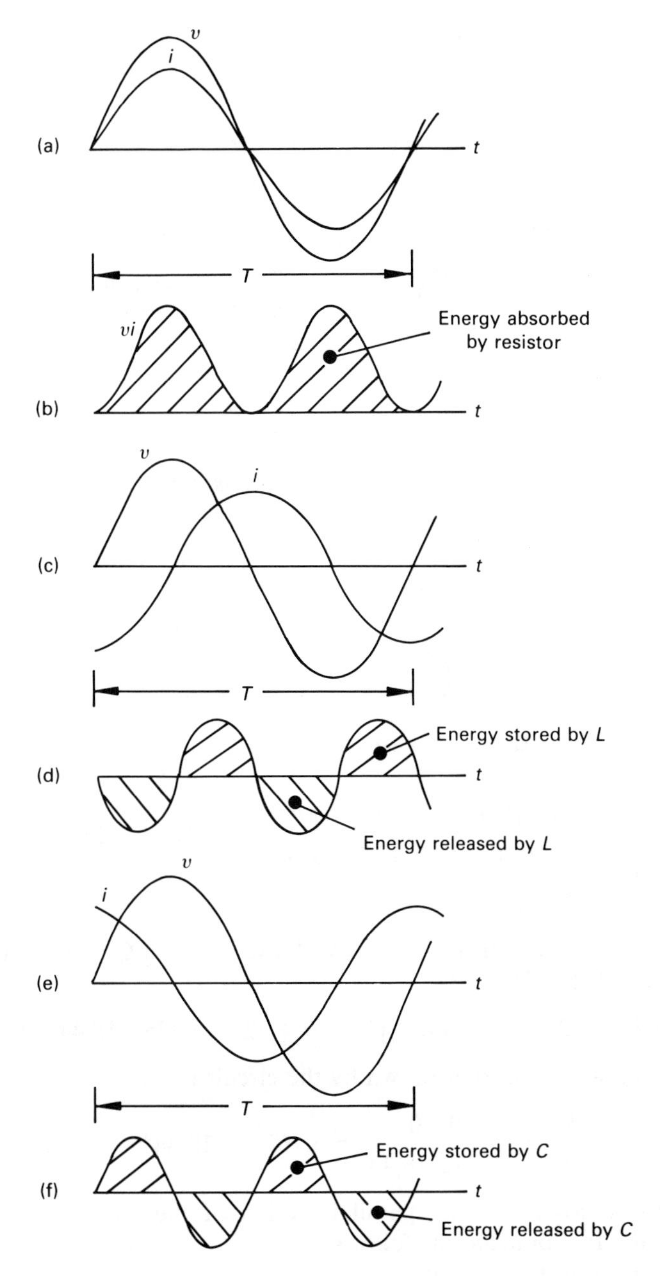

Figure 5.6 v, i *and* p *waveforms (a) and (b) for a pure resistor, (c) and (d) for a pure inductor and (e) and (f) for a pure capacitor.*

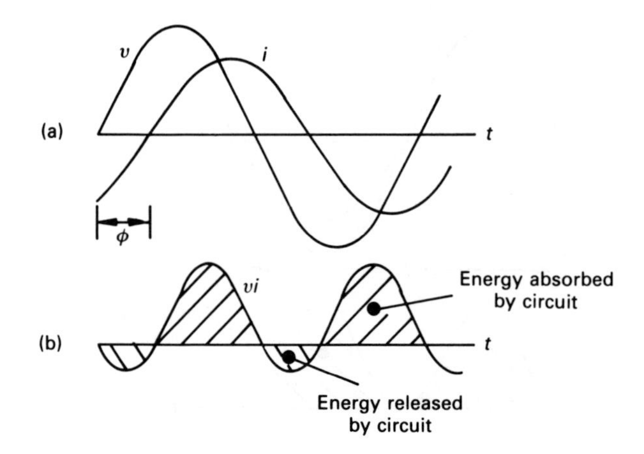

Figure 5.7 *Waveforms for* v, i *and* v–i *product when* i *lags behind* v *by* ϕ.

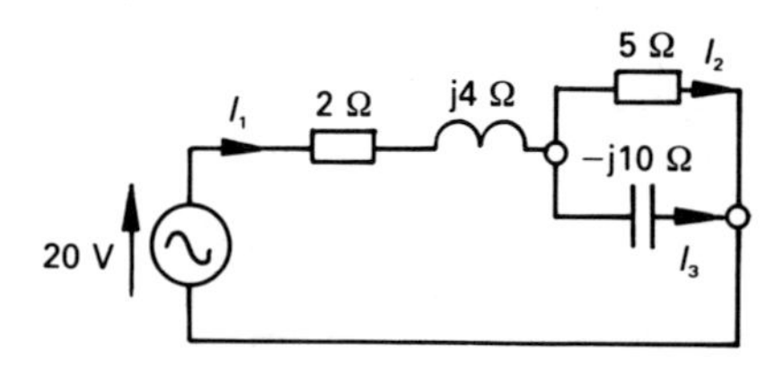

Figure 5.8 *Figure for the example in section 5.16.*

$$\boldsymbol{Z}_{\mathbf{P}} = \frac{5(-\mathrm{j}10)}{5 - \mathrm{j}10} = \frac{50\angle -90^\circ}{11.18\angle -63.43^\circ} = 4.472\angle -26.57^\circ$$

$$= 4 - \mathrm{j}2\ \Omega$$

Since the parallel circuit is in series with the (2 + j4) Ω section, the total impedance of the circuit is

$$\boldsymbol{Z}_{\mathbf{E}} = (2 + \mathrm{j}4) + (4 - \mathrm{j}2) = 6 + \mathrm{j}2 = 6.325\angle 18.44^\circ\ \Omega$$

By Ohm's law, the current drawn by the circuit is

$$\boldsymbol{I}_1 = \frac{\boldsymbol{V}_{\mathbf{S}}}{\boldsymbol{Z}_{\mathbf{E}}} = \frac{20\angle 0^\circ}{6.325\angle 18.44^\circ} = 3.162\angle -18.44^\circ = 3 - \mathrm{j}1\ \mathrm{A}$$

This current flows into the parallel section of the circuit, and divides between the two branches as follows.

The current in the capacitor is

$$\boldsymbol{I}_3 = \boldsymbol{I}_1 \frac{\boldsymbol{Z}_{\mathbf{P}}}{\boldsymbol{Z}_{\mathbf{C}}} = 3.162\angle -18.44^\circ \frac{4.472\angle -26.57^\circ}{10\angle -90^\circ}$$

$$= 1.414\angle 45^\circ = 1 + \mathrm{j}1\ \mathrm{A}$$

Applying KCL to the parallel circuit gives

$$\boldsymbol{I}_2 = \boldsymbol{I}_1 - \boldsymbol{I}_3 = (3 - \text{j}1) - (1 + \text{j}1) = 2 - \text{j}2$$
$$= 2.828\angle -45° \text{ A}$$

The power consumed by the 2 Ω resistor is

$$I_1^2 R = 3.162^2 \times 2 = 20 \text{ W}$$

and the power consumed by the 5 Ω resistor is

$$I_1^2 R = 2.828^2 \times 5 = 40 \text{ W}$$

Neither of the resistors consumes reactive VA. Also, since the inductor and the capacitor are both pure elements, they do not consume any power. The reactive VA consumed by the inductor is

$$Q_L = I_1^2 X_L = 3.162^2 \times 4 = 40 \text{ VAr lagging}$$

and that consumed by the capacitor is

$$Q_C = I_3^2 X_C = 1.414^2 \times 10 = 20 \text{ VAr leading}$$

It should be pointed out here that the current through the inductor lags behind the voltage across it by 90°, and that the current through the capacitor leads the voltage across it by 90°. That is, Q_L and Q_C have opposite mathematical signs! If we assign a positive sign to Q_L, then we must assign a negative sign to Q_C, and the total VAr consumed by the circuit is

$$Q = 40 - 20 = 20 \text{ VAr (lagging)}$$

The mathematical sign associated with Q_L and Q_C is discussed further in section 5.17.

5.17 Complex power

If a voltage $\boldsymbol{V} = V\angle\alpha$ results in a current $\boldsymbol{I} = I\angle\beta$ in a circuit, then the average power P absorbed is

$$P = VI \cos(\alpha - \beta)$$

Using complex number nomenclature, we may say that

$$P = VI \,\text{Re}[\text{e}^{\text{j}(\alpha - \beta)}] = \text{Re}[V\text{e}^{\text{j}\alpha} I\text{e}^{-\text{j}\beta}]$$

The first term within the final set of brackets can be recognised as the phasor voltage. However, since the current in the circuit is

$$\boldsymbol{I} = I\angle\beta = I\text{e}^{\text{j}\beta}$$

the second term in the square brackets is not the current phasor. To correct this, we make use of conjugate notation (see chapter 15 for details) as follows

$$\boldsymbol{I}^* = I\mathrm{e}^{-\mathrm{j}\beta}$$

where $\boldsymbol{I}^*$ is the *complex conjugate* of the current. That is

$$P = \mathrm{Re}[\boldsymbol{VI}^*]$$

and we can let the apparent power become complex by defining it as

$$\boldsymbol{S} = \boldsymbol{VI}^* = VI\mathrm{e}^{\mathrm{j}(\alpha - \beta)} = VI\mathrm{e}^{\mathrm{j}\phi} = P + \mathrm{j}Q$$

where P is the power absorbed by the circuit, Q is the reactive power, and the apparent power is $|S|$. The power factor is given by $\cos \phi$.

If the current lags behind the voltage, ϕ has a negative sign, with the result that the complex part of $\boldsymbol{VI}^*$ has a positive value. By international agreement, *complex lagging VA is given a positive sign* and leading VA is given a negative sign.

We will now apply this method of calculation to the problem in section 5.16 in which $\boldsymbol{V}_S = 20\angle 0°$ V, $\boldsymbol{I}_1 = 3.162\angle -18.44°$ A, $\boldsymbol{I}_2 = 2.828\angle -45°$ A and $\boldsymbol{I}_3 = 1.414\angle 45°$ A. The complex power absorbed by the circuit is

$$\begin{aligned}\boldsymbol{S} &= \boldsymbol{V}_S\boldsymbol{I}_1^* = 20\angle 0° \times 3.162\angle 18.44° = 63.24\angle 18.44° \\ &= 60 + \mathrm{j}20 \text{ VA}\end{aligned}$$

That is, the circuit absorbs 63.24 VA, 60 W and 20 VAr (lagging). Referring to the solution in section 5.16, the reader will note that the 2 Ω resistor consumes 20 W, and the 5 Ω resistor consumes 40 W, giving a total power consumption of 60 W. Moreover, the inductor receives 40 VAr (lagging), and the capacitor receives 20 VAr (leading). In the notation adopted in this method of calculation these respectively correspond to +40 VAr and −20 VAr, giving a total reactive power consumption of +20 VAr, as calculated above.

Unworked problems

5.1. A sinusoidal voltage waveform has a value of −45 V at $t = 0$ and is rising; it reaches zero volts at $t = 0.104$ ms, and has its first positive peak at $t = 0.4165$ ms. Calculate the frequency and the periodic time of the wave, and also its maximum value. Write down an expression for the voltage waveform. If this voltage produces a current of 10 cos $(\omega t - 135°)$ A, calculate the phase angle of $i(t)$; with respect to $v(t)$. [800 Hz; 1.25×10^{-3} s; 90 V; 90 sin $(1600\pi t - 30°)$ V; $v(t)$ leads $i(t)$ by 15°]

5.2. The voltage induced in a coil of an electrical generator increases at a constant rate from zero volts at zero radians to V_m at α radians, then remains constant from α to $(\pi - \alpha)$, after which it decreases at a constant rate until it reaches zero volts at π radians. This is repeated with the reverse polarity in the second half of the wave. Calculate the r.m.s. value and the average value of the wave for a value of α of 0, $\pi/6$ and $\pi/2$.

α(rad)	0	$\pi/6$	$\pi/2$
r.m.s.	V_m	$V_m\sqrt{(7/9)}$	$V_m\sqrt{(1/3)}$
average	V_m	$5V_m/6$	$V_m/2$

5.3. A sinusoidal current of 20 A (r.m.s.) at a frequency of 50 Hz is added to a current of the same magnitude whose frequency is 60 Hz. Write down an expression for the instantaneous current in the circuit, and calculate its value at $t = 10$ ms.
[28.28(sin 100 πt + sin 120 πt); -16.62 A]

5.4. Convert the following values into their rectangular complex form:
(a) $20\angle 150°$,
(b) $9\angle 10° - 10\angle 110°$,
(c) $(5 - j3) \times (6 + j2)$,
(d) $5\angle 200°/7\angle 90°$,
(e) $8.7\angle 10° + (8 - j12)$.
[(a) $-17.32 + j10$; (b) $12.28 - j\,7.83$; (c) $36 - j8$;
(d) $-0.244 + j\,0.671$; (e) $16.57 - j\,10.49$]

5.5. Convert the following complex values into polar form: (a) $-6 - j8$, (b) $(7 + j6) + (-8 - j10)$, (c) $(-7 + j9)/(5 + j3)$, (d) $4\angle -170° + (2 - j3)$.
[(a)$10\angle -126.9°$; (b)$4.123\angle -104°$; (c)$1.96\angle -96.9°$;
(d)$4.173\angle -117.7°$]

5.6. A two-terminal black box contains (a) a 50 Ω resistor, (b) a 50 μF capacitor, (c) a 0.5 H inductor. If the current flowing into the box is $5e^{j(1000t - 30°)}$ A, calculate the voltage between the terminals of the box in each case.
[(a) $250e^{j(1000t - 30°)}$ V; (b) $250e^{j(1000t - 120°)}$ V; $2500e^{j(1000t + 60°)}$ V]

5.7. Three wires are connected to a node in a circuit; the wires carry currents i_1, i_2 and i_3, respectively. (a) If $i_1(t) = 10 \sin(300t + 40°)$ A and $i_2(t) = 15 \cos(300t - 60°)$ A, calculate $i_3(t)$; (b) if $\boldsymbol{I}_1 = 15\angle -20°$ A and $\boldsymbol{I}_3 = 20\angle 40°$ A, calculate $\boldsymbol{I}_2$.
((a) $24.9 \sin(300t - 146°)$ A; (b) $30.42\angle -165.3°$ A]

5.8. If $v_{AB}(t) = 50 \sin(50t - 30°)$ mV, $v_{BC}(t) = 25 \sin(50t + 40°)$ mV and $v_{AD}(t) = 90 \sin(50t - 10°)$ mV, calculate $\boldsymbol{V}_{CD}$ and $\boldsymbol{V}_{BD}$.

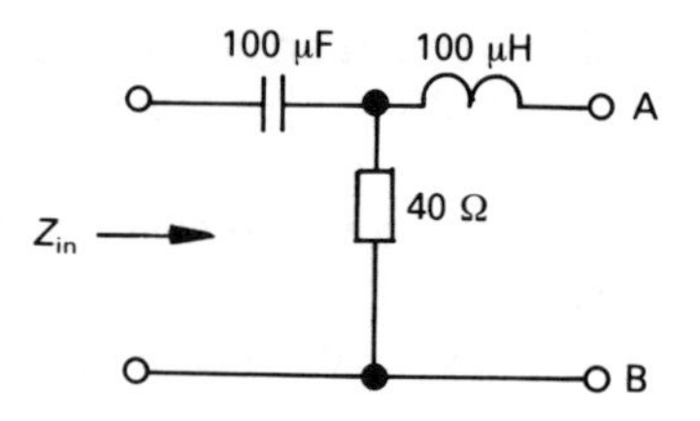

Figure 5.9

[19.1∠165.6° mV (*note*: $\boldsymbol{V}_{DC}$ = 19.1∠−14.4° mV); 32.73∠−168.3° mV]

5.9. If the supply frequency to the circuit in figure 5.9 is 400 rad/s, calculate the complex input impedance $\boldsymbol{Z}_{in}$ if terminals AB are (a) open-circuited, (b) short-circuited, (c) connected by a 50 Ω resistor. [(a) 47.17∠−32° Ω; (b) 20.62∠−14° Ω; (c) 31.17∠−36.2° Ω]

5.10. A 1 H inductance, a 1 μF capacitance and a resistor *R* are connected in parallel to a supply voltage of frequency 500 rad/s. Express the real and imaginary parts of the impedance of the parallel circuit in terms of *R*.
[Real part = $R \times 10^6/(10^6 + 2.25R^2)$; imaginary part = $1500R/(10^6 + 2.25R^2)$]

5.11. What value of inductance must be connected in series with a device whose internal circuit is equivalent to a resistance of 5000 Ω in series with a capacitance of 0.1 μF, to give an overall circuit admittance of (51.42 + j87.4) μS. The supply frequency is 1000 rad/s.
[1.5 H]

5.12. A series circuit contains a resistance of 5 Ω, an inductor of reactance 10 Ω and a capacitor of reactance 6 Ω. If the r.m.s. voltage of the supply is 100 V, calculate (a) the complex impedance and admittance of the circuit, (b) the current drawn by the circuit, (c) the phase angle and power factor of the circuit and (d) the voltage across each element in the circuit.
[(a) 6.4∠38.66° Ω, 0.156∠−38.66° S; (b) 15.63∠−38.66° A; (c) $\boldsymbol{I}$ lags $\boldsymbol{V}$ by 38.66°, power factor = 0.781 (lagging); (d) $\boldsymbol{V}_R$ = 78.13∠−38.66° V; $\boldsymbol{V}_L$ = 156.3°∠51.3° V, $\boldsymbol{V}_C$ = 93.75∠−128.7° V]

5.13. In problem 5.12, calculate the apparent power, the power and the VAr consumed. Calculate also the VAr consumed by the inductor and the capacitor.
[1563 VA; 1220 W; 976.4 VAr (lagging); 2443 VAr (lagging); 1466.6 VAr (leading)]

5.14. A resistor and a coil are connected in series to a 240 V supply. If the current drawn from the supply lags the supply voltage by 37°, and its r.m.s. value is 3 A, calculate (given that the r.m.s. voltage across the coil is 171.4 V) the resistance and reactance of the inductor, and the resistance of the resistor. Calculate also the power and VAr consumed by the circuit.
[30.74 Ω, 48.15 Ω; 33.26 Ω; 575 W; 433 VAr (lagging)]

5.15. A 3-branch parallel circuit contains a 4 kΩ resistor in one branch, a 5 kΩ inductive reactance in the second branch and a 7 kΩ capacitive reactance in the third branch. Calculate the effective admittance and impedance of the circuit. If it is energised by a 200 V r.m.s. supply, calculate the current in each branch, the total current drawn by the circuit and the phase angle of the circuit. Determine also the apparent power, the power and the VAr consumed, together with the power factor of the circuit.
[$0.2564\angle -12.87°$ mS, $3.9\angle 12.87°$ kΩ; $\boldsymbol{I}_1 = 0.05\angle 0°$ A; $\boldsymbol{I}_2 = 0.04\angle -90°$ A; $\boldsymbol{I}_3 = 0.0286\angle 90°$ A; −12.87°; 10.26 VA; 10 W; 2.29 VAr (lagging); 0.975 (lagging)]

5.16. A two-branch parallel circuit contains impedances of (8 − j7) Ω and (5 + j6) Ω, respectively. If the current in the (8 − j7) Ω branch is $9.41\angle 41.2°$ A, calculate the current in the other branch, and the supply voltage. Also evaluate the complex admittance of the circuit. What is the phase angle of the circuit? What apparent power, power and VAr are consumed?
[$12.8\angle -50.1°$ A; $100\angle 0°$ V; $0.157\angle -13.41°$ S; 13.41° (lagging); 1570 VA; 1527.2 W; 364.1 VAr (lagging)]

6

Sinusoidal Steady-state Analysis

6.1 Introduction

This chapter not only deals with a.c. applications of techniques dealt with in the work on d.c. circuits, but deals with other circuit theorems particularly appropriate to a.c. circuit analysis. In particular, the application of complex numbers to electric circuits; the theory of complex numbers is covered in chapter 15, where the reader can study this and other mathematical techniques appropriate to electrical and electronic engineering.

We commence by analysing the circuit in figure 6.1(a), which is specified in terms of time-domain data. Following the concepts developed in earlier chapters, the frequency-domain equivalent circuit is drawn in diagram (b).

We shall retain ohmic values in kΩ, so that the resulting currents are in mA. The impedance of the series branch is

$$\mathbf{Z_S} = 2 + \mathrm{j}1\ \mathrm{k}\Omega$$

and the impedance of the parallel branch of the circuit is

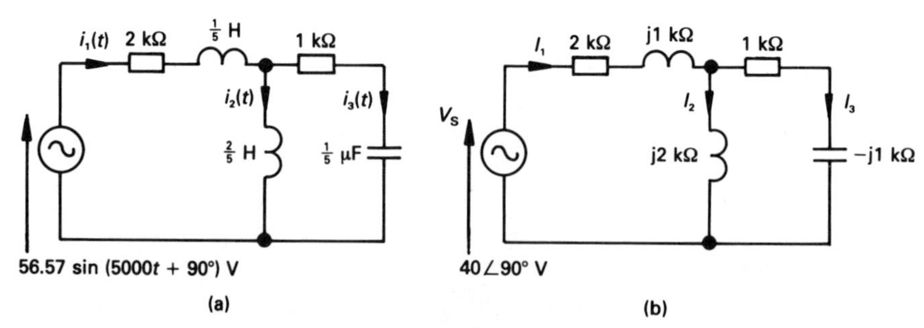

Figure 6.1 *The circuit in diagram (a) contains time-domain data which, in diagram (b), is converted into its corresponding frequency-domain equivalent.*

$$Z_P = \frac{j2(1-j1)}{j2+(1-j1)} = \frac{j2(1-j1)}{1+j1} = \frac{2\angle 90° \times 1.414\angle -45°}{1.414\angle 45°}$$
$$= 2\angle 0°\ \text{k}\Omega$$

Since the two branches are in series with one another, the effective impedance of the complete circuit is

$$Z_E = Z_S + Z_P = (2+j1) + (2+j0) = 4+j1$$
$$= 4.123\angle 14.04°\ \text{k}\Omega$$

By Ohm's law, the current I_1 drawn by the circuit is

$$I_1 = V_S/Z_E = 40\angle 90°/4.123\angle 14.04° = 9.7\angle 75.96°\ \text{mA}$$

The current in each branch of the parallel circuit may be evaluated by any of the methods described earlier, typical calculations being

$$I_3 = I_1 \frac{Z_2}{Z_2+Z_3} = 9.7\angle 75.96° \times \frac{j2}{j2+(1-j1)}$$
$$= 9.7\angle 75.96° \times \frac{2\angle 90°}{1.414\angle 45°} = 13.72\angle 120.96°\ \text{mA}$$

and

$$I_2 = I_1 \frac{Z_3}{Z_2+Z_3} = 9.7\angle -14.04°\ \text{mA}$$

The power, VA and VAr in the various sections of the circuit can be calculated by any one of the methods outlined in chapter 5.

Since types of a.c. analysis deal with circuit equations involving complex numbers, we will be concerned with the solution of simultaneous equations involving complex quantities. Generally speaking, their solution is no more difficult than solving conventional simultaneous equations, with the exception that we also need to deal with phase angles as well as magnitude. That is, the solution takes a little longer, and needs a little more care. Consider the solution of the following simultaneous equations

$$50\angle 0° = 11.18\angle -26.56° X + 5\angle 90° Y \qquad (6.1)$$
$$0 = 5\angle 90° X + 3.162\angle -18.43° Y \qquad (6.2)$$

where X and Y are unknowns. To eliminate variable Y from the equations, we need to make the coefficient of Y have the same value in both equations; that is *it must have the same magnitude and phase in both equations*. We then add (or subtract) the equations, leaving only the variable X, whose value can then be determined. Full details of the theory of complex number manipulation is given in chapter 15.

If we multiply equation (6.2) by the complex value $\frac{5\angle 90°}{3.162\angle -18.43°} = 1.581\angle 108.43°$, it makes the coefficient Y in the new equation $5\angle 90°$. When the new equation is subtracted from equation (6.1), Y is eliminated as follows

$$50\angle 0° = 11.18\angle -26.56° X + 5\angle 90° Y \qquad \text{((6.1) repeated)}$$

$$\underline{0 \quad = \quad 7.91\angle 198.43° X + 5\angle 90° Y} \qquad ((6.2) \times 1.581\angle 108.43°)$$

$$50\angle 0° = 17.68\angle -8.21° X \qquad \text{(SUBTRACT the equations)}$$

To perform the subtraction it is first necessary to convert both equations into rectangular form, as described in chapter 15. The value of X is

$$X = \frac{50\angle 0°}{17.68\angle -8.21°} = 2.83\angle 8.12°$$

Substituting this value into equation (6.2) gives

$$Y = \frac{-5\angle 90° \times 2.83\angle 8.12°}{3.162\angle -18.43°} = \frac{-14.15\angle 98.12°}{3.162\angle -18.43°}$$

$$= -4.47\angle 116.55° = 4.47\angle -63.45°$$

Alternatively, the equations can be solved by determinants which, in the opinion of the author, is an easier and more reliable routine. As with other mathematical processes, this is dealt with in chapter 15.

Yet another method of solving a pair of simultaneous equations with complex values in polar form is given by the computer program in listing 6.1.

The main program is in lines 10 to 610, inclusive; the input of data is complete by line 290, and lines 310 to 590, inclusive, make computations on each of the determinants. Two subroutines are used; the one commencing at line 1000 calculates the magnitude and phase angle of the determinants, and that commencing at line 2000 keeps the final phase angle within the limits ±180°.

Listing 6.1
Solution of two simultaneous equations having complex values.

```
10  CLS
20  PRINT TAB(3); "Solution of two complex simultaneous equations"
30  PRINT TAB(18); "of the form"
40  PRINT TAB(15); "V1 = A*X + B*Y"
50  PRINT TAB(15); "V2 = C*X + D*Y"
60  PRINT TAB(3); "Where V1 and V2 are complex values,"
70  PRINT TAB(3); "A,B,C and D are complex coefficients,"
80  PRINT TAB(3); "and X and Y are the variables."
```

```
90  PRINT
100 INPUT "Magnitude of V1 = ", V1
110 INPUT "Phase of V1 (degrees) = ", P1
120 REM ** Convert degrees to radians **
130 P1 = P1 * .01745
140 INPUT "Magnitude of A = ", A
150 INPUT "Phase of A (degrees) = ", PA
160 PA = PA * .01745
170 INPUT "Magnitude of B = ", B
180 INPUT "Phase of B (degrees) = ", PB
190 PB = PB * .01745
200 PRINT
210 INPUT "Magnitude of V2 = ", V2
220 INPUT "Phase of V2 (degrees) = ", P2
230 P2 = P2 * .01745
240 INPUT "Magnitude of C = ", C
250 INPUT "Phase of C (degrees) = ", PC
260 PC = PC * .01745
270 INPUT "Magnitude of D = ", D
280 INPUT "Phase of D (degrees) = ", PD
290 PD = PD * .01745
300 PRINT
310 REM ** Calculations for main determinant **
320 Mag = A * D: PH = PA + PD
330 Re1 = Mag * COS(PH): Im1 = Mag * SIN(PH)
340 Mag = C * B: PH = PC + PB
350 Re2 = Mag * COS(PH): Im2 = Mag * SIN(PH)
360 GOSUB 1000
370 Det = Mag: DetP = PH * 57.3066
380 REM ** Calculations for determinant X **
390 Mag = V1 * D: PH = P1 + PD
400 Re1 = Mag * COS(PH): Im1 = Mag * SIN(PH)
410 Mag = V2 * B: PH = P2 + PB
420 Re2 = Mag * COS(PH): Im2 = Mag * SIN(PH)
430 GOSUB 1000
440 DetX = Mag: DetXP = PH * 57.3066
450 REM ** Calculations for determinant Y **
460 Mag = A * V2: PH = PA + P2
470 Re1 = Mag * COS(PH): Im1 = Mag * SIN(PH)
480 Mag = V1 * C: PH = P1 + PC
490 Re2 = Mag * COS(PH): Im2 = Mag * SIN(PH)
500 GOSUB 1000
510 DetY = Mag: DetYP = PH * 57.3066
520 REM ** There is no solution of Det = 0 **
530 IF Det = 0 THEN PRINT TAB(3); "The equations cannot be solved": END
540 REM ** Calculate and print the value of the variables **
550 PRINT TAB(3); "X = "; DetX / Det
560 Phi = DetXP - DetP: GOSUB 2000
```

```
570 PRINT TAB(9); "at angle "; Phi; " degrees."
580 PRINT TAB(3); "Y = "; DetY / Det
590 Phi = DetYP - DetP: GOSUB 2000
600 PRINT TAB(9); "at angle "; Phi; " degrees."
610 END
1000 REM ** Calculate magnitude and phase of determinant **
1010 Re = Re1 - Re2: Im = Im1 - Im2
1020 Mag = SQR(Re ^ 2 + Im ^ 2): pi = 3.141593
1030 IF Re = 0 AND Im > 0 THEN PH = pi / 2: RETURN
1040 IF Re = 0 AND Im < 0 THEN PH = -pi / 2: RETURN
1050 PH = ATN(Im / Re)
1060 REM ** Compensation for angular calculation **
1070 IF Re < 0 AND Im < 0 THEN PH = PH - pi
1080 IF Re < 0 AND Im > 0 THEN PH = PH + pi
1090 IF Re < 0 AND Im = 0 THEN PH = pi
1100 RETURN
2000 REM ** Modify phase angle to within 180 deg **
2010 IF Phi > 180 THEN Phi = Phi - 360: RETURN
2020 IF Phi < -180 THEN Phi = Phi + 360: RETURN
2030 RETURN
```

6.2 Nodal, mesh and loop analysis

The technique involved for each of these methods of analysis is generally similar to that described in chapter 2 for resistive circuits, with the exception that complex values are used.

We will illustrate nodal and mesh analysis using simple problems. When a circuit contains a mixture of voltage and current sources it may be necessary to modify the circuit equations using supermeshes (when using mesh analysis) or supernodes (when using nodal analysis). Where dependent sources are involved the techniques are, once again, similar to those outlined for resistive circuits.

As with d.c. circuits, loop analysis can be used to evaluate loop currents in any circuit, but has the disadvantage that it is less systematic than mesh analysis, and is not generally so convenient for computer solution.

Worked example 6.2.1

In the first example in this section, we apply *nodal analysis* to the solution of the circuit in figure 6.2.

Solution

In this case we take node 0 to be the reference node. Applying KCL to node 1 we get

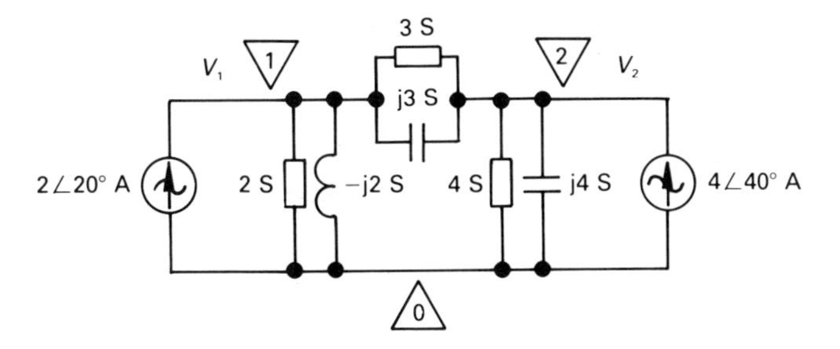

Figure 6.2 *Circuit for worked example 6.2.1.*

$$\begin{aligned} 2\angle 20° &= ((2 + 3) + j(-2 + 3))\boldsymbol{V}_1 - (3 + j3)\boldsymbol{V}_2 \\ &= (5 + j1)\boldsymbol{V}_1 - (3 + j3)\boldsymbol{V}_2 \\ &= 5.1\angle 11.31°\boldsymbol{V}_1 - 4.24\angle 45°\boldsymbol{V}_2 \end{aligned} \tag{6.3}$$

The application of KCL to node 2 yields

$$\begin{aligned} 4\angle 40° &= -(3 + j3)\boldsymbol{V}_1 + ((4 + 3) + j(3 + 4))\boldsymbol{V}_2 \\ &= -(3 + j3)\boldsymbol{V}_1 + (7 + j7)\boldsymbol{V}_2 \\ &= -4.242\angle 45°\boldsymbol{V}_1 + 9.9\angle 45°\boldsymbol{V}_2 \end{aligned} \tag{6.4}$$

$\boldsymbol{V}_2$ can be eliminated between these equations by multiplying equation (6.3) by $9.9\angle 45°/4.24\angle 45° = 2.335\angle 0°$, and adding it to equation (6.4) as follows.

$$\begin{array}{rll} 4.67\angle 20° & = 11.91\angle 11.31°\boldsymbol{V}_1 - 9.9\angle 45°\boldsymbol{V}_2 & ((6.3) \times 2.335) \\ 4\angle 40° & = -4.242\angle 45°\boldsymbol{V}_1 + 9.9\angle 45°\boldsymbol{V}_2 & ((6.4)\text{ repeated}) \\ \hline 8.45\angle 29.22° & = 8.7\angle -4.37°\boldsymbol{V}_1 & (\text{ADD the equations}) \end{array}$$

Hence

$$\boldsymbol{V}_1 = \frac{8.45\angle 29.22°}{8.7\angle -4.37°} = 0.982\angle 33.6°\text{ V}$$

Substituting this value into equation (6.3) gives

$$\boldsymbol{V}_2 = \frac{(5.1\angle 11.31° \times 0.982\angle 33.6°) - 2\angle 20°}{4.24\angle 45°}$$

$$= 0.778\angle 14.72°\text{ V}$$

This calculation involves several polar-to-rectangular and rectangular-to-polar conversions, which are not shown. A simpler method may be to use determinants, or the computer program in listing 6.1.

In examples which follow, details of the solution of simultaneous equations is omitted.

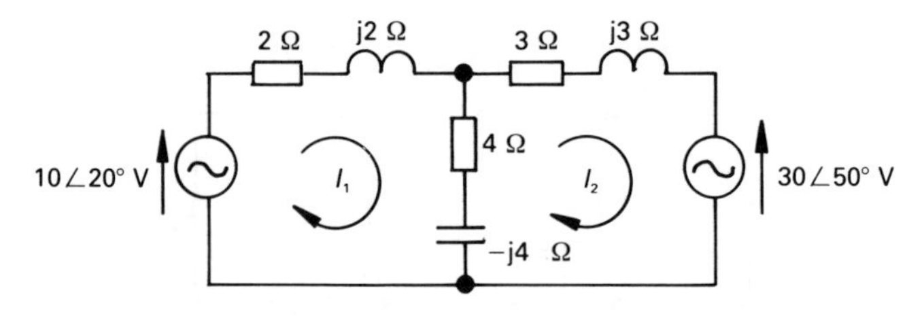

Figure 6.3 *Mesh current analysis.*

Worked example 6.2.2

In this problem we will use *mesh analysis* to determine the currents in figure 6.3.

Solution

Applying KVL to mesh 1 yields

$$\begin{aligned}10\angle 20^\circ &= ((2 + \mathrm{j}2) + (4 - \mathrm{j}4))\boldsymbol{I}_1 - (4 - \mathrm{j}4)\boldsymbol{I}_2 \\ &= (6 - \mathrm{j}2)\boldsymbol{I}_1 - (4 - \mathrm{j}4)\boldsymbol{I}_2 \\ &= 6.325\angle -18.44^\circ \boldsymbol{I}_1 - 5.657\angle -45^\circ \boldsymbol{I}_2\end{aligned}$$

and to mesh 2 gives

$$\begin{aligned}-30\angle 50^\circ &= -(4 - \mathrm{j}4)\boldsymbol{I}_1 + ((3 + \mathrm{j}3) + (4 - \mathrm{j}4))\boldsymbol{I}_2 \\ &= -(4 - \mathrm{j}4)\boldsymbol{I}_1 + (7 - \mathrm{j}1)\boldsymbol{I}_2 \\ &= -5.657\angle -45^\circ \boldsymbol{I}_1 + 7.071\angle -8.13^\circ \boldsymbol{I}_2\end{aligned}$$

Solving these equations in the manner described earlier gives

$$\boldsymbol{I}_1 = 2.39\angle 163.4^\circ \text{ A}$$

and

$$\boldsymbol{I}_2 = 3.69\angle -148.6^\circ \text{ A}$$

6.3 Principle of superposition

The principle of superposition can be applied to linear circuits, including a.c. circuits. Consider the calculation of the voltage $\boldsymbol{V}_1$ in the circuit in figure 6.4. The value of this voltage is dependent on the two sources in the circuit, one being a current source and the other a voltage source.

When the circuit is energised by the current source alone (in the meantime, the voltage source is replaced by its internal impedance, that is, zero), the voltage at the left-hand node is $\boldsymbol{V}_{1\mathrm{A}}$. In this case, $2\angle 0^\circ$ A flows

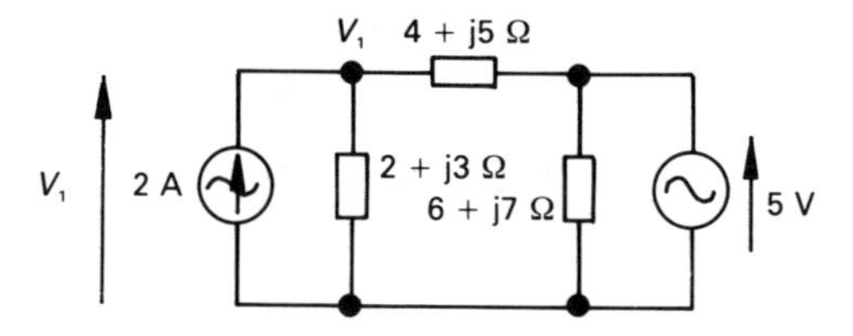

Figure 6.4 *Principle of superposition.*

into a two-branch parallel circuit comprising (2 + j3) Ω in parallel with (4 + j5) Ω, and V_{1A} is

$$V_{1A} = 2\angle 0° \frac{(2 + j3)(4 + j5)}{(2 + j3) + (4 + j5)} = 2.68 + j3.76 \text{ V}$$

When the circuit is energised by the voltage source alone (in the meantime, the current source is replaced by its internal impedance, that is, infinity), the voltage at the left-hand node is now V_{1B}. This voltage is the proportion of the $5\angle 0°$ V source appearing across the (2 + j3) Ω impedance, which is

$$V_{1B} = 5\angle 0° \frac{2 + j3}{(2 + j3) + (4 + j5)} = 1.81 + j0.1 \text{ V}$$

Hence, by superposition

$$V_1 = V_{1A} + V_{1B} = 4.49 + j3.86 = 5.92\angle 40.8° \text{ V}$$

6.4 Thévenin's theorem and Norton's theorem

Thévenin's and *Norton's theorems* (see also chapter 3) may be used for frequency-domain circuits, the only difference being that we need to replace the term *resistance* by *impedance*, and *conductance* by *admittance*. We will illustrate their use in a.c. circuits by way of examples.

Worked example 6.4.1

Determine Thévenin's equivalent circuit with respect to terminals A and B of the network in figure 6.5. Hence calculate the current which would flow in a (6 − j5) Ω impedance connected between A and B.

Solution

Firstly, we must replace the source by its internal impedance (zero in this case), and determine the impedance between terminals A and B as follows.

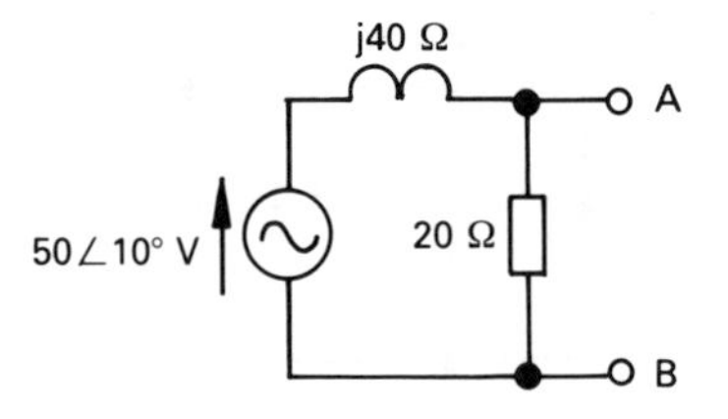

Figure 6.5 *Thévenin's theorem.*

$$Z_{TH} = \frac{20 \times j40}{20 + j40} = 16 + j8\ \Omega$$

Next we calculate the voltage V_{AB}, which is the Thévenin source voltage as follows

$$V_{TH} = V_{AB} = 50\angle 10° \frac{20\angle 0°}{20 + j40} = 22.36\angle -53.44°\ V$$

That is, the Thévenin source between terminals A and B comprises a voltage $V_{TH} = V_{AB} = 22.36\angle -53.44°$ V having an internal impedance of $(16 + j8)\ \Omega$. When the external load is connected, the total impedance in the circuit is

$$Z_E = Z_{TH} + (6 - j5) = 22 + j3 = 22.2\angle 7.77°\ \Omega$$

and the current in the load is

$$I = \frac{V_{TH}}{Z_E} = \frac{22.3\angle -53.44°}{22.2\angle 7.77°} = 1.007\angle -61.21°\ A$$

Worked example 6.4.2

Repeat worked example 6.4.1 using Norton's theorem.

Solution

We can avoid some work simply by saying that the Norton internal admittance is equal to the reciprocal of the Thévenin internal impedance, that is

$$Y_N = 1/Z_{TH} = 1/(16 + j8) = 0.0559\angle -26.56°$$
$$= 0.05 - j0.025\ S$$

The value of the Norton internal source current is equal to the current which flows from A to B when these terminals are short-circuited. That is

$$I_N = 50\angle 10°/j40 = 1.25\angle -80°\ A$$

The Norton equivalent source current could, alternatively, have been deduced by saying

$$\begin{aligned} I_N &= V_{TH}/Z_{TH} = 22.36\angle -53.44°/(16 + j8) \\ &= 1.25\angle -80° \text{ A} \end{aligned}$$

The Norton source current divides between the internal admittance of the source and the admittance of the load, the latter value being

$$Y_L = 1/(6 - j5) = 0.098 + j0.082 \text{ S}$$

The effective admittance of the Norton generator and the load in parallel is

$$\begin{aligned} Y_E &= Y_N + Y_L = (0.05 - j0.025) + (0.098 + j0.082) \\ &= 0.148 + j0.057 = 0.159\angle 21.06° \text{ S} \end{aligned}$$

and the current in the load is

$$I = I_N \frac{Y_L}{Y_E} = 1.25\angle -80° \frac{0.128\angle 39.8°}{0.159\angle 21.06°} = 1\angle -61.26° \text{ A}$$

When comparing the values of I obtained by the two methods of calculation, minor discrepancies are due to rounding errors.

6.5 Millman's theorem

Other circuit theorems, subject to any special conditions discussed earlier, apply equally well in the frequency domain; Millman's theorem (also known as the Parallel Generator theorem) is no exception. This theorem is widely used in electronics, and is well suited to the solution of unbalanced 3-phase, 3-wire power systems (see also chapter 7) as is illustrated below.

Worked example 6.5.1

Consider the system in figure 6.6, in which an *unbalanced load* (see chapter 7 for a description of the term unbalanced) is supplied by an unbalanced three-phase supply. We need to calculate the current in each line, together with the voltage V_{SN}.

Solution

Using the equation developed in chapter 3, the voltage V_{SN} is

$$V_{SN} = \frac{V_{AN}Y_{AS} + V_{BN}Y_{BS} + V_{CN}Y_{CS}}{Y_{AS} + Y_{BS} + Y_{CS}}$$

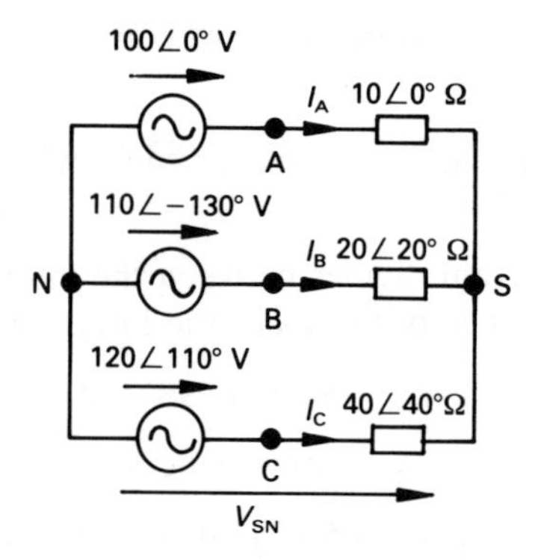

Figure 6.6 *Millman's theorem applied to an unbalanced 3-phase, 3-wire power system.*

$$= \frac{\dfrac{100\angle 0^\circ}{10\angle 0^\circ} + \dfrac{110\angle -130^\circ}{20\angle 20^\circ} + \dfrac{120\angle 110^\circ}{40\angle -40^\circ}}{\dfrac{1}{10\angle 0^\circ} + \dfrac{1}{20\angle 20^\circ} + \dfrac{1}{40\angle -40^\circ}}$$

$$= \frac{2.92\angle -25.35^\circ}{0.166\angle 0.36^\circ} = 17.6\angle -25.71^\circ$$

$$= 15.86 - j7.64 \text{ V}$$

The current in line A is

$$\boldsymbol{I}_A = \frac{\boldsymbol{V}_{AS}}{\boldsymbol{Z}_{AS}} = \frac{\boldsymbol{V}_{AN} - \boldsymbol{V}_{SN}}{\boldsymbol{Z}_{AS}} = \frac{100\angle 0^\circ - 17.6\angle -25.71^\circ}{10\angle 0^\circ}$$

$$= 8.45\angle 5.19^\circ \text{ A}$$

Similarly, it may be shown that

$$\boldsymbol{I}_B = 5.79\angle -158.4^\circ \text{ A and } \boldsymbol{I}_C = 3.33\angle 155.4^\circ \text{ A}$$

6.6 Rosen's, star–delta and delta–star theorems

Rosen's theorem or the general star–mesh transformation is as applicable in the frequency domain as it is to d.c. circuits, as are the star–delta and delta–star transformations.

Worked example 6.6.1

Three impedances are connected in delta as follows. An impedance of (6 + j3) Ω is connected between lines 1 and 2, (7 − j5) Ω is connected between lines 2 and 3, and an impedance of (6 + j0) Ω is connected between lines 3 and 1. Determine the impedances in the equivalent star-connected network.

Solution

Since each of the star-connected impedances is connected to the star-point S, then

$$\mathbf{Z}_{1S} = \frac{\mathbf{Z}_{12}\mathbf{Z}_{13}}{\Sigma \mathbf{Z}}$$

where

$$\Sigma \mathbf{Z} = \mathbf{Z}_{12} + \mathbf{Z}_{23} + \mathbf{Z}_{23}$$

$$= (6 + j3) + (7 - j5) + (6 + j0) = 19 - j2$$

$$= 19.1\angle -6^\circ\ \Omega$$

Hence

$$\mathbf{Z}_{1S} = \frac{(6 + j3)(6 + j0)}{19.1\angle -6^\circ} = \frac{40.25\angle 26.57^\circ}{19.1\angle -6^\circ}$$

$$= 2.11\angle 32.57^\circ\ \Omega$$

Also

$$\mathbf{Z}_{2S} = \mathbf{Z}_{21}\mathbf{Z}_{23}/\Sigma\mathbf{Z} = 3.02\angle -2.96^\circ\ \Omega$$

$$\mathbf{Z}_{3S} = \mathbf{Z}_{31}\mathbf{Z}_{32}/\Sigma\mathbf{Z} = 2.7\angle -29.53^\circ\ \Omega$$

6.7 Maximum power transfer theorem

The maximum power transfer theorem as applied to a.c. circuits is more involved than is the case in d.c. circuits because of the existence of complex impedances and complex voltages and currents. There are four general conditions for maximum power transfer in a.c. networks, and these are given below without proof:

1. A *pure resistance load* extracts maximum power from a circuit when the load resistance is equal to the magnitude of the internal impedance of the source to which it is connected.

2. A *constant reactance, variable resistance load* extracts maximum power when the resistance of the load is equal to the magnitude of the internal impedance of the network plus the reactance of the load.
3. A *constant power factor, variable impedance load* extracts maximum power when the magnitude of the load impedance is equal to the magnitude of the internal impedance of the network.
4. A *variable resistance, variable reactance load* extracts maximum power from a circuit when the impedance of the load is equal to the complex conjugate of the impedance of the network.

While we cannot illustrate each version of the theorem here, we will take a look at the final version.

Worked example 6.7.1

If a load having independently variable resistance and reactance is connected to terminals A and B of the circuit in worked example 6.4.1 (figure 6.5) and extracts maximum power from it, calculate the value of the impedance of the load and the power absorbed.

Solution

The Thévenin equivalent circuit of worked example 6.4.1 is a voltage source of $22.36\angle -53.44°$ V having an internal impedance of $(16 + j8)\ \Omega$. Maximum power is absorbed when the impedance of the load is the complex conjugate of this value, namely

$$\mathbf{Z}_L = \mathbf{Z}_{TH}^* = 16 - j8\ \Omega$$

The effective impedance of the complete circuit is then

$$\mathbf{Z}_E = \mathbf{Z}_{TH} + \mathbf{Z}_L = (16 + j8) + (16 - j8) = 32\ \Omega$$

and the current in the load is

$$\boldsymbol{I} = \mathbf{V}_{TH}/\mathbf{Z}_E = 22.36\angle -53.44°/32 = 0.7\angle -53.44°\ \text{A}$$

The maximum power absorbed by the load is

$$P_{max} = |\mathbf{I}^2|\ R_L = 0.7^2 \times 16 = 7.84\ \text{W}$$

6.8 a.c. circuits with dependent sources

So far we have only looked at circuits containing independent sources. The techniques involved in circuits with dependent sources are generally the same as those in d.c. circuits, with the exception that we are dealing with complex quantities, as shown in the following worked example.

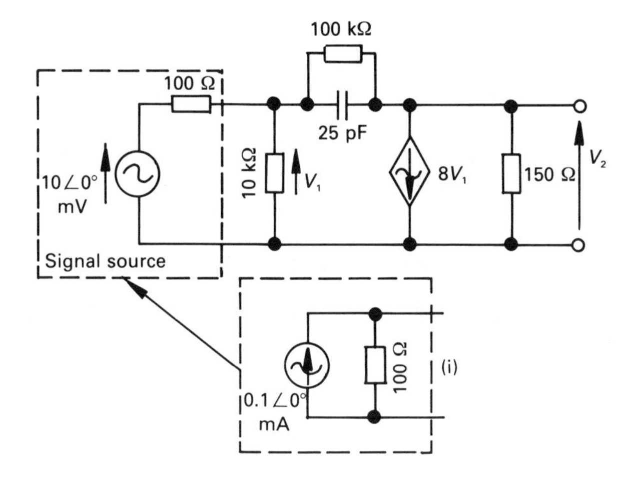

Figure 6.7 *Figure for worked example 6.8.1.*

Worked example 6.8.1

The circuit in figure 6.7 is a simplified circuit of a transistor amplifier. Calculate the voltage gain (V_2/V_1) at a frequency of (a) 1 kHz, (b) 200 kHz.

Solution

One method of solving this problem is by nodal analysis as follows. Initially, the 10 mV source and its associated 100 Ω internal resistor (shown in the broken box in figure 6.7 are converted into a 0.1∠0° mA source which is shunted by a 100 Ω resistor, as shown in insert (i). The 100 Ω resistor is seen to be in parallel with the 10 kΩ resistor, the two being equivalent to a 99 Ω resistor (or a 0.0101 S conductance). The nodal equations for the circuit are as follows

Node 1:

$$0.1 \times 10^{-3}\angle 0^\circ = \left(0.0101 + \frac{1 + \text{j}100 \times 10^3\omega C}{100 \times 10^3} \right) \boldsymbol{V}_1 - \frac{1 + \text{j}100 \times 10^3\omega \text{C}}{100 \times 10^3} \boldsymbol{V}_2$$

$$= (0.01011 + \text{j}\omega C)\boldsymbol{V}_1 - (1 \times 10^{-5} + \text{j}\omega C)\boldsymbol{V}_2$$

Node 2:

$$0 = \left(8 - \frac{1 + j100 \times 10^3 \omega C}{100 \times 10^3}\right) V_1 +$$

$$\left(6.667 \times 10^{-3} + \frac{1 + j100 \times 10^3 \omega C}{100 \times 10^3}\right) V_2$$

$$= (7.9999 - j\omega C)V_1 + (6.677 \times 10^{-3} + j\omega C)V_2$$

(a) When $f = 1$ kHz, $\omega C = 2\pi \times 1000 \times 25 \times 10^{-12} = 1.57 \times 10^{-7}$, and the equations are

$$0.1 \times 10^{-3} + j0 = (0.01011 + j1.57 \times 10^{-7})V_1 - (1 \times 10^{-5} + j1.57 \times 10^{-7})V_2$$

$$0 = (7.9999 - j1.57 \times 10^{-7})V_1 + (6.677 \times 10^{-3} + j1.57 \times 10^{-7})V_2$$

Solving by one of the methods outlined earlier

$$V_2 = 5.423\angle 179.5° \text{ V}$$

hence

$$\text{voltage gain} = V_2/V_1 = 542.3\angle 179.5°$$

(b) When $f = 200$ kHz, $\omega C = 2\pi \times 200 \times 10^3 \times 25 \times 10^{-12} = 3.142 \times 10^{-5}$, and the equations are

$$0.1 \times 10^{-3} + j0 = (0.01011 + j3.142 \times 10^{-5})V_1 - (1 \times 10^{-5} + j3.142 \times 10^{-5})V_2$$

$$0 = (7.9999 - j3.142 \times 10^{-5})V_1 + (6.677 \times 10^{-3} + j3.142 \times 10^{-5})V_2$$

Solving gives $V_2 = 2.741\angle 120.4°$ V
and voltage gain $= V_2/V_1 = 274.1\angle 120.4°$

The reduction in gain (and the change in phase shift) when the frequency is increased from 1 kHz to 200 kHz is due to the reduction in the reactance of the capacitor with increase in frequency. Frequency response is dealt with more fully in chapter 11.

Unworked problems

6.1. A series circuit containing a 10 Ω resistor and an 8 H inductor is energised by a voltage $v(t) = 8 \cos 2t + 6 \sin 2t$ V. Determine an expression for the current in the circuit. What is the phase relationship between the voltage and the current?
[0.53 sin $(2t - 4.87°)$ A; $i(t)$ lags $v(t)$ by 4.87°]

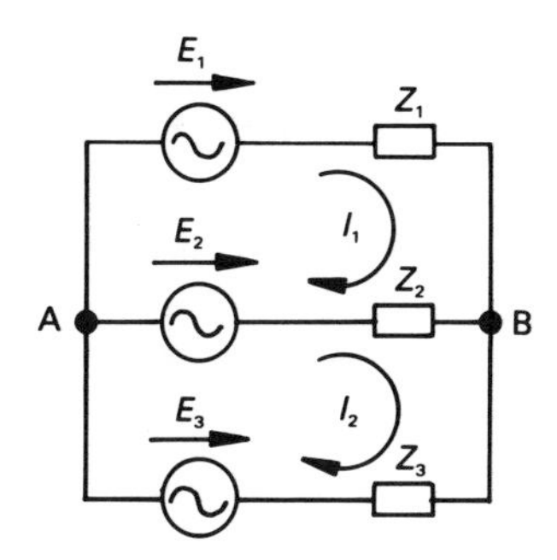

Figure 6.8

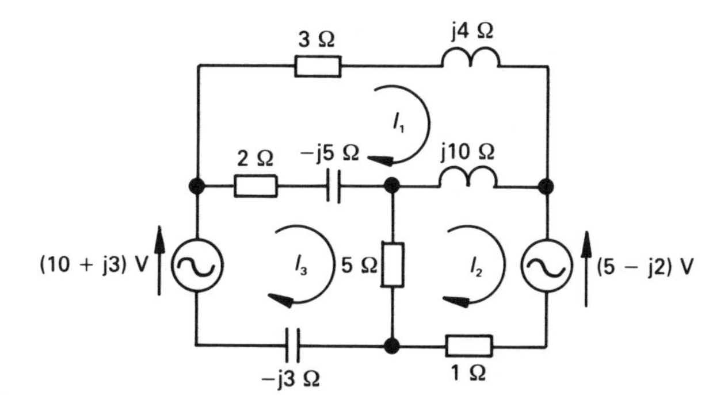

Figure 6.9

6.2. A voltage of 5 sin ($5t - \pi/4$) V is applied to a series circuit containing a coil of 10 Ω resistance and 4 H inductance, and a 0.04 F capacitor. Calculate the voltage across the coil. [6.2 sin ($5t$ + 74.74°) V]

6.3. For the circuit in figure 6.8, if $\boldsymbol{E}_1 = 10\angle 0°$ V, $\boldsymbol{E}_2 = 0$, $\boldsymbol{E}_3 = 6\angle 0°$ V, $\boldsymbol{Z}_1 = \text{j}4\ \Omega$, $\boldsymbol{Z}_2 = 4\ \Omega$, $\boldsymbol{Z}_3 = -\text{j}2\ \Omega$, use mesh analysis to calculate $\boldsymbol{I}_1$ and $\boldsymbol{I}_2$. [$\boldsymbol{I}_1 = 2.262\angle{-96.4°}$ A; $\boldsymbol{I}_2 = 2.594\angle{-101.3°}$ A]

6.4. If in figure 6.8, $\boldsymbol{E}_1 = (20 + \text{j}7)$ V, $\boldsymbol{E}_2 = (10 + \text{j}0)$ V, $\boldsymbol{E}_3 = 0$, $\boldsymbol{Z}_1 = (1 + \text{j}2)\ \Omega$, $\boldsymbol{Z}_2 = (5 - \text{j}6)\ \Omega$ and $\boldsymbol{Z}_3 = \text{j}8\ \Omega$, calculate $\boldsymbol{I}_1$ and $\boldsymbol{I}_2$ using mesh analysis. [$\boldsymbol{I}_1 = 1.21\angle{-50°}$ A; $\boldsymbol{I}_2 = 2.31\angle{-69.9°}$ A]

6.5. Write down, but do not solve, the mesh equations for the circuit in figure 6.9.

$$\begin{bmatrix} 0 & = & (5+\text{j}9)\boldsymbol{I}_1 & - & \text{j}10\boldsymbol{I}_2 & - & (2-\text{j}5)\boldsymbol{I}_3 \\ -(5-\text{j}2) & = & -\text{j}10\boldsymbol{I}_1 & + & (6+\text{j}10)\boldsymbol{I}_2 & - & 5\boldsymbol{I}_3 \\ (10+\text{j}3) & = & -(2-\text{j}5)\boldsymbol{I}_1 & - & 5\boldsymbol{I}_2 & + & (7-\text{j}8)\boldsymbol{I}_3 \end{bmatrix}$$

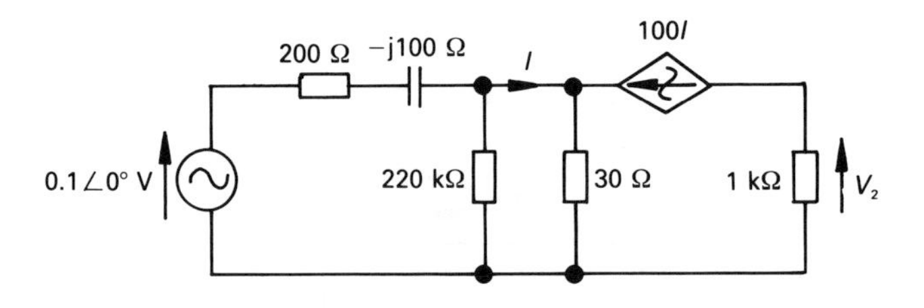

Figure 6.10

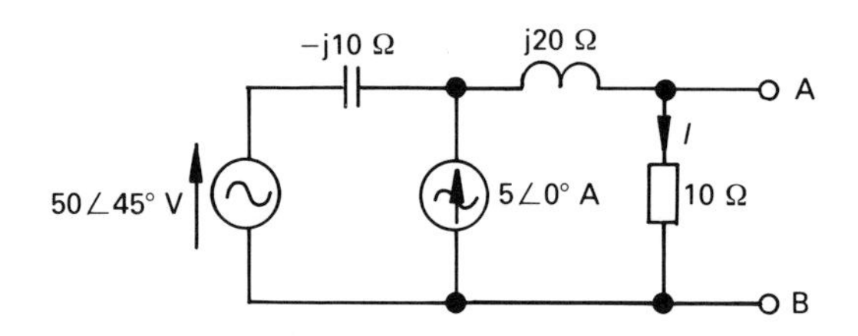

Figure 6.11

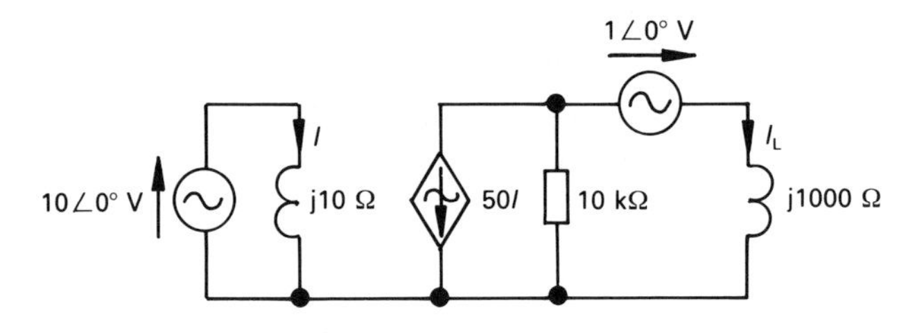

Figure 6.12

6.6. Solve the circuit in figure 6.9 for $\boldsymbol{I}_1$, $\boldsymbol{I}_2$ and $\boldsymbol{I}_3$.
[$\boldsymbol{I}_1 = 1.05\angle 57°$ A; $\boldsymbol{I}_2 = 1.8\angle 63.4°$ A; $\boldsymbol{I}_3 = 2.07\angle 75.8°$ A]

6.7. Figure 6.10 shows a simplified equivalent circuit of a common-base amplifier. Calculate the value of $\boldsymbol{V}_2$.
[$3.09\angle -178.2°$ V]

6.8. Using the superposition theorem, calculate the current $\boldsymbol{I}$ in figure 6.11.
[$2.704\angle -67.5°$ A]

6.9. Using superposition, calculate the current $\boldsymbol{I}_L$ in figure 6.12.
[$4.97\angle 84.3°$ A]

6.10. The network in figure 6.13 is the small-signal hybrid parameter equivalent circuit of a transistor. Determine the Thévenin equiva-

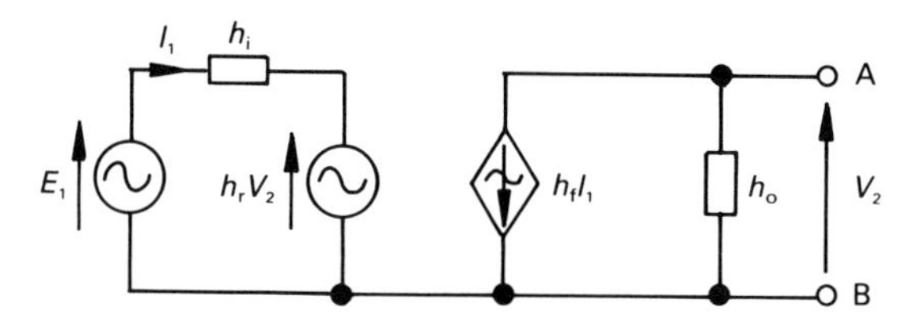

Figure 6.13

lent circuit with respect to terminals A and B; parameters h_r and h_f are dimensionless, h_i is a resistance and h_o is a conductance.
[$E_{TH} = -h_f E_1/(h_o h_i - h_r h_f)$; $Z_{TH} = 1/(h_o - h_r h_f/h_i)$ Ω)]

6.11. Deduce Thévenin's equivalent circuit with respect to terminals A and B for the circuit in problem 6.8.
[$\boldsymbol{E}_{TH} = 27.04\angle{-67.5°}$ V; $\boldsymbol{Z}_{TH} = 7.071\angle 45°$ Ω]

6.12. Deduce the Norton equivalent circuit with respect to terminals A and B of problem 6.3.
[$\boldsymbol{I}_N = 0.502\angle 90°$ A; $\boldsymbol{Y}_N = 0.354\angle 45°$ S]

6.13. Use Millman's theorem to solve problems 6.3 and 6.4.

6.14. An impedance of (2 + j4) Ω is connected between nodes A and B, an impedance (3 − j5) Ω is connected between nodes B and C, and an impedance (2 + j0) Ω is connected between nodes A and C. Determine the equivalent star-connected circuit.
[$\boldsymbol{Z}_{AS} = 1.26\angle 71.57°$ Ω; $\boldsymbol{Z}_{BS} = 3.69\angle 12.53°$ Ω; $\boldsymbol{Z}_{CS} = 1.65\angle{-50.9°}$ Ω]

6.15. The load in problem 6.14 is energised by a 3-phase, 3-wire system, whose neutral point is node N. If node A is energised by a voltage of $100\angle 10°$ V with respect to node N, node B is energised by a voltage of $120\angle{-120°}$ V with respect to node N, and node C is energised by a voltage of $90\angle 100°$ V with respect to node N, use Millman's theorem to calculate the current flowing into nodes A, B and C, respectively.
[$\boldsymbol{I}_A = 111.9\angle{-29.4°}$ A; $\boldsymbol{I}_B = 16.3\angle{-166.2°}$ A; $\boldsymbol{I}_C = 100.7\angle 144.2°$ A]

7

Polyphase Circuits

7.1 Introduction

Polyphase supply systems have a number of advantages over single-phase systems including the fact that, for a given amount of electrical power transmitted, the total power loss is lower in the polyphase system; moreover, the total volume of conductor material needed in the cable is less. For these and other reasons, power is transmitted across the nation by a three-phase system. Another feature is that the torque produced by a single-phase motor is pulsating rather than rotating; that is, an ideal single-phase motor has no starting torque! In order to cause the rotor of a single-phase motor to begin to rotate, it is necessary (at starting) to convert it to a two-phase motor. On the other hand, polyphase motors produce a smooth torque, and their speed can be controlled using relatively straightforward methods.

Industry uses many other types of multi-phase system. For example, many control systems use two-phase supplies to drive servomotors, and many rectifier systems use a six-, twelve- or twentyfour-phase supply. In this chapter we shall concentrate on three-phase systems.

The US notation for phase markings has been adopted, so that the phases in a three-phase system are called A, B and C rather than R, Y and B, respectively.

7.2 Three-phase generation

The usual method of producing a three-phase set of voltages is to have three separate, but identical, windings on the rotor of an alternator, each winding being physically displaced from the next winding by 120°. Since the windings are displaced in space by 120°, and are moving through the same magnetic field system, the net result is three single-phase voltages (referred

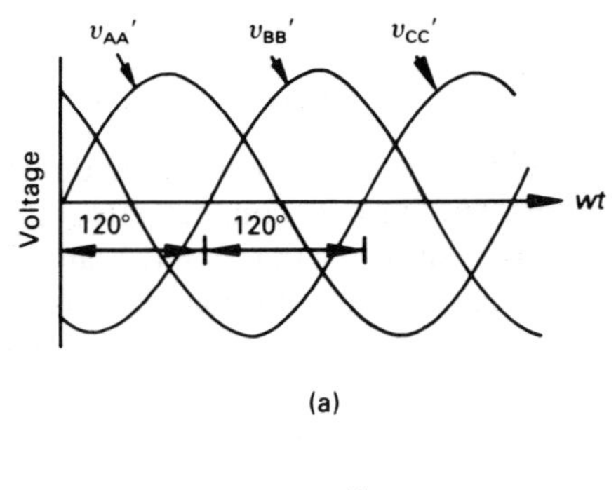

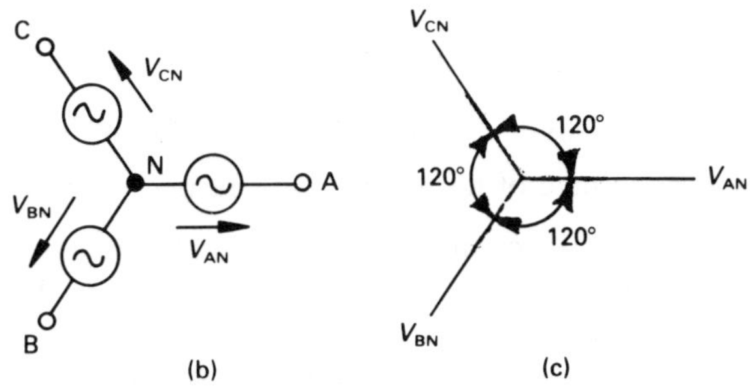

Figure 7.1 *(a) Three-phase voltage waveforms, (b) star-connected voltages and (c) a typical phasor diagram.*

to as the *phase* voltages) having an angular displacement of 120° between them.

Alternatively, the voltages can be produced electronically, the phase displacement being controlled either by *R–C* circuits and operational amplifiers, or by integrated circuits.

While 3-phase voltage sources are widely used, the reader should note that 3-phase current sources are very uncommon.

7.3 Star-connection or Y-connection

Imagine an alternator with three separate windings with ends marked A and A′, B and B′, and C and C′, respectively having the time-varying voltage $V_{AA'}$, $V_{BB'}$ and $V_{CC'}$ induced in them, the respective waveform diagrams being shown in figure 7.1(a). The voltages can be written down in the form

$$
\begin{aligned}
V_{AA'} &= V_{m(AA')} \sin \omega t \text{ V} \\
V_{BB'} &= V_{m(BB')} \sin (\omega t - 120°) \\
&= V_{m(BB')} \sin (\omega t + 240°) \text{ V}
\end{aligned}
$$

$$V_{CC'} = V_{m(CC')} \sin(\omega t - 240°)$$
$$= V_{m(CC')} \sin(\omega t + 120°) \text{ V}$$

If the ends A′, B′ and C′ (each of which can be thought of as the 'start' end of the corresponding winding) are connected together at node N, then the connection diagram is as shown in diagram (b), and the phasor diagram at $\omega t = 0$ is illustrated in diagram (c). Once again, it is pointed out that the magnitude of each phasor is the r.m.s. value of the associated sinusoidal waveform.

Moreover, we use the double-subscript notation for three-phase voltages as follows

$\boldsymbol{V}_{AN}$ = voltage of terminal A with respect to N

$\boldsymbol{V}_{BN}$ = voltage of terminal B with respect to N

$\boldsymbol{V}_{CN}$ = voltage of terminal C with respect to N

In many systems, point N is connected to earth, that is, to a point of neutral potential, and is known as the *neutral point* of the supply. Where it is not earthed, it is sometimes called the *star point*.

The reader should note that, in the above analysis, we have taken $\boldsymbol{V}_{AN}$ to lie in the reference direction. Although this will generally be the case, it may not always be so for any given problem.

7.4 Phase sequence

As explained in chapter 5, a sinusoidal waveform can be thought of as the vertical component of the tip of a radial line which rotates at constant speed in a counterclockwise direction. A three-phase set of voltages can therefore be thought of as the vertical component of three radial lines which are separated from one another by 120°, and which rotate at a constant speed in a counterclockwise direction.

The *phase sequence* of the system in question is the order in which they pass a fixed point in space. In the case of the phasors in figure 7.1(c), this is the sequence ABC or *positive phase sequence* (PPS). PPS supplies are generally associated with electrical power systems, and are the subject of the majority of the work in this chapter.

The windings can be reconnected so that the phase sequence is reversed, that is, the sequence ACB, which is known as *negative phase sequence* (NPS). Power supply systems normally generate PPS supplies but, under abnormal conditions such as certain types of electrical fault, NPS components may be produced. NPS systems are discussed more fully in section 7.17.

7.5 Balanced and unbalanced systems

A *balanced polyphase supply* is one having phase voltages which are *equal in magnitude*, and which are phase displaced from one another by an *equal angle*; in a three-phase system, this angle is 120°. If either of the conditions is not met, the supply is said to be *unbalanced.*

A *balanced polyphase load* is one in which the impedance of each phase of the load has the same magnitude and the same phase angle. If either condition is not met, the load is said to be *unbalanced.*

In the majority of large industrial loads, both the supply and the load are balanced, but in domestic situations and small industries, we meet with unbalanced supplies connected to unbalanced loads.

While we have already met with many analytical tools which allow us to deal with most polyphase circuits, one of the most powerful techniques (*symmetrical components*) has yet to be introduced (see section 7.17).

7.6 Phase and line voltages in a star-connected system

As explained earlier, a *phase voltage* is the voltage induced in a winding or phase of an alternator. In the case of the phasor diagram of the star-connected supply in figure 7.2, the phase voltages are $\boldsymbol{V}_{AN}$, $\boldsymbol{V}_{BN}$ and $\boldsymbol{V}_{CN}$.

A *line voltage* or *line-to-line voltage* is the voltage between a pair of lines. For example, the line voltage between lines A and B in figure 7.2 is

$$\boldsymbol{V}_{AB} = \text{voltage of line A with respect to line B}$$
$$= \boldsymbol{V}_{AN} - \boldsymbol{V}_{BN}$$

Similarly

$$\boldsymbol{V}_{BC} = \text{voltage of line B with respect to line C}$$
$$= \boldsymbol{V}_{BN} - \boldsymbol{V}_{CN}$$

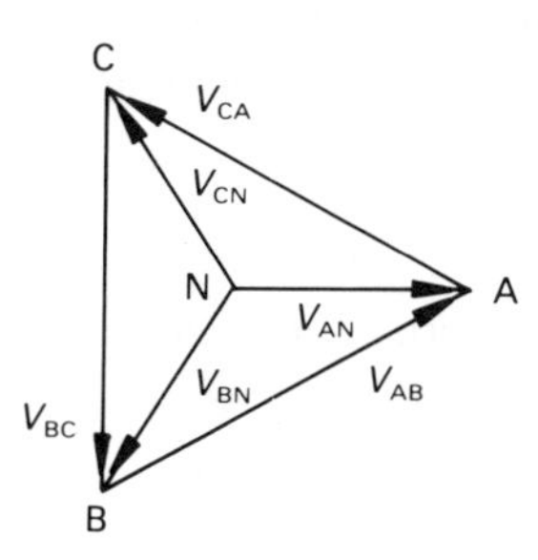

Figure 7.2 *Phase and line voltages.*

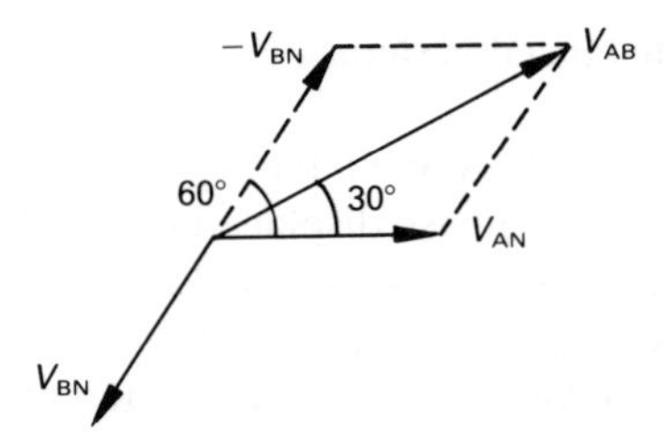

Figure 7.3 *Calculation of the line voltage* $\mathbf{V}_{AB}$.

and

$$\mathbf{V}_{CA} = \text{voltage of line C with respect to line A}$$
$$= \mathbf{V}_{CN} - \mathbf{V}_{AN}$$

In the case of a *balanced system*, the phase voltages are equal in magnitude, and each is called V_p. The way in which the line voltage $\mathbf{V}_{AB}$ can be related to V_p is shown below (see also figure 7.3).

$$\mathbf{V}_{AB} = \mathbf{V}_{AN} - \mathbf{V}_{BN}$$

In a balanced system

$$\mathbf{V}_{AB} = V_p\angle 0° - V_p\angle{-120°} = \sqrt{3}\, V_p\angle 30° \text{ V}$$

That is the *magnitude*, V_L, of the voltage between a pair of lines (the *line voltage*) in a *balanced star-connected supply system* is

$$V_L = \sqrt{3}\, V_p$$

Referring to figures 7.2 and 7.3 we see that

$$\mathbf{V}_{AB} = \sqrt{3}\, V_p\angle 30° \text{ V}$$
$$\mathbf{V}_{CA} = \sqrt{3}\, V_p\angle 150° \text{ V}$$
$$\mathbf{V}_{BC} = \sqrt{3}\, V_p\angle{-90°} \text{ V}$$

That is, *the line voltages in a balanced 3-phase supply are equal in magnitude, and are phase displaced from one another by 120°.*

Worked example 7.6.1

Calculate complex expressions for phase and line voltages in a three-phase system in which $V_p = 250$ V.

Solution

$$V_p = 250 \text{ V, hence } V_L = \sqrt{3}\, V_p = 433 \text{ V}$$

Assuming that V_{AN} lies in the reference direction then

$$V_{AN} = 250\angle 0° = 250 + j0\ \text{V}$$

$$V_{BN} = 250\angle -120° = -125 - j216.5\ \text{V}$$

$$V_{CN} = 250\angle 120° = -125 + j216.5\ \text{V}$$

and

$$V_{AB} = 433\angle 30° = 375 + j216.5\ \text{V}$$

$$V_{CA} = 433\angle 150° = -375 + j216.5\ \text{V}$$

$$V_{BC} = 433\angle -90° = 0 - j433\ \text{V}$$

Worked example 7.6.2

If $V_{AN} = 400\angle 20°$ V, $V_{BN} = 350\angle -130°$ V and $V_{CN} = 450\angle 110°$ V, determine a complex expression for each line voltage.

Solution

In this case we are dealing with an unbalanced set of voltages and, moreover, V_{AN} is not in the reference direction.

$$V_{AB} = V_{AN} - V_{BN} = 400\angle 20° - 350\angle -130°$$

$$= (375.88 + j136.8) - (-224.98 - j268.12)$$

$$= 600.86 + j404.92 = 724.6\angle 33.98°\ \text{V}$$

There are two points the reader should note. Firstly, it is necessary to convert the polar complex values to rectangular complex values before they can be subtracted. Secondly, the answer is given in polar form, which is more usual form for practical purposes since most engineering instruments give readings in this form rather than in rectangular complex form.

$$V_{CA} = V_{CN} - V_{AN} = 450\angle 110° - 400\angle 20°$$

$$= (-153.9 + j422.86) - (375.88 + j136.8)$$

$$= -529.78 + j286.06 = 602.1\angle 151.6°\ \text{V}$$

$$V_{BC} = V_{BN} - V_{CN} = 350\angle -130° - 450\angle 110°$$

$$= (-224.98 - j268.12) - (-153.9 + j422.9)$$

$$= -71.08 - j691.02 = 694.02\angle -95.9°\ \text{V}$$

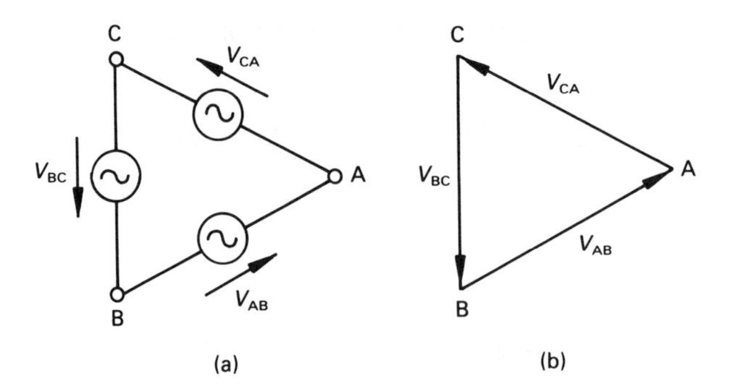

Figure 7.4 *(a) A 3-phase delta-connected generator, and (b) its phasor diagram.*

7.7 Delta-connected or mesh-connected three-phase source

If the three separate windings on the rotor of a 3-phase alternator are interconnected so that the end of one winding is connected to the start of the next (see figure 7.4(a)), the result is the *delta-* or *mesh-connection*, having the phasor diagram in figure 7.4(b).

Since each phase of the generator is connected between a pair of lines, then the magnitude of the line voltage is equal to the phase voltage. That is, in a *balanced delta-connected system*

$$V_L = V_P$$

or

$$\mathbf{V_{AB}} = V_L \angle 30° \text{ V}$$

$$\mathbf{V_{CA}} = V_L \angle 150° \text{ V}$$

$$\mathbf{V_{BC}} = V_L \angle -90° \text{ V}$$

The reader will observe that in a balanced system, the line voltages are equal in magnitude, and are phase displaced from one another by 120°.

7.8 Three-phase, four-wire, star–star system

A typical system is illustrated in figure 7.5; the neutral point (N) of the supply is connected to the star point (S) of the load by a line of zero resistance.

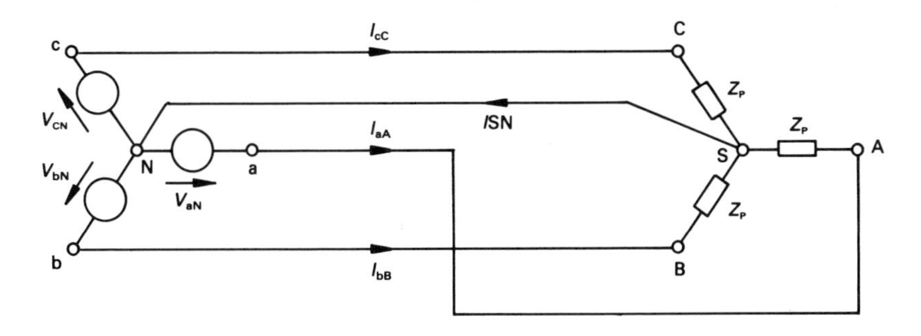

Figure 7.5 *A three-phase, four-wire, star–star system.*

7.8.1 Balanced supply, balanced load

In this case each phase voltage has the same value, and each is separated from the next phase voltage by 120°. Also, a balanced load is one in which the magnitude of the impedance in each phase of the load has the same value, and each has the same phase angle. That is for each load

$$\mathbf{Z_p} = Z_p\angle\phi\ \Omega$$

The current $\boldsymbol{I}_{\mathbf{aA}}$ flowing from node a of the supply to node A of the load is

$$\boldsymbol{I}_{\mathbf{aA}} = \frac{\boldsymbol{V}_{\mathbf{aN}}}{\mathbf{Z_p}} = \frac{V_p\angle 0^\circ}{Z_p\angle\phi} = \frac{V_p}{Z_p}\angle\,\phi = \frac{V_p}{Z_p}\,(\cos\phi + \mathrm{j}\sin\phi)$$

also

$$\boldsymbol{I}_{\mathbf{bB}} = \frac{\boldsymbol{V}_{\mathbf{bN}}}{\mathbf{Z_p}} = \frac{V_p\angle{-120^\circ}}{Z_p\angle\phi} = \frac{V_p}{Z_p}\angle(-120^\circ + \phi)$$

$$= \frac{V_p}{Z_p}\,(\cos(-120^\circ + \phi) + \mathrm{j}\sin(-120^\circ + \phi))$$

$$= \frac{V_p}{Z_p}\,((-0.5\cos\phi + 0.866\sin\phi) +$$

$$\mathrm{j}(-0.866\cos\phi - 0.5\sin\phi))$$

and

$$\boldsymbol{I}_{\mathbf{cC}} = \frac{\boldsymbol{V}_{\mathbf{cN}}}{\mathbf{Z_p}} = \frac{V_p\angle 120^\circ}{Z_p\angle\phi} = \frac{V_p}{Z_p}\angle(120^\circ - \phi)$$

$$= \frac{V_P}{Z_P}\,((-0.5 \cos \phi - 0.866 \sin \phi)$$

$$+ j(0.866 \cos \phi + 0.5 \sin \phi))$$

The current in the neutral wire, $\boldsymbol{I}_{SN}$ (usually described as the *neutral wire current* or *neutral current*, $\boldsymbol{I}_N$) is

$$\boldsymbol{I}_N = \boldsymbol{I}_{aA} + \boldsymbol{I}_{bB} + \boldsymbol{I}_{cC} = 0 + j0 \text{ A}$$

That is, *in a balanced star-connected 4-wire system the current in the neutral current is zero*. In such a system, there is no need for a neutral wire, and a three-phase three-wire supply can be used (see also section 7.9). This is the case in most industrial systems (but not in most domestic supply systems).

7.8.2 Three-phase, four-wire, star-connected system with an unbalanced load

With an unbalanced load, it is generally the case that $\boldsymbol{I}_N$ is non-zero, and is illustrated in the following example,

Worked example 7.8.1

A 502.3 V, 3-phase, 4-wire supply is connected to a star-connected load having the following load impedances

$$\boldsymbol{Z}_{AS} = 10\angle 0° \ \Omega, \quad \boldsymbol{Z}_{BS} = 10\angle 20° \ \Omega, \quad \boldsymbol{Z}_{CS} = 10\angle{-40°} \ \Omega$$

Calculate the current in each line and in the neutral wire.

Solution

The phase voltage is

$$V_P = V_L/\sqrt{3} = 502.3/\sqrt{3} = 290 \text{ V}$$

The circuit is generally as shown in figure 7.5, and the current in line A is

$$\boldsymbol{I}_A = \boldsymbol{V}_{AN}/\boldsymbol{Z}_{AS} = 290\angle 0°/10\angle 0° = 29\angle 0° = 29 + j0 \text{ A}$$

in line B is

$$\boldsymbol{I}_B = \boldsymbol{V}_{BN}/\boldsymbol{Z}_{BS} = 290\angle{-120°}/10\angle 20° = 29\angle{-140°}$$
$$= -22.22 - j18.64 \text{ A}$$

and

$$\boldsymbol{I}_C = \boldsymbol{V}_{CN}/\boldsymbol{Z}_{CS} = 290\angle 120°/10\angle{-40°} = 29\angle 160°$$
$$= -27.25 + j9.92 \text{ A}$$

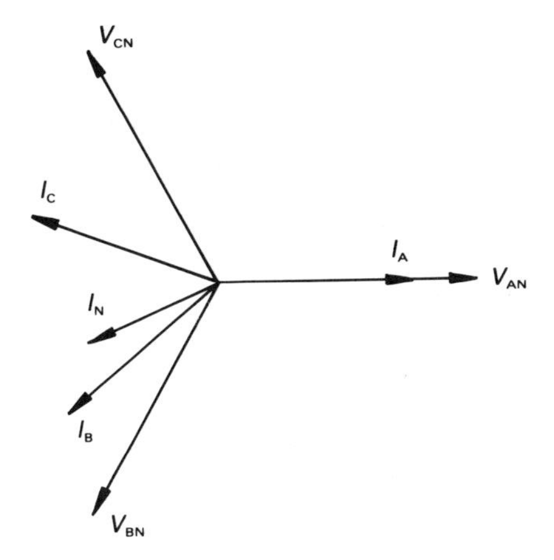

Figure 7.6 *Phasor diagram for worked example 7.8.1.*

The current in the neutral wire is

$$\begin{aligned} \boldsymbol{I}_N &= \boldsymbol{I}_A + \boldsymbol{I}_B + \boldsymbol{I}_C \\ &= (29 + j0) + (-22.22 - j18.64) + (-27.25 + j9.92) \\ &= 22.25\angle{-156.9°} \text{ A} \end{aligned}$$

The corresponding phasor diagram is shown in figure 7.6.

7.9 Three-phase, three-wire, star–star system

In many industrial systems, both the supply and the load are balanced, so that the neutral wire current is zero; in this case the neutral wire may be removed without any effect on the voltages and currents in the system.

However, if either the supply or the load is unbalanced, or both are unbalanced, then it is usually the case that the neutral point of the supply is not at the same potential as the star point of the load. This is illustrated in the following example, in which both the supply and the load are unbalanced.

Worked example 7.9.1

Calculate the star-to-neutral voltage, the voltage across each phase of the load, and the current in each phase of the load in the following 3-phase, 3-wire, star–star system.

$$\boldsymbol{V}_{AN} = 200\angle 10° \text{ V}, \quad \boldsymbol{V}_{BN} = 220\angle{-140°} \text{ V}, \quad \boldsymbol{V}_{CN} = 180\angle 100° \text{ V}$$

$$\boldsymbol{Z}_{AS} = 10\angle 0° \ \Omega, \quad \boldsymbol{Z}_{BS} = 15\angle 10° \ \Omega, \quad \boldsymbol{Z}_{CS} = 5\angle{-20°} \ \Omega$$

Solution

In this case we can use Millman's theorem to calculate the voltage between the star point of the load and the neutral point of the supply as follows.

$$V_{SN} = \frac{\dfrac{V_{AN}}{Z_{AS}} + \dfrac{V_{BN}}{Z_{BS}} + \dfrac{V_{CN}}{Z_{CS}}}{\dfrac{1}{Z_{AS}} + \dfrac{1}{Z_{BS}} + \dfrac{1}{V_{CS}}}$$

$$= \frac{\dfrac{200\angle 10^\circ}{10\angle 0^\circ} + \dfrac{220\angle -140^\circ}{15\angle 10^\circ} + \dfrac{180\angle 100^\circ}{5\angle -20^\circ}}{\dfrac{1}{10\angle 0^\circ} + \dfrac{1}{15\angle 10^\circ} + \dfrac{1}{5\angle -20^\circ}}$$

$$= \frac{29.44\angle 111.9^\circ}{0.358\angle 9^\circ} = 82.23\angle 102.9^\circ$$

$$= -18.36 + j80.15 \text{ V}$$

That is to say, a voltage exists between the neutral point of the supply and the star point of the load.

Using this value, we can calculate the voltage across each phase of the load as follows.

$$V_{AS} = V_{AN} - V_{SN} = 200\angle 10^\circ - 82.23\angle 102.9^\circ$$
$$= 220\angle -11.9^\circ \text{ V}$$

$$V_{BS} = V_{BN} - V_{SN} = 220\angle -140^\circ - 82.23\angle 102.9^\circ$$
$$= 267.63\angle -124.1^\circ \text{ V}$$

and

$$V_{CS} = V_{CN} - V_{SN} = 180\angle 100^\circ - 82.23\angle 102.9^\circ$$
$$= 97.97\angle 97.57^\circ \text{ V}$$

The current in each line can be calculated as follows.

$$I_A = \frac{V_{AS}}{Z_{AS}} = \frac{220\angle -11.9^\circ}{10\angle 0^\circ} = 22\angle -11.9^\circ \text{ A}$$

$$I_B = \frac{V_{BS}}{Z_{BS}} = \frac{267.63\angle -124.1^\circ}{15\angle 10^\circ} = 17.84\angle -134.1^\circ \text{ A}$$

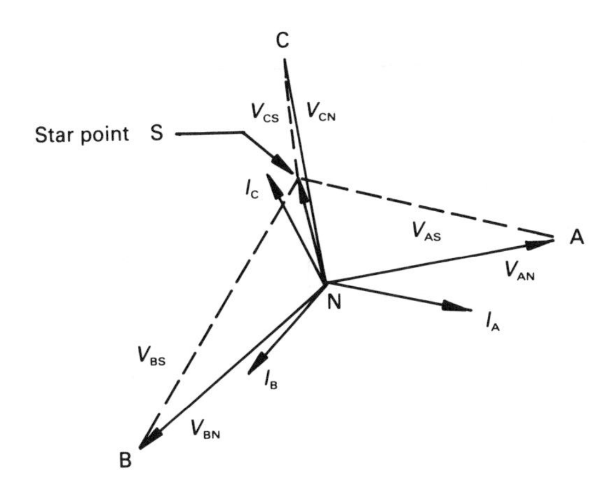

Figure 7.7 *Phasor diagram for worked example 7.9.1.*

and

$$\boldsymbol{I}_C = \frac{\boldsymbol{V}_{CS}}{\boldsymbol{Z}_{CS}} = \frac{97.97\angle 97.57°}{5\angle -20°} = 19.59\angle 117.6° \text{ A}$$

The corresponding phasor diagram is shown in figure 7.7.
Note: The problem can, alternatively, be solved either by using mesh current or node voltage analysis.

7.10 Delta-connected systems

Many three-phase loads are delta-connected rather than Y-connected, the principal reason being that if the load is unbalanced, it can either be connected or removed without affecting the voltage distribution at the load end (in a 3-phase, 3-wire system, the addition or removal of an unbalanced Y-connected load almost invariably produces a change in the individual phase voltages at the load).

The reader will observe that, in figure 7.8, each phase of the load is connected to a corresponding phase of the supply. Consequently, we can (to a large extent) regard a delta-connected system as three single-phase systems, in which a pair of phases use one line in common to carry the line current. Arising from these observations, we may say that

$$\boldsymbol{I}_{AB} = \frac{\boldsymbol{V}_{AB}}{\boldsymbol{Z}_{AB}} \qquad \boldsymbol{I}_{CA} = \frac{\boldsymbol{V}_{CA}}{\boldsymbol{Z}_{CA}} \qquad \boldsymbol{I}_{BC} = \frac{\boldsymbol{V}_{BC}}{\boldsymbol{Z}_{BC}}$$

Applying KCL at node A shows that

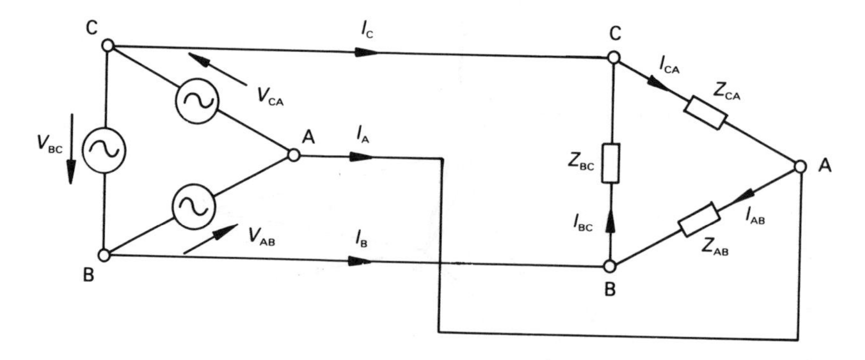

Figure 7.8 *A delta–delta connected system.*

$$\boldsymbol{I}_{\mathrm{A}} = \boldsymbol{I}_{\mathrm{AB}} - \boldsymbol{I}_{\mathrm{CA}}$$

Applying it to node B yields

$$\boldsymbol{I}_{\mathrm{B}} = \boldsymbol{I}_{\mathrm{BC}} - \boldsymbol{I}_{\mathrm{AB}}$$

and at node C gives

$$\boldsymbol{I}_{\mathrm{C}} = \boldsymbol{I}_{\mathrm{CA}} - \boldsymbol{I}_{\mathrm{BC}}$$

The above equations are generally applicable to any delta-connected system, whether balanced or unbalanced.

7.10.1 Balanced delta-connected load with a balanced supply

In this case

$$V_{\mathrm{L}} = |\boldsymbol{V}_{\mathrm{AB}}| = |\boldsymbol{V}_{\mathrm{BC}}| = |\boldsymbol{V}_{\mathrm{CA}}|$$

and

$$Z_{\mathrm{p}} = |\boldsymbol{Z}_{\mathrm{AB}}| = |\boldsymbol{Z}_{\mathrm{BC}}| = |\boldsymbol{Z}_{\mathrm{CA}}|$$

The magnitude of the phase current, I_{p}, flowing in each phase of the load is

$$I_{\mathrm{p}} = \frac{V_{\mathrm{P}}}{Z_{\mathrm{P}}} = \frac{V_{\mathrm{L}}}{Z_{\mathrm{P}}}$$

That is, each phase current has the same value. Now (see also figure 7.4(b))

$$\boldsymbol{I}_{\mathrm{AB}} = \frac{\boldsymbol{V}_{\mathrm{AB}}}{\boldsymbol{Z}_{\mathrm{AB}}} = \frac{V_{\mathrm{L}}\angle 30^\circ}{Z_{\mathrm{p}}\angle\phi} = \frac{V_{\mathrm{L}}}{Z_{\mathrm{p}}}\angle(30^\circ - \phi)$$

$$I_{BC} = \frac{V_{BC}}{Z_{BC}} = \frac{V_L\angle -90°}{Z_p\angle\phi} = \frac{V_L}{Z_p}\angle(-90° - \phi)$$

and

$$I_{CA} = \frac{Z_{CA}}{Z_{CA}} = \frac{V_L\angle 150°}{Z_p\angle\phi} = \frac{V_L}{V_p}\angle(150° - \phi)$$

The line current I_A is calculated from

$$\begin{aligned} I_A &= I_{AB} - I_{CA} = \frac{V_L}{Z_p}(\angle(30° - \phi) - \angle(150° - \phi)) \\ &= \sqrt{3}\,\frac{V_L}{Z_p}(\cos\phi + j\sin\phi) = \sqrt{3}\frac{V_L}{Z_p}\angle\phi \\ &= \sqrt{3}\,I_p\angle\phi \end{aligned}$$

Similarly it may be shown that the magnitude of the other line currents (I_B and I_C) is $\sqrt{3}I_p$, and the phase angle between each phase current and the associated phase voltage at the load (which is one of the line voltages) is ϕ.

Worked example 7.10.1

A balanced 500 V, 3-phase source supplies a balanced 3-phase delta-connected load of (4 + j5) Ω per phase. Calculate the current in each phase of the load and in each line. Draw the phasor diagram.

Solution

The impedance in each phase of the load is

$$4 + j5 = 6.4\angle 51.34°\ \Omega$$

and the modulus of the current in each phase of the load is

$$I_p = 500/6.4 = 78.125\ \text{A}$$

and the magnitude of the line current is

$$I_L = \sqrt{3}\,I_p = \sqrt{3} \times 78.125 = 135.32\ \text{A}$$

Since the load is balanced, each phase current lags behind the associated phase voltage by 51.34°. The corresponding phasor diagram is shown in figure 7.9. The reader should note that, since each node of the delta is separate, it is sometimes convenient (though not strictly true) to draw all the phasors from a common centre point.

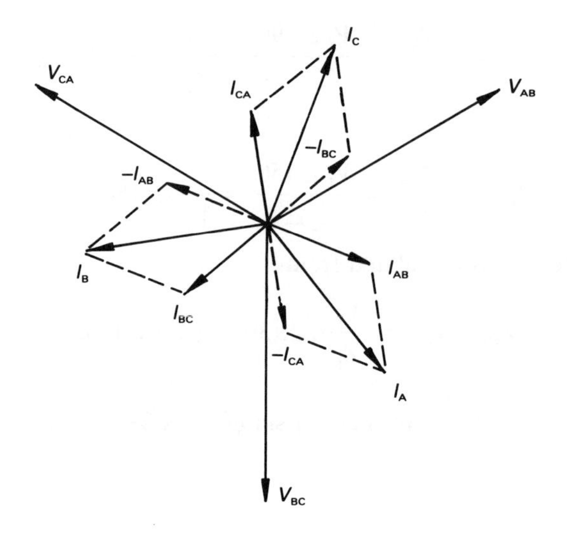

Figure 7.9 *Figure for worked example 7.10.1.*

7.10.2 Unbalanced delta-connected load with a balanced supply

If the load is unbalanced, it is necessary to calculate individually the phase currents in the load, and then implement the equations at the beginning of this section to calculate the line currents, as outlined in worked example 7.10.2.

Worked example 7.10.2

The following impedances are connected in delta to a balanced 3-phase, 500 V supply

$$\mathbf{Z}_{AB} = 10\angle 25°\ \Omega, \quad \mathbf{Z}_{BC} = 15\angle{-30°}\ \Omega, \quad \mathbf{Z}_{AC} = 20\angle 0°\ \Omega$$

Solution

The phase currents are

$$\boldsymbol{I}_{AB} = \frac{\mathbf{V}_{AB}}{\mathbf{Z}_{AB}} = \frac{500\angle 30°}{10\angle 25°} = 50\angle 5° = 49.8 + \text{j}4.36\ \text{A}$$

$$\boldsymbol{I}_{BC} = \frac{\mathbf{V}_{BC}}{\mathbf{Z}_{BC}} = \frac{500\angle{-90°}}{15\angle{-30°}} = 33.33\angle{-60°} = 16.67 - \text{j}28.86\ \text{A}$$

$$I_{CA} = \frac{V_{CA}}{Z_{CA}} = \frac{500\angle 150°}{20\angle 0°} = 25\angle 150° = -21.65 + j12.5 \text{ A}$$

and the corresponding line currents are

$$I_A = I_{AB} - I_{CA} = (49.8 + j4.36) - (-21.65 + j12.5)$$
$$= 71.45 - j8.14 = 71.91\angle -6.5° \text{ A}$$

$$I_B = I_{BC} - I_{AB} = (16.67 + j28.14) - (49.8 + j4.36)$$
$$= -33.13 - j33.22 = 46.92\angle -134.9° \text{ A}$$

$$I_C = I_{CA} - I_{BC} = (-21.65 + j12.5) - (16.67 - j28.14)$$
$$= -38.32 + j41.36 = 56.38\angle 132.8° \text{ A}$$

7.10.3 Unbalanced delta-connected load with an unbalanced supply

The solution of this type of circuit is generally similar to that outlined in worked example 7.10.2, with the exception that the unbalanced line voltages are used in the calculation.

7.11 Delta-connected supply and a star-connected load

This situation generally presents no problem if the supply and load are both balanced, since it can be dealt with as though the load is supplied by a 3-phase, 4-wire supply (even though the neutral wire is absent). The reason is that, since the load is balanced, the neutral wire current is zero and the voltage across each phase of the load is $V_L/\sqrt{3}$.

If either the supply or the load is unbalanced, a convenient method of dealing with the problem is to convert the load into its equivalent delta network (see chapter 6), and deal with the circuit as described in section 7.10.

7.12 Star-connected supply and delta-connected load

Once again, the situation is fairly straightforward, because we can calculate the line voltages (which may either be balanced or unbalanced), and then treat the problem as outlined in section 7.10.

7.13 Summary of balanced star- and delta-connected systems

	Phase voltage	*Line voltage*	*Phase current*	*Line current*
Star	$V_P = V_L/\sqrt{3}$	$V_L = \sqrt{3}V_P$	$I_P = I_L$	$I_L = I_P$
Delta	$V_P = V_L$	$V_L = V_P$	$I_P = I_L/\sqrt{3}$	$I_L = \sqrt{3}I_P$

7.14 Power consumed in a three-phase system

The power consumed by *one phase* of a polyphase system is

$$P_P = V_P I_P \cos \phi$$

where $\cos \phi$ is the power factor of the load in that phase.

Unbalanced load

The total power consumed is the sum of the power in each of the three phases. That is, the total power P_T is

$$P_T = P_A + P_B + P_C$$

where P_A, P_B and P_C are the power consumed, respectively, in the A-, B- and C-phases.

Balanced load, balanced supply

Once again, the total power is the sum of the individual value of power in the three phases but, in this case, an equal value of power is consumed by each phase so that

$$P_T = 3V_P I_P \cos \phi$$

In a *Y*-connected load, $V_L = \sqrt{3}V_P$ and $I_L = I_P$ and

$$P_T = \sqrt{3}V_L I_L \cos \phi$$

In a *delta-connected load*, $V_L = V_P$ and $I_L = \sqrt{3}I_P$, hence

$$P_T = \sqrt{3}V_L I_L \cos \phi$$

That is, in *any balanced 3-phase load* the total power consumed is

$$P_T = \sqrt{3}V_L I_L \cos \phi$$

Moreover, in a balanced system, each phase consumes not only the same value of VA but also the same value of VAr, so that

$$S_T = \sqrt{3}V_L I_L$$
$$P_T = \sqrt{3}V_L I_L \cos\phi$$
$$Q_T = \sqrt{3}V_L I_L \sin\phi$$

Worked example 7.14.1

Calculate the VA, the power and the VAr consumed by the balanced delta-connected load in worked example 7.10.1.

Solution

We are dealing here with a system with a balanced 500 V supply, whose line current is 135.32 A and the phase angle of the load is −51.34°. Hence

$$S_T = \sqrt{3}V_L I_L = \sqrt{3} \times 500 \times 135.32 \text{ W} = 117.19 \text{ kVA}$$
$$P_T = \sqrt{3}V_L I_L \cos\phi = 117.19 \times 10^3 \times 0.6247 \text{ W}$$
$$= 73.2 \text{ kW}$$
$$Q_T = \sqrt{3}V_L I_L \sin\phi = 117.19 \times 10^3 \times 0.781 \text{ VAr}$$
$$= 91.51 \text{ kVAr lagging}$$

Alternatively, we may say that in the A-phase of the load

$$\mathbf{V}_{AB} = 500 \angle 30° \text{ V and } \mathbf{I}_{AB} = 78.125 \angle (30° - 51.34°)$$
$$= 78.125 \angle -21.34° \text{ A, so that}$$
$$\mathbf{S}_{AB} = \mathbf{V}_{AB}\mathbf{I}^*_{AB} = 500 \angle 30° \times 78.125 \angle 21.34°$$
$$= 39.06 \angle 51.34° \text{ kVA}$$

Hence

$$\mathbf{S}_T = 3 \times 39.06 \angle 51.34° = 117.18 \angle 51.34°$$
$$= (73.2 + \text{j}91.5) \text{ kVA}$$

or

$$S_T = 117.18 \text{ kVA}$$
$$P_T = 73.2 \text{ kW}$$
$$Q_T = 91.5 \text{ kVAr (lagging)}$$

Worked example 7.14.2

Calculate the power consumed by the unbalanced 3-phase, 3-wire, star-connected load in worked example 7.9.1.

Solution

The relevant data are (see also worked example 7.9.1)

$$\boldsymbol{V}_{AS} = 220\angle -11.9^\circ \text{ V}, \quad \boldsymbol{I}_A = 22\angle -11.9^\circ \text{ A}$$

$$\boldsymbol{V}_{BS} = 267.63\angle -124.1^\circ \text{ V}, \quad \boldsymbol{I}_B = 17.84\angle -134.1^\circ \text{ A}$$

$$\boldsymbol{V}_{CS} = 97.97\angle 97.57^\circ \text{ V}, \quad \boldsymbol{I}_B = 19.59\angle 117.6^\circ \text{ A}$$

The power consumed by phase A is

$$P_A = \text{Re}(\boldsymbol{V}_{AS}\boldsymbol{I}_A^*) = \text{Re}(220\angle -11.9^\circ \times 22\angle 11.9^\circ)$$

$$= \text{Re}(4840\angle 0^\circ) = 4840 \text{ W}$$

and by phase B is

$$P_B = \text{Re}(\boldsymbol{V}_{BS}\boldsymbol{I}_B^*) = \text{Re}(267.63\angle -124.1^\circ \times 17.84\angle 134.1^\circ)$$

$$= \text{Re}(4774.5\angle 10^\circ) = 4702 \text{ W}$$

and by phase C

$$P_C = \text{Re}(\boldsymbol{V}_{CS}\boldsymbol{I}_C^*) = \text{Re}(97.97\angle 97.57^\circ \times 19.57\angle -117.6^\circ)$$

$$= \text{Re}(1919.2\angle -20.03^\circ) = 1803 \text{ W}$$

Hence

$$P_T = P_A + P_B + P_C = 11\,345 \text{ W}$$

Note: The power in each phase could, alternatively, have been calculated on the basis of

$$(\text{phase current})^2 \times \text{resistance per phase}$$

7.15 Power measurement in three-phase systems

The majority of power measurement is carried out using analogue wattmeters, and these have one coil to sense the voltage applied to a circuit, and another to sense the current flowing in the circuit. The instrument produces a deflecting torque proportional to the *average power* consumed by the circuit.

In the special case of a balanced load, it is theoretically only necessary to

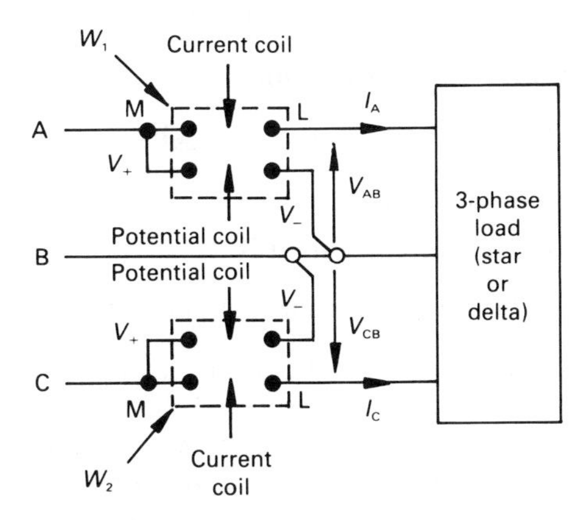

Figure 7.10 *The two-wattmeter method of power measurement.*

have one instrument, which measures the power consumed by one phase. The total power is three times the reading of the instrument.

In general (both for balanced and unbalanced loads) it is necessary to *connect correctly* $(N - 1)$ wattmeters, where N is the number of lines used to supply power to the load. That is, a 3-phase, 4-wire system requires $(4 - 1) = 3$ wattmeters to measure the total power, and a 3-phase, 3-wire system needs two wattmeters to measure the total power.

7.16 The two-wattmeter method of power measurement

The total power consumed by a three-phase, three-wire system can be measured by two wattmeters *connected in any two lines*. For example, they can be connected as shown in figure 7.10.

The markings on the wattmeter terminals are as follows. When current flows from terminal M to terminal L, AND terminal V_+ is positive with respect to terminal V_-, the wattmeter produces a positive deflection. If the current flows in the reverse direction, OR the voltage polarity is reversed, the wattmeter produces a negative torque. If both the current and polarity are reversed, the wattmeter produces a positive deflection. For power measurement, the usual connection between the potential coil and current coil is as shown in figure 7.10, that is V_+ is linked to M.

The instantaneous total power consumption by the load in figure 7.10 is

$$p_T = v_{AS}i_A + v_{BS}i_B + v_{CS}i_C$$

where v_{AS} is the instantaneous voltage between line A and the star point of the load (assuming for the moment that the load is star-connected), i_A is the current in line A, etc. Since the load is supplied by a 3-wire system

$$i_B = -(i_A + i_C)$$

that is, the instantaneous power consumed is

$$p_T = (v_{AS} - v_{BS})i_A + (v_{CS} - v_{BS})i_C$$
$$= v_{AB}i_A + v_{CB}i_C$$

and the total average power consumed is

$$P_T = V_{AB}I_A \cos \alpha + V_{CB}I_C \cos \beta$$

where α is the phase angle between $\boldsymbol{V}_{AB}$ and $\boldsymbol{I}_A$, and β is the angle between $\boldsymbol{V}_{CB}$ and $\boldsymbol{I}_C$; P_T is the sum of the readings of the wattmeters.

The reader is asked to note that we have not applied any conditions to the load, so that *it may either be balanced or unbalanced*. There are, of course, two other possible ways in which the wattmeters may be connected to measure power, and a similar analysis will show that in either case the sum of the readings of the two wattmeters is equal to the total power consumed.

Alternatively, using complex quantities for figure 7.10, we may say that

$$P_T = \text{Re}(\boldsymbol{V}_{AB}\boldsymbol{I}_A^*) + \text{Re}(\boldsymbol{V}_{CB}\boldsymbol{I}_C^*)$$

The analysis above assumed that the load was star-connected. It will prove an interesting exercise for the reader to show that the analysis is also correct if the load is delta-connected.

7.17 Introduction to symmetrical components

An unbalanced set of 3-phase voltages or currents can be analysed into three sets of balanced components, namely:

1. A *positive phase sequence* (PPS) balanced system having the same phase sequence as the original unbalanced system (say ABC).
2. A *negative phase sequence* (NPS) balanced system having a phase sequence opposite to that of the original unbalanced system (say ACB).
3. A *zero phase sequence* (ZPS) system whose elements are all in phase with one another, and have the same magnitude.

If either the supply or load is unbalanced, either two or all three sets of the symmetrical component elements exist.

If a 'healthy' balanced system has a fault on it (other than a symmetrical short-circuit), either two or all three sets of symmetrical components exist. If only for this fact, the determination of the symmetrical components or voltage and current is a useful weapon in the armoury of an electrical engineer.

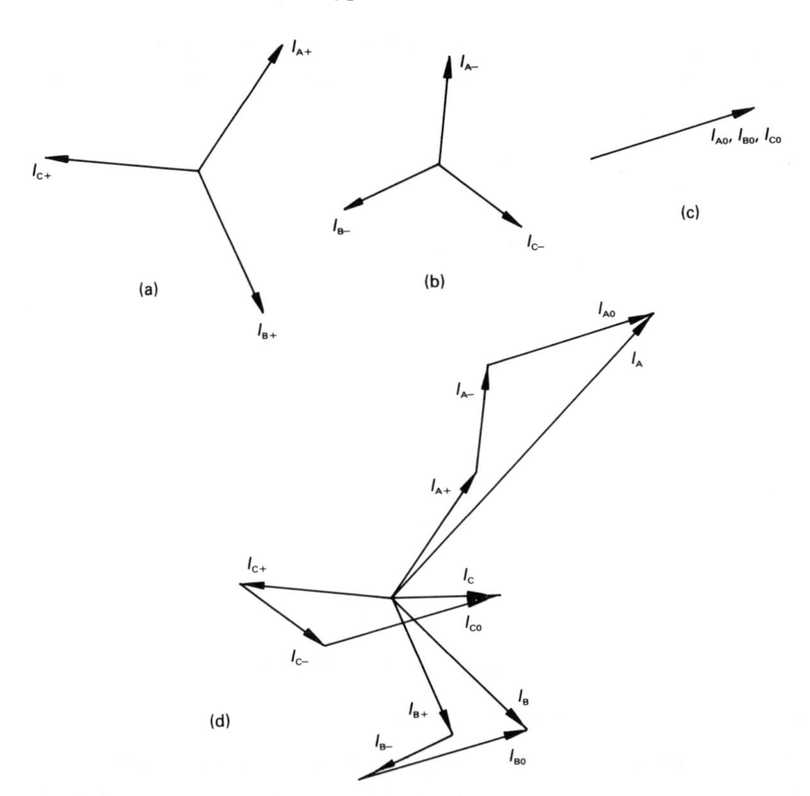

Figure 7.11 *Symmetrical components of current.*

7.18 Analysis of unbalanced conditions

If $\boldsymbol{I}_A$, $\boldsymbol{I}_B$ and $\boldsymbol{I}_C$ are a set of 3-phase unbalanced currents, then

$$\boldsymbol{I}_A = \boldsymbol{I}_{A+} + \boldsymbol{I}_{A-} + \boldsymbol{I}_{A0}$$
$$\boldsymbol{I}_B = \boldsymbol{I}_{B+} + \boldsymbol{I}_{B-} + \boldsymbol{I}_{B0}$$
$$\boldsymbol{I}_C = \boldsymbol{I}_{C+} + \boldsymbol{I}_{C-} + \boldsymbol{I}_{C0}$$

where $\boldsymbol{I}_{A+}$, $\boldsymbol{I}_{B+}$ and $\boldsymbol{I}_{C+}$ are the PPS components of the currents, $\boldsymbol{I}_{A-}$, $\boldsymbol{I}_{B-}$ and $\boldsymbol{I}_{C-}$ are the NPS components of the currents, and $\boldsymbol{I}_{A0}$, $\boldsymbol{I}_{B0}$ and $\boldsymbol{I}_{C0}$ are the ZPS components of the currents.

Figure 7.11 illustrates the use of these equations. The PPS, NPS and ZPS components of a set of unbalanced currents are shown in diagrams (a), (b) and (c), respectively, and the way in which the unbalanced currents are synthesised from them using the above equations is shown in diagram (d).

For a set of unbalanced voltages, there is a similar set of equations, that is

$$\boldsymbol{V}_A = \boldsymbol{V}_{A+} + \boldsymbol{V}_{A-} + \boldsymbol{V}_{A0}, \text{ etc.}$$

The phase sequence components can be calculated from the original unbalanced components as follows. If the complex operator h is defined as $h = 1 \angle 120°$, then

$$\boldsymbol{I}_{B+} = h^2\boldsymbol{I}_{A+} \qquad \boldsymbol{I}_{C+} = h\boldsymbol{I}_{A+}$$
$$\boldsymbol{I}_{B-} = h\boldsymbol{I}_{A-} \qquad \boldsymbol{I}_{C-} = h^2\boldsymbol{I}_{A-}$$
$$\boldsymbol{I}_{A0} = \boldsymbol{I}_{B0} = \boldsymbol{I}_{C0}$$

and

$$h + h^2 + h^3 = h + h^2 + 1 = 0$$

hence

$$\boldsymbol{I}_A = \boldsymbol{I}_{A+} + \boldsymbol{I}_{A-} + \boldsymbol{I}_{A0}$$
$$\boldsymbol{I}_B = h^2\boldsymbol{I}_{B+} + h\boldsymbol{I}_{B-} + \boldsymbol{I}_{B0}$$
$$\boldsymbol{I}_C = h\boldsymbol{I}_{C+} + h^2\boldsymbol{I}_{C-} + \boldsymbol{I}_{C0}$$

or

$$\boldsymbol{I}_{A+} = (\boldsymbol{I}_A + h\boldsymbol{I}_B + h^2\boldsymbol{I}_C)/3$$
$$\boldsymbol{I}_{A-} = (\boldsymbol{I}_A + h^2\boldsymbol{I}_B + h\boldsymbol{I}_C)/3$$
$$\boldsymbol{I}_{A0} = (\boldsymbol{I}_A + \boldsymbol{I}_B + \boldsymbol{I}_C)/3$$

The reader should note that in a 3-phase, 4-wire system, the current in the neutral wire is $(\boldsymbol{I}_A + \boldsymbol{I}_B + \boldsymbol{I}_C) = 3\boldsymbol{I}_{A0}$; that is to say, the current in the neutral wire is entirely ZPS current. Similarly, we may conclude that in a 'healthy' 3-phase, 3-wire system, none of the lines carries ZPS current. However, should a fault occur on the system (such as, for example, an earth fault when the system has a supply with an earthed neutral point), then ZPS current may flow in the supply lines.

In an unbalanced but 'healthy' delta-connected system, ZPS current does not flow in the supply lines, *but may flow* inside the mesh. This situation can arise, for example, in the case of a delta–star connected transformer when current flows in the neutral wire of the secondary circuit. A corresponding component of ZPS current flows around the closed delta-connected primary winding in order to maintain m.m.f. balance between the two windings.

7.18.1 Power consumed by symmetrical components

The total average power consumed in a system is the sum of the individual powers due to the PPS, NPS and ZPS components. No average power is associated with the voltage from one phase sequence and the current from another phase sequence.

Worked example 7.18.1

A 3-phase, 4-wire balanced supply of line voltage 502.3 V supplies the following line currents

$$\boldsymbol{I}_A = 29 \angle 0° \text{ A}, \quad \boldsymbol{I}_B = 29 \angle -140° \text{ A}, \quad \boldsymbol{I}_C = 29 \angle 160° \text{ A}$$

Calculate the PPS, NPS and ZPS components of the line current, and determine the associated power consumption.

Solution

From the equations derived earlier

$$\begin{aligned}
\boldsymbol{I}_{A+} &= (\boldsymbol{I}_A + h\boldsymbol{I}_B + h^2\boldsymbol{I}_C)/3 \\
&= (29 \angle 0° + [1 \angle 120° \times 29 \angle -140°] \\
&\qquad\qquad + [1 \angle 120° \times 29 \angle 160°])/3 \\
&= 29(1 \angle 0° + 1 \angle -20° + 1 \angle 40°)/3 = 26.32 \angle 6.3° \text{ A} \\
\boldsymbol{I}_{A-} &= (\boldsymbol{I}_A + h^2\boldsymbol{I}_B + h\boldsymbol{I}_C)/3 \\
&= (29 \angle 0° + [1 \angle 240° \times 29 \angle -140°] \\
&\qquad\qquad + [1 \angle -240° \times 29 \angle 160°])/3 \\
&= 29(1 \angle 0° + 1 \angle 100° + 1 \angle -80°)/3 = 9.67 \angle 0° \text{ A} \\
\boldsymbol{I}_{A0} &= (\boldsymbol{I}_A + \boldsymbol{I}_B + \boldsymbol{I}_C)/3 \\
&= 29(1 \angle 0° + 1 \angle -140° + 1 \angle 160°)/3 = 7.42 \angle -157° \text{ A}
\end{aligned}$$

Since the supply is balanced, $\boldsymbol{V}_{A-} = 0$ and $\boldsymbol{V}_{A0} = 0$. The total power is therefore supplied by the PPS component of the current as follows. Now

$$\boldsymbol{V}_{A+} = V_P \angle 0° = (502.3/\sqrt{3}) \angle 0° = 290 \angle 0° \text{ V}$$

hence

$$\begin{aligned}
P_T &= 3 \text{ Re}(\boldsymbol{V}_{A+}\boldsymbol{I}_{A+}^*) = 3 \text{ Re}(290 \angle 0° \times 26.32 \angle -6.3°) \\
&= 22\ 788 \text{ W}
\end{aligned}$$

Unworked problems

7.1. The voltage $\boldsymbol{V}_{AB}$ in a balanced 3-phase system is $173.2 \angle 50°$ V. Determine the value of $\boldsymbol{V}_{CN}$.
[$100 \angle 140°$ V]

7.2. If the source in problem 7.1 is a negative phase sequence source, calculate $\boldsymbol{V}_{CN}$.
[$100 \angle -40°$ V]

7.3. Calculate the active (in-phase) and reactive current components in each phase of a Y-connected, 12 kV, 3-phase alternator supplying a 5 MW load with a power factor of 0.7 lagging. If the magnitude of the line current remains unchanged, and the load power factor is raised to 0.9 lagging, calculate the new output power from the alternator.
[In-phase component = 240.6 A, quadrature component = 245.5 A; 6.43 MW]

7.4. A balanced Y-connected load of $(6 + j8)\ \Omega$ per phase is connected to a balanced 3-phase, 500 V supply. Calculate the magnitude of the line current, the power factor, and the total VA, power and VAr consumed.
[28.9 A; 0.6; 25 kVA; 15 kW; 20 kVAr]

7.5. A *phase sequence indicator* is an instrument which can be used to determine the phase sequence of a polyphase supply. A simple indicator comprises the star-connected combination of a 19 μF capacitor connected between phase A and the star point, an electrical lamp of resistance 200 Ω connected between phase B and star, and an identical lamp connected between phase C and star (the star point of the indicator *is not connected to the neutral point of the supply*). If the line voltage is 200 V, 50 Hz, calculate the voltage across the lamp connected to the B-phase when the supply is (a) PPS, (b) NPS.
[(a) $182.1 \angle -104.5°$ V; $51.36 \angle 152.4°$ V]

7.6. A balanced mesh-connected load of $(8 + j6)\ \Omega$ per phase is supplied by a balanced 500 V, 3-phase supply. Calculate the modulus of the phase current, its power factor, and determine the total apparent power, power and reactive volt-amperes.
[50 A; 0.8 (lagging); 43.3 kVA; 34.64 kW; 25.98 kVAr]

7.7. The currents in branches AB, BC and CA of a mesh-connected system supplied by a symmetrical source of phase sequence ABC are as follows:

AB: 50 A at a power factor of 0.8 lagging
BC: 60 A at a power factor of 0.7 leading
CA: 40 A at unity power factor.

Calculate the current in each line.
[$\boldsymbol{I}_A = 30 \angle -90°$ A; $\boldsymbol{I}_B = 72.9 \angle 88.4°$ A; $\boldsymbol{I}_C = 42.9 \angle -92.6°$ A]

7.8. An unbalanced set of phase voltages are

$V_{AN} = 200 \angle 0° \text{ V}, \quad V_{BN} = 200 \angle -90° \text{ V}, \quad V_{CN} = 200 \angle 90° \text{ V}$

and the corresponding line currents are

$I_A = 30 \angle 0° \text{ A}, \quad I_B = 40 \angle -145° \text{ A}, \quad I_C = 40 \angle 125° \text{ A}$

Calculate the symmetrical components of voltage and current, together with the power supplied by each set of symmetrical components, and the total power consumed.
[$V_{A0} = 66.67 \angle 0°$ V; $V_{A+} = 182.14 \angle 0°$ V; $V_{A-} = 48.8 \angle 180°$ V; $I_{A0} = 9.17 \angle 159°$ A; $I_{A+} = 35.65 \angle -7.2°$ A; $I_{A-} = 3.42 \angle 20.5°$ A; $P_{ZPS} = -1712$ W; $P_{PPS} = 19\,326$ W; $P_{NPS} = -469$ W; $P_T = 17\,145$ W]

7.9. Three delta-connected impedances of (40 − j20) Ω per phase are connected in parallel with three Y-connected impedances of (20 + j40) Ω per phase. If the line voltage is 500 V, calculate the magnitude and phase angle of the line current, and determine the total apparent power, the power and the reactive power consumed.
[20.42 A; −8.13°; 5.9 kVA; 5.84 kW; 0.83 kVAr]

7.10. A 300 V, 3-phase supply of sequence ABC is connected to the load in figure 7.12. Neglecting instrument losses, determine the power indicated by the wattmeter.
[187.3 W]

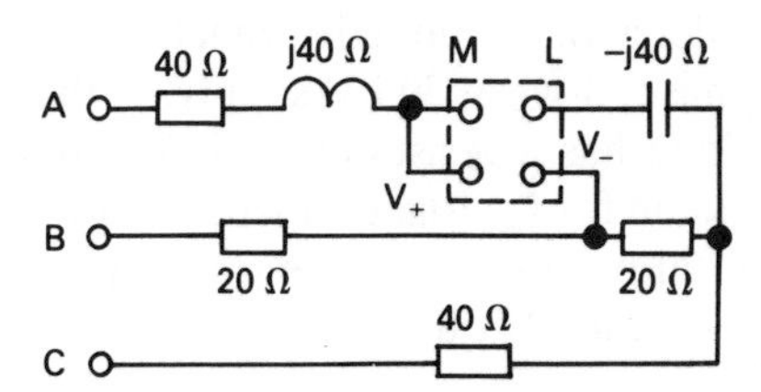

Figure 7.12

7.11. The circuit in figure 7.13 is energised by a 200 V, 3-phase supply of sequence ABC. Calculate the reading of the wattmeter.
[493 W]

7.12. A symmetrical 500 V, 3-phase, 50 Hz supply is connected to the circuit in figure 7.14. Determine the magnitude of the voltage between points XY, YZ and ZX for the phase sequence (a) ABC, (b) ACB.
[(a) 0 V; (b) 500 V]

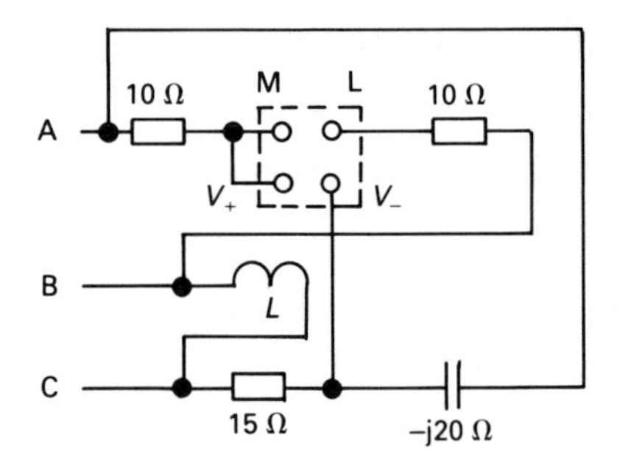

Figure 7.13

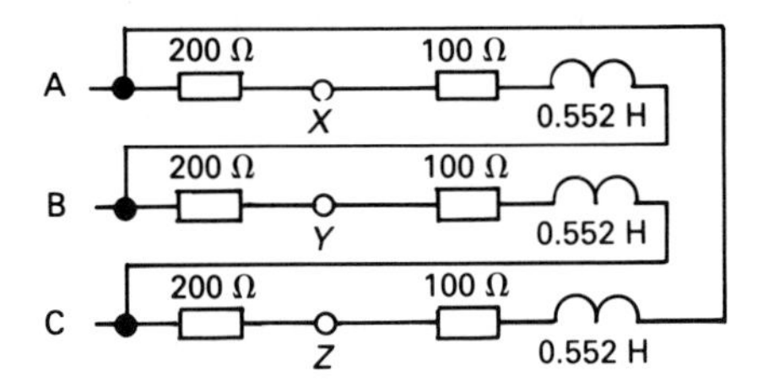

Figure 7.14

7.13. A balanced star-connected load, each phase containing a 15 Ω resistor in parallel with an inductive reactance of 5 Ω, is supplied by a balanced 3-phase, 3-wire, 200 V supply. If the line connecting the supply to the load has a resistance of 0.5 Ω per phase, calculate a complex expression for $\boldsymbol{I}_A$, together with the power consumed by the load and the total power loss in the transmission line.
[23.47 ∠ −66° A; 2480 W; 826 W]

7.14. Repeat problem 7.13 but for a delta-connected load having an impedance in each phase consisting of a 15 Ω resistance in parallel with a −j5 Ω reactance.
[23.47 ∠ 66° A; 2480 W; 826 W]

8

Two-port Networks

8.1 Introduction

A *port* is a pair of terminals in a network where a signal may either enter or leave. A *two-port network* is a class of *multi-port networks*, which have more than one port. Figure 8.1 shows a general block diagram of a two-port network, which has voltage $\boldsymbol{V}_1$ between the terminals of one port, and $\boldsymbol{V}_2$ between the terminals of the other port. A condition *which must be satisfied* in all two-port networks is $\boldsymbol{I}_1 = \boldsymbol{I}_3$ and $\boldsymbol{I}_2 = \boldsymbol{I}_4$.

Two-port networks are important building blocks in a wide range of applications, including electronics, automatic control systems, communication circuits, transmission and distribution systems, etc.

There is a wide range of parameters which can be used to define the operation of a two-port network, and are selected according to the ease with which they can be applied to a particular situation. In this chapter we will be looking at *admittance parameters*, *impedance parameters*, *hybrid parameters* and *transmission parameters*.

The parameters described in this chapter have particular application to electrical and electronic engineering; for example, admittance parameters have special use in high-frequency circuits (transistor amplifiers, RF applications, etc.), hybrid parameters are particularly useful in describing transistor characteristics, transmission parameters are important in transmission line calculations, etc.

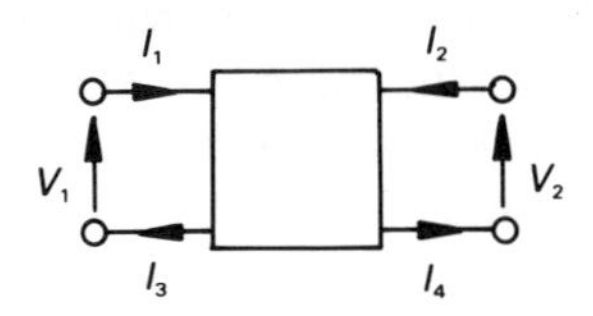

Figure 8.1 *A two-port network.*

Whenever we deal with most two-terminal networks, it is the usual convention to assume that *current flows into the upper terminal of both ports*. However, in the case of the transmission parameters, which are largely concerned with transmission lines, the current $\boldsymbol{I}_2$ flows out of the upper terminal of the right-hand port.

In every case, we assume that the two-terminal network is composed of linear elements and contains no independent sources; dependent sources are permissible. We may therefore assume that $\boldsymbol{I}_1$ and $\boldsymbol{I}_2$ are produced by the superposition of two components, one produced by $\boldsymbol{V}_1$ and the other by $\boldsymbol{V}_2$.

It is usual to think of $\boldsymbol{V}_1$ and $\boldsymbol{V}_2$ (or $\boldsymbol{I}_1$ and $\boldsymbol{I}_2$) as being produced by ideal sources, but this is not necessarily the case. Moreover, a network may be terminated at either end by another 2-port network, which may further be connected to another 2-port network, etc.

8.2 Input impedance, output impedance, voltage gain, current gain and power gain

Input impedance and admittance

When dealing with the input impedance of a 2-port network, we are concerned only with the *V–I* relationship at the input port. It is therefore only necessary to think at this stage about a simple one-port network. Quite simply, we can define the input impedance of the port as

$$\boldsymbol{Z}_{\text{in}} = \frac{\text{input voltage}}{\text{input current}} = \frac{\boldsymbol{V}_1}{\boldsymbol{I}_1}$$

Similarly, we may define the input admittance as

$$\boldsymbol{Y}_{\text{in}} = \frac{\text{input current}}{\text{input voltage}} = \frac{\boldsymbol{I}_1}{\boldsymbol{V}_1}$$

Consider the 2-port passive network enclosed in the broken lines in figure 8.2. A load is usually connected to the output terminals of the network before it can be used, and we will assume that a 2 Ω load is connected. In order to write the mesh equations for the network, we will assume that a current I_3 circulates in a clockwise direction around the mesh containing the 10 Ω, 20 Ω and 5 Ω resistors. It should be pointed out that we have adopted the 2-terminal network convention that current *flows into* the output terminals. This is taken account of in the resulting mesh equations for the circuit (including the load), which are

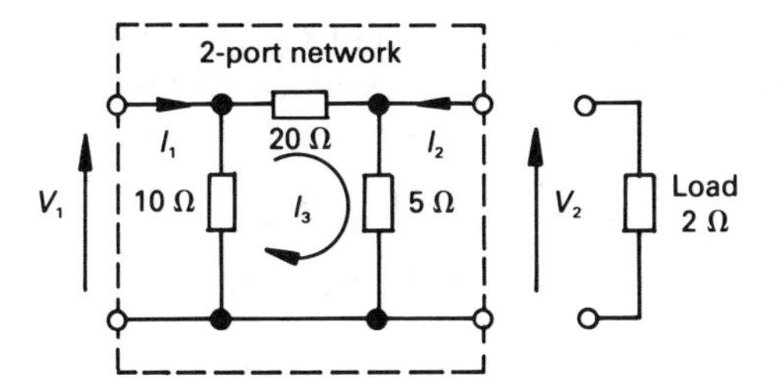

Figure 8.2 *A passive two-port network.*

$$V_1 = 10I_1 - 10I_3$$
$$0 = -10I_1 + 35I_3 + 5(-I_2)$$
$$0 = - 5I_3 - 7(-I_2)$$

If we let $V_1 = 1$ V, and solve the equations for I_1, we get $I_1 = 0.1467$ A. Accordingly

$$Z_{in} = V_1/I_1 = 1/0.1467 = 6.82\ \Omega$$

and

$$Y_{in} = 1/Z_{in} = 0.1467\ \text{S}$$

Output impedance

The output impedance of a network is simply the Thévenin equivalent impedance between the output terminals. There are a number of ways of calculating $\mathbf{Z}_{out}$, one of which (after removing the load from the circuit) is to replace the input supply source by its internal impedance, and drive the output terminals by a $1 \angle 0°$ A current. The potential difference between the output terminals is equal to $\mathbf{Z}_{out}$.

In the case of the network in figure 8.2, we can obtain the output impedance by observation as follows. Since the network is energised by a voltage source, the output impedance of the network is seen to be a 5 Ω resistor in parallel with a 20 Ω resistor (the 10 Ω resistor being short-circuited by the internal impedance of the voltage source), that is

$$Z_{out} = 5 \times 20/(5 + 20) = 4\ \Omega$$

If, on the other hand, the network had been energised at the input by an ideal current source, we would have open-circuited the input terminals when evaluating the output impedance, giving

$$Z_{out} = 5 \times (20 + 10)/(5 + [20 + 10]) = 4.29\ \Omega$$

Yet another method of calculating the output impedance is to energise the input of the network and determine (i) the open-circuit output voltage

V_{2OC}, (ii) the short-circuit output current I_{2SC}; the output impedance is the ratio of these two values. If I_{2SC} is assumed to *flow out* of the network, then

$$Z_{out} = V_{2OC}/I_{2SC}$$

and if I_{2SC} is assumed to *flow into* the network, then

$$Z_{out} = V_{2OC}/(-I_{2SC})$$

We illustrate the latter approach using nodal analysis applied to the network in figure 8.2. If the input is energised by a $1 \angle 0°$ A current source, the nodal equations with the output terminal open-circuited ($I_2 = 0$) are

$$I_1 = 1 = 0.15V_1 \quad - 0.05V_{2OC}$$

$$I_2 = 0 = -0.05V_1 + 0.25V_{2OC}$$

Solving gives $V_{2OC} = 1.429$ V.

The short-circuit ($V_2 = 0$) output current when $I_1 = 1 \angle 0°$ A can be evaluated from the following nodal equations

$$I_1 = 1 = 0.15V_1 - 0 \qquad I_{2SC} = -0.05V_1 + 0$$

giving $I_{2SC} = -0.3333$ A, hence

$$Z_{out} = V_{2OC}/(-I_{2SC}) = 4.29\ \Omega$$

Finally, another method of evaluating Z_{out} is to replace the input source by its internal impedance and, with the load disconnected, drive the output terminals with a current of $1 \angle 0°$ A; the voltage (in volts) between the output terminals is equal to Z_{out} (in Ω).

Voltage gain

The voltage gain, G_V, of a network is given under normal operating conditions *with the load connected* by the ratio

$$G_V = V_2/V_1$$

The reader will note that we use G for *gain*; it is a convention to use G in this way, and the reader should not confuse it with G for conductance.

To determine its value for the passive network in figure 8.2, let us drive the input with a current of $1 \angle 0°$ A, and determine the voltage from the nodal equations, which are

$$I_1 = 1 = \quad 0.15V_1 - 0.05V_2$$

and

$$0 = -0.05V_1 + 0.75V_2$$

From the second equation we see that

$$G_V = V_2/V_1 = 0.05/0.75 = 0.06667$$

Although, in this case, $\boldsymbol{V}_2$ is less than $\boldsymbol{V}_1$ we, none-the-less, refer to the ratio as the voltage gain.

Current gain

The current gain, $\boldsymbol{G}_{\text{I}}$, of a four-terminal network is calculated under normal operating conditions (*with the load connected*) from the ratio

$$\boldsymbol{G}_{\text{I}} = \boldsymbol{I}_2/\boldsymbol{I}_1$$

If we energise the passive network in figure 8.2 by a voltage of $1 \angle 0°$ V, and write down the mesh equations we obtain

$$\begin{aligned} \boldsymbol{V}_1 = 1 &= \quad 10\boldsymbol{I}_1 - 10\boldsymbol{I}_3 \\ 0 &= -10\boldsymbol{I}_1 + 35\boldsymbol{I}_3 + 5(-\boldsymbol{I}_2) \\ 0 &= \qquad\quad - \; 5\boldsymbol{I}_3 - 7(-\boldsymbol{I}_2) \end{aligned}$$

where $\boldsymbol{I}_2$ flows into the network; solving for $\boldsymbol{I}_1$ and $\boldsymbol{I}_2$ yields

$$\boldsymbol{I}_1 = 0.147 \angle 0° \text{ A}$$

$$\boldsymbol{I}_2 = -0.033 \angle 0° \text{ A (flowing into the network)}$$

and the current gain of the network is

$$\boldsymbol{G}_{\text{I}} = \boldsymbol{I}_2/\boldsymbol{I}_1 = -0.033/0.147 = -0.224$$

The reader will have observed that we could have calculated $\boldsymbol{I}_1$ from

$$\boldsymbol{I}_1 = \boldsymbol{V}_1/\boldsymbol{Z}_{\text{in}} = 1/6.82 = 0.147 \text{ A}$$

Power gain

The power gain, G_{P}, of a network is given by the ratio

$$G_{\text{P}} = \frac{\text{output power}}{\text{input power}} = \frac{P_{\text{out}}}{P_{\text{in}}}$$

Assuming sinusoidal excitation, the power gain is

$$G_{\text{P}} = \frac{\text{Re}[-\boldsymbol{V}_2\boldsymbol{I}_2^*]}{\text{Re}[\boldsymbol{V}_1\boldsymbol{I}_1^*]}$$

The negative sign in the numerator arises from the fact that the output current is assumed to flow into the output terminals of the 2-port network.

There are several ways of calculating the power gain, one of the simplest being

$$G_{\text{P}} = |G_{\text{V}}G_{\text{I}}|$$

In the case of the network in figure 8.2, the power gain is

$$G_P = |0.0667 \times (-0.224)| = 0.0149$$

Although the values above were calculated for a passive network, the techniques involved are applicable to networks containing dependent sources.

8.3 Admittance parameters or *y*-parameters

In this case the network is defined by the equations

$$I_1 = y_{11}V_1 + y_{12}V_2$$

$$I_2 = y_{21}V_1 + y_{22}V_2$$

The *y*-parameters are simply constants of proportionality, and have dimensions of siemens (S). We may write the equations in general matrix form as follows.

$$\begin{bmatrix} I_1 \\ I_2 \end{bmatrix} = \begin{bmatrix} y_{11} & y_{12} \\ y_{21} & y_{22} \end{bmatrix} \begin{bmatrix} V_1 \\ V_2 \end{bmatrix}$$

The parameter y_{11} can be evaluated by measuring V_1 and I_1 when the output terminals are short-circuited ($V_2 = 0$). That is

$$I_1 = y_{11}V_1 + 0$$

or

$$y_{11} = \left.\frac{I_1}{V_1}\right|_{V_2 = 0}$$

Similarly, we may calculate the other *y*-parameters as follows

$$y_{12} = \left.\frac{I_1}{V_2}\right|_{V_1 = 0}$$

$$y_{21} = \left.\frac{I_2}{V_1}\right|_{V_2 = 0}$$

$$y_{22} = \left.\frac{I_2}{V_2}\right|_{V_1 = 0}$$

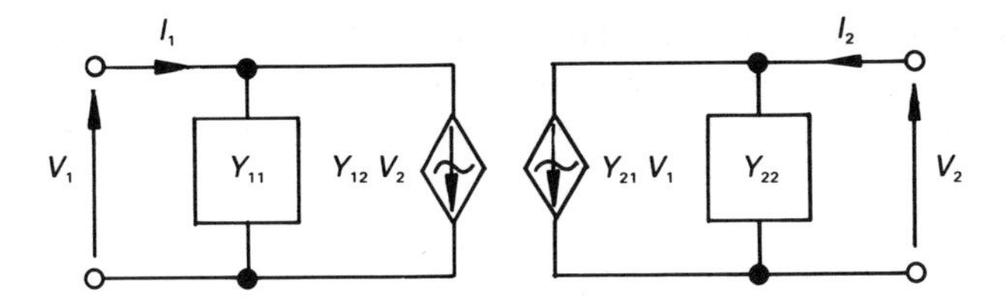

Figure 8.3 *A two-port,* y-*parameter equivalent circuit.*

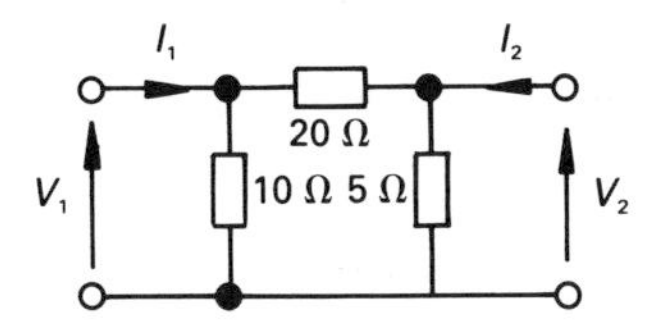

Figure 8.4 *Calculation of* y-*parameters.*

Because each parameter has the dimensions of admittance, and each is obtained by short-circuiting either the input or output port, the y-parameters are also known as the *short-circuit admittance parameters*. More specifically, the parameters are known respectively as

$\mathbf{y}_{11}$ is the *short-circuit input admittance*

$\mathbf{y}_{22}$ is the *short-circuit output admittance*

$\mathbf{y}_{12}$ and $\mathbf{y}_{21}$ are the *short-circuit transfer admittances*

There are several equivalent circuits corresponding to the y-parameter equations, one circuit being shown in figure 8.3.

It is pointed out here that the parameters y_{11}, y_{12}, y_{21} and y_{22} are sometimes known as parameters y_i, y_r, y_f and y_o, respectively, corresponding to input, reverse, forward and output parameters.

As a simple example, we will calculate the y-parameters for the passive 2-port network in figure 8.4.

Now, $\mathbf{y}_{11}$ is the input admittance with the output terminals shorted, that is its value is (1/10 + 1/20) S, or

$$\mathbf{y}_{11} = 0.15 \text{ S}$$

To calculate $\mathbf{y}_{12}$, we short-circuit the input terminals, and apply 1 V to the output terminals. The current which flows in the short-circuit in the direction of $\mathbf{I}_1$ is equal to the value of $\mathbf{y}_{12}$. Inspecting figure 8.4, we see that the current which *flows out* of the input terminals is 1/20 A. However, we define $\mathbf{I}_1$ as the current *flowing into the input terminals*, hence

$$\mathbf{y}_{12} = -0.05 \text{ S}$$

Similarly, to calculate $\boldsymbol{y}_{21}$, we short-circuit the output terminals and apply 1 V to the input terminals. The current which flows in the short-circuit in the direction of $\boldsymbol{I}_2$ is equal to $\boldsymbol{y}_{21}$. By a similar argument to that above, we see that

$$\boldsymbol{y}_{21} = -0.05 \text{ S}$$

The parameter $\boldsymbol{y}_{22}$ is the admittance between the output terminals with the input short-circuited which, by observation, is

$$\boldsymbol{y}_{22} = (1/5 + 1/20) = 0.25 \text{ S}$$

The y-parameter equations for the 2-port network in figure 8.4 are therefore

$$\boldsymbol{I}_1 = 0.15\boldsymbol{V}_1 - 0.05\boldsymbol{V}_2$$

$$\boldsymbol{I}_2 = -0.05\boldsymbol{V}_1 + 0.25\boldsymbol{V}_2$$

and

$$[y] = \begin{bmatrix} 0.15 & -0.05 \\ -0.05 & 0.25 \end{bmatrix}$$

It is not a coincidence that $y_{12} = y_{21}$ in the above calculation. Elements such as resistors, inductors and capacitors (other than, perhaps, electrolytic capacitors) can be connected in a circuit in either direction, and the result is the same. These elements are known as *bilateral elements*. A network containing only bilateral elements is known as a *bilateral network*, and for such a two-port it can be shown that $y_{12} = y_{21}$. Some two-ports contain non-bilateral elements, such as dependent sources; these networks also have this property. A two-port for which $y_{12} = y_{21}$ is known as a *reciprocal network*.

Worked example 8.3.1

Determine the y-parameters of the network in figure 8.5, which is a small-signal linear equivalent circuit of a transistor in the common-emitter mode with resistive feedback between the collector and base. Calculate the

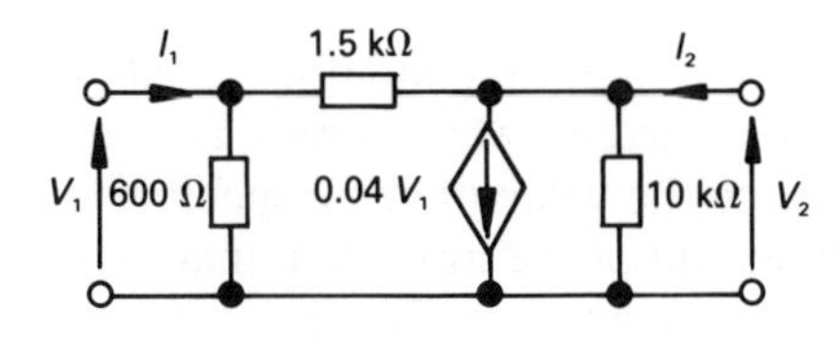

Figure 8.5 *Figure for worked example 8.3.1.*

voltage gain and the current gain of the 2-port network when a 2 kΩ resistor is connected to the output terminals.

Solution

Whatever circuit we analyse with the y-parameters, we must strictly adhere to the parameter definitions. The parameter $\boldsymbol{y}_{11}$ is the input admittance measured with the output terminals short-circuited. Since, in this case, the dependent source in figure 8.5 is shorted, then

$$\boldsymbol{y}_{11} = (1/600 + 1/1500) = 2.333 \text{ mS}$$

The parameter $\boldsymbol{y}_{12}$ is equal to the current which flows in a short-circut between the input terminals in the direction of $\boldsymbol{I}_1$ when 1 V is applied to the output terminals. Since $\boldsymbol{V}_1 = 0$, no current flows in the dependent source, and

$$\boldsymbol{y}_{12} = -1/1500 \text{ S} = -0.6667 \text{ mS}$$

The parameter $\boldsymbol{y}_{21}$ is the current flowing in a short-circuit between the output terminals in the direction of $\boldsymbol{I}_2$ when we apply 1 V to the input terminals. Since $\boldsymbol{V}_2 = 0$, the current flowing through the 1.5 kΩ resistor towards the output is 1/1500 A. The current in the current source is $0.04\boldsymbol{V}_1 = 0.04$ A, hence the current flowing in the short-circuit between the output terminals gives the value

$$\boldsymbol{y}_{21} = (0.04 - 1/1500) \text{ S} = 39.33 \text{ mS}$$

Finally, $\boldsymbol{y}_{22}$ is the admittance between the output terminals when the input terminals are shorted. Since $\boldsymbol{V}_1 = 0$, no current flows through the dependent source, and the 600 Ω resistor is short-circuited. Hence

$$\boldsymbol{y}_{22} = 1/10\,000 + 1/1500 = 0.767 \text{ mS}$$

The y-parameter equations which apply to figure 8.5 are therefore

$$\boldsymbol{I}_1 = 2.333\boldsymbol{V}_1 - 0.6667\boldsymbol{V}_2 \text{ mA}$$

$$\boldsymbol{I}_2 = 39.33\boldsymbol{V}_1 + 0.767\boldsymbol{V}_2 \text{ mA}$$

In order to calculate the voltage and current gain, we need to connect the 2 kΩ load, and supply a current to the input; let this be 1 mA. That is

$$\boldsymbol{I}_1 = 1 \text{ mA} \quad \text{and} \quad \boldsymbol{V}_2 = -2000\boldsymbol{I}_2 \text{ A} = -2\boldsymbol{I}_2 \text{ mA}$$

or

$$\boldsymbol{I}_2 = -0.5\boldsymbol{V}_2 \text{ mA}$$

Substituting these values into the y-parameter equations for the network gives

$$\begin{aligned} \boldsymbol{I}_1 &= 1 \times 10^{-3} = 2.333 \times 10^{-3}\boldsymbol{V}_1 - 0.6667 \times 10^{-3}\boldsymbol{V}_2 \\ \boldsymbol{I}_1 &= 0 \qquad\quad\;\; = 39.33 \times 10^{-3}\boldsymbol{V}_1 + 1.267 \times 10^{-3}\boldsymbol{V}_2 \end{aligned}$$

From the second of these equations we get

$$G_V = \frac{V_2}{V_1} = -\frac{1.348}{0.0434} = -31.06$$

and from the first of the equations

$$1 \times 10^{-3} = 2.333 \times 10^{-3} \times \left[-\frac{V_2}{31.06} \right] - 0.6667 \times 10^{-3} V_2$$

$$= -0.472 V_2$$

or $V_2 = -1.35$ V

hence $I_2 = -0.5 V_2 = 0.675$ mA

therefore $G_I = I_2/I_1 = 0.675/1 = 0.675$

8.4 Impedance parameters or z-parameters

Once again, input and output voltages and currents are assigned as in section 8.1. The impedance parameters are specified by the following equations

$$V_1 = z_{11} I_1 + z_{12} I_2$$
$$V_2 = z_{21} I_1 + z_{22} I_2$$

where

$$z_{11} = \left. \frac{V_1}{I_1} \right|_{I_2 = 0}$$

$$z_{12} = \left. \frac{V_1}{I_2} \right|_{I_1 = 0}$$

$$z_{21} = \left. \frac{V_2}{I_1} \right|_{I_2 = 0}$$

$$z_{22} = \left. \frac{V_2}{I_2} \right|_{I_1 = 0}$$

Since the parameter values are calculated for zero current at either input or output, the z-parameters are known as *open-circuit impedance parameters*, all having dimensions of ohms. More specifically

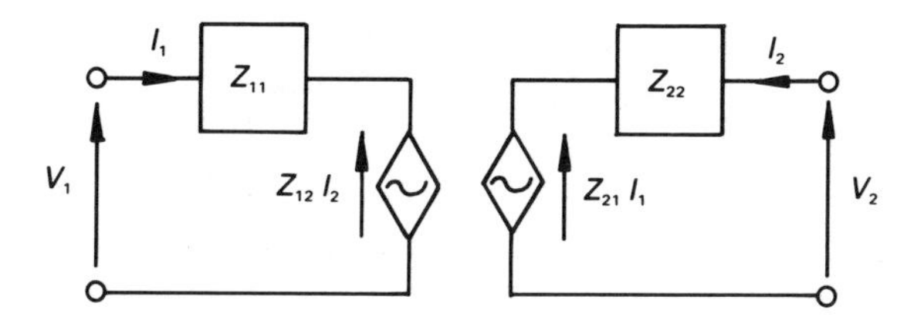

Figure 8.6 *A z-parameter equivalent circuit.*

z_{11} is the *open-circuit input impedance*
z_{22} is the *open-circuit output impedance*
z_{12} and z_{21} are the *open-circuit transfer impedances*

The above equations establish the equivalent circuit of the network, one version being shown in figure 8.6.

As with the y-parameters, we often use z_i, z_r, z_f and z_o to represent z_{11}, z_{12}, z_{21} and z_{22} respectively.

We will now calculate the z-parameters for the passive 2-port network in figure 8.4. The parameter z_{11} is the input impedance with the output terminals open-circuited, leaving 10 Ω in parallel with (20 + 5) Ω, or

$$z_{11} = 10 \times (20 + 5)/(10 + (20 + 5)) = 7.143\ \Omega$$

To calculate z_{12} we open-circuit the input terminals, and cause a current of $1\angle 0°$ A to flow into the upper output terminal: z_{12} is equal to the voltage between the input terminals (the upper terminal being assumed positive).

With the input terminals open-circuited, the current which flows in the 10 Ω resistor (making the upper input terminal positive) when a current of 1 A enters the upper output terminal is

$$1 \times 5/(5 + (20 + 10)) = 0.1428\ \text{A}$$

Hence

$$z_{12} = 10 \times 0.1428 = 1.428\ \Omega$$

Similarly

$$z_{21} = 5 \times 10/(10 + (20 + 5)) = 1.428\ \Omega$$

and

$$\begin{aligned} z_{22} &= (5 \times (20 + 10))/(5 + (20 + 10) \\ &= 4.286\ \Omega \end{aligned}$$

Once again, it will be seen that $z_{12} = z_{21}$ for a reciprocal network.

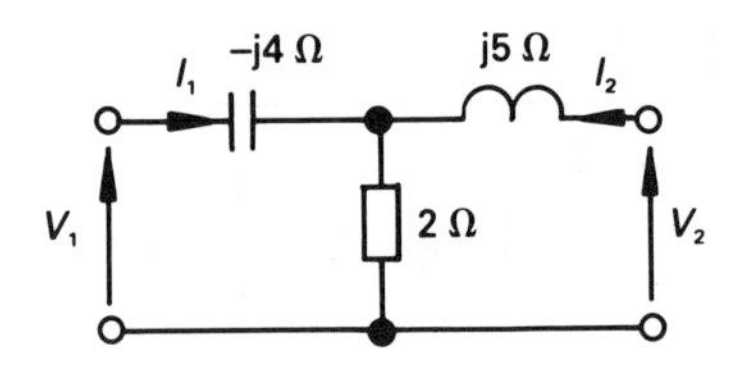

Figure 8.7 *Circuit for worked example 8.4.1.*

Worked example 8.4.1

Calculate the z-parameters of the 2-port network in figure 8.7. Hence calculate the current and voltage gain when a 10 Ω resistor is connected to the output port.

Solution

Many of the values can, in this case, be obtained by observation. The parameter z_{11} is the input impedance when the output terminals are open-circuited. By observation

$$z_{11} = 2 - \mathrm{j}4\ \Omega$$

Since z_{12} is calculated with $I_1 = 0$, then V_1 is equal to the voltage across the 2 Ω resistor, that is

$$V_1 = 2I_2$$

and

$$z_{12} = V_1/V_2 = 2\ \Omega$$

Similarly, z_{21} is calculated with $I_2 = 0$; V_2 is equal to the voltage across the 2 Ω resistor, hence

$$V_2 = 2I_1$$

and

$$z_{21} = V_2/I_1 = 2\ \Omega$$

Since the circuit contains only bilateral elements, it is a reciprocal network, and one would expect that $z_{12} = z_{21}$.

The parameter z_{22} is the impedance between the output terminals when the input is open-circuited. That is

$$z_{22} = 2 + \mathrm{j}5\ \Omega$$

The z-parameter equations which apply to figure 8.6 are therefore

$$V_1 = (2 - j4)I_1 + 2I_2$$
$$V_2 = 2I_1 + (2 + j5)I_2$$

To determine the current and voltage gain of the complete circuit, we must connect the 10 Ω load and apply a voltage of, say, 1 V to the input port. In this case

$$V_1 = 1\angle 0° \text{ V} \qquad V_2 = -10I_2$$

The equations which apply to the complete circuit therefore are

$$V_1 = 1 = (2 - j4)I_1 + 2I_2$$
$$V_2 = -10I_2 = 2I_1 + (2 + j5)I_2$$

The second equation yields

$$G_I = \frac{I_2}{I_1} = -\frac{2}{12 + j5} = 0.154\angle 157.4°$$

and the first gives

$$1 = \frac{(2 - j4)I_2}{0.154\angle 157.4°} + 2I_2$$

or $$I_2 = 0.036\angle -136.5° \text{ A}$$

Hence $$V_2 = -10I_2 = 0.36\angle 43.5° \text{ V}$$

therefore

$$G_V = V_2/V_1 = 0.36\angle 43.5°$$

8.5 Hybrid parameters or *h*-parameters

Hybrid parameter or mixed parameter networks are well suited to bipolar transistor circuits, because the parameters are relatively easy to measure experimentally. The hybrid parameter equations for a 2-port network are written in the form

$$V_1 = h_{11}I_1 + h_{12}V_2$$
$$I_2 = h_{21}I_1 + h_{22}V_2$$

By the nature of the equations, it is evident that the parameters have different dimensions, two of them being dimensionless.

The parameters h_{11} and h_{21} are obtained by letting $V_2 = 0$, while h_{12} and h_{22} are determined by letting $I_1 = 0$, as follows

$$\mathbf{h}_{11} = \left.\frac{\mathbf{V}_1}{\mathbf{I}_1}\right|_{\mathbf{V}_2 = 0} \quad \text{(ohms)}$$

$$\mathbf{h}_{12} = \left.\frac{\mathbf{V}_1}{\mathbf{V}_2}\right|_{\mathbf{I}_1 = 0} \quad \text{(dimensionless)}$$

$$\mathbf{h}_{21} = \left.\frac{\mathbf{I}_2}{\mathbf{I}_1}\right|_{\mathbf{V}_2 = 0} \quad \text{(dimensionless)}$$

$$\mathbf{h}_{22} = \left.\frac{\mathbf{I}_2}{\mathbf{V}_2}\right|_{\mathbf{I}_1 = 0} \quad \text{(siemens)}$$

The parameters are respectively known as

$\mathbf{h}_{11}$ = *short-circuit input impedance*
$\mathbf{h}_{12}$ = *open-circuit reverse voltage gain*
$\mathbf{h}_{21}$ = *short-circuit forward current gain*
$\mathbf{h}_{22}$ = *open-circuit output admittance*

In transistor analysis, the parameters $\mathbf{h}_{11}$, $\mathbf{h}_{12}$, $\mathbf{h}_{21}$ and $\mathbf{h}_{22}$ become $\mathbf{h}_i$, $\mathbf{h}_r$, $\mathbf{h}_f$ and $\mathbf{h}_o$, respectively. In addition, the subscripts b, e and c are added to indicate whether the transistor is operating in the common-base, common-emitter or common-collector mode. Thus, in the common-emitter mode, we have the parameters $\mathbf{h}_{ie}$, $\mathbf{h}_{re}$, $\mathbf{h}_{fe}$ and $\mathbf{h}_{oe}$.

Once again, the equations lead us directly to the equivalent circuit of the network, shown in figure 8.8.

We will analyse the passive 2-port in figure 8.2 to illustrate the application of *h*-parameters. The value of $\mathbf{h}_{11}$ is equal to the input impedance of the network with the output short-circuited. By observation, this is 10 Ω in parallel with 20 Ω, giving

$$\mathbf{h}_{11} = 10 \times 20/(10 + 20) = 6.667\ \Omega$$

The value $\mathbf{h}_{12}$ is calculated from the ratio of $\mathbf{V}_1$ to $\mathbf{V}_2$ with the input open-circuited, and a voltage of $1\angle 0°$ V applied to the output terminals.

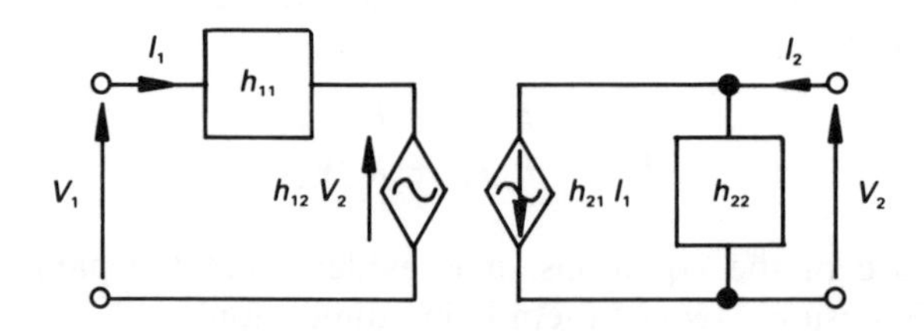

Figure 8.8 *A 2-port* h-*parameter equivalent circuit.*

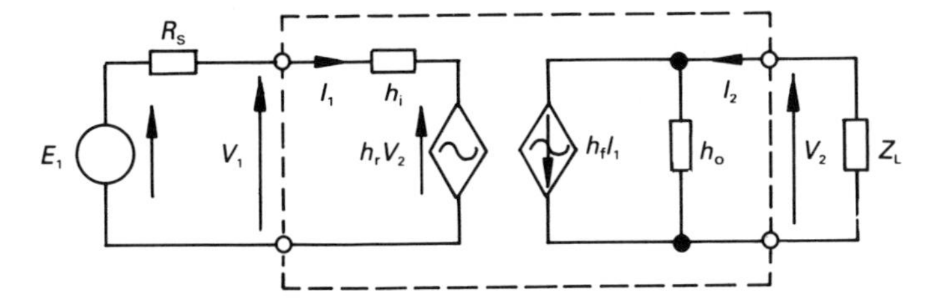

Figure 8.9 *Figure for worked example 8.5.1.*

This is fairly easy to calculate in this case because $\boldsymbol{V}_1$ is the voltage across the 10 Ω resistor in a chain containing a 10 Ω and a 20 Ω resistor between the output terminals. That is

$$\boldsymbol{h}_{12} = 1 \times 10/(10 + 20) = 0.333$$

$\boldsymbol{h}_{21}$ is the short-circuit forward current gain, and can be calculated by short-circuiting the output terminals, and applying a current of $1\angle 0°$ A to the input terminals. The value of $\boldsymbol{h}_{21}$ is the ratio of the current in the short-circuit which flows in the direction of $\boldsymbol{I}_2$ in figure 8.2 to the input current. With 1 A at the input, the current in the short-circuit is

$$-\boldsymbol{I}_2 = 1 \times 10/(10 + 20) = 0.3333 \text{ A}$$

hence

$$\boldsymbol{h}_{21} = \boldsymbol{I}_2/\boldsymbol{I}_1 = -0.3333$$

Finally, we calculate $\boldsymbol{h}_{22}$ by open-circuiting the input terminals and evaluate the admittance between the output terminals, that is

$$\boldsymbol{h}_{22} = \frac{1}{5} + \frac{1}{20 + 10} = 0.2333 \text{ S}$$

Worked example 8.5.1

For the hybrid parameter equivalent circuit in figure 8.9, determine an expression for (a) the current gain $\boldsymbol{I}_2/\boldsymbol{I}_1$, (b) the voltage gain $\boldsymbol{V}_2/\boldsymbol{V}_1$

Solution

The diagram in figure 8.9 satisfies the small-signal operation of a transistor in any mode (provided that the parameters for the correct operating mode are supplied), and the solutions are applicable to any mode of operation.
(a) *Current gain*
Applying the rule for current division in the output circuit we get

$$I_2 = h_f I_1 \frac{1/h_o}{Z_L + 1/h_o} = I_1 \frac{h_f}{1 + h_o Z_L}$$

hence

$$G_I = \frac{I_2}{I_1} = \frac{h_f}{1 + h_o Z_L}$$

(b) *Voltage gain*
Applying KVL to the input circuit yields

$$I_1 = \frac{E_1 - h_r V_2}{R_S + h_i}$$

When KCL is applied to the output circuit, we see that

$$I_2 = h_f I_1 + h_o V_2$$

however

$$I_2 = -V_2/Z_L$$

or

$$\frac{-V_2}{Z_L} = h_f I_1 + h_o V_2$$

Substituting I_1 from above gives

$$\frac{-V_2}{Z_L} = h_f \left[\frac{V_1 - h_r V_2}{R_S + h_i}\right] + h_o V_2$$

or

$$V_2 \left[\frac{h_f h_r}{R_S + h_i} - h_o - \frac{1}{Z_L}\right] = \frac{h_f V_1}{R_S + h_i}$$

That is

$$G_V = \frac{V_2}{V_1} = \frac{-h_f Z_L}{(R_S + h_i)(1 + Z_L h_o) - h_f h_r Z_L}$$

The negative sign associated with G_V implies that there is a phase shift of 180° between the input and the output. This value of phase shift will, of course, be modified at differing values of frequency if Z_L contains a reactive element (see also chapters 11 and 12).

8.6 Transmission parameters

These parameters are also known as the *ABCD*-parameters or *t*-parameters or *a*-parameters. As the name implies, they are largely concerned

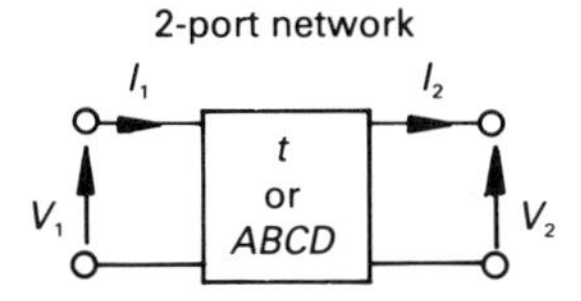

Figure 8.10 *Transmission parameters.*

with transmission lines (but are also useful in other applications including control systems). Consequently, we are concerned with current which *leaves* the output port (figure 8.10), and we define the parameters as follows

$$\begin{aligned} V_1 &= t_{11}V_2 + t_{12}I_2 \\ I_1 &= t_{21}V_2 + t_{22}I_2 \end{aligned}$$

A positive value of I_2 means that current leaves the network. The t-parameters are related to the $ABCD$-parameters as follows

$$A = t_{11}, \quad B = t_{12}, \quad C = t_{21}, \quad D = t_{22}$$

The parameters are calculated as follows

$$t_{11} = \left.\frac{V_1}{V_2}\right|_{I_2 = 0} \quad \text{(dimensionless)}$$

$$t_{12} = \left.\frac{V_1}{I_2}\right|_{V_2 = 0} \quad \text{(ohms)}$$

$$t_{21} = \left.\frac{I_1}{V_2}\right|_{I_2 = 0} \quad \text{(S)}$$

$$t_{22} = \left.\frac{I_1}{I_2}\right|_{V_2 = 0} \quad \text{(dimensionless)}$$

The reciprocity theorem can be used to find the relationship between the t-parameters of a *passive two-port* as follows. If we apply voltage V_T to the input terminals of the two-port, as shown in figure 8.11(a) then, since the output terminals are short-circuited ($V_2 = 0$), the circuit equations are

$$V_T = t_{12}I_{2SC} \tag{8.1}$$

$$I_{1SC} = t_{22}I_{2SC}$$

When V_T is applied to the 'input' terminals, as shown in figure 8.11(b), and the 'output' terminals are short-circuited, then $V_1 = 0$ and the direction of the current reverses. The equations for figure 8.11(b) are

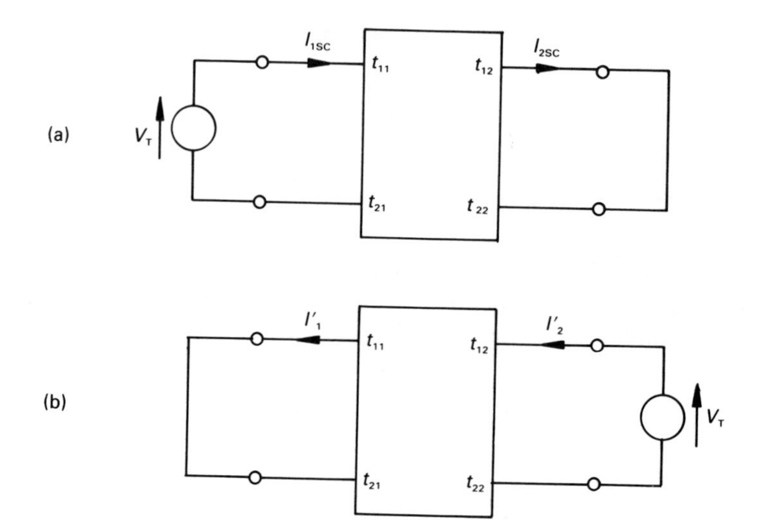

Figure 8.11 *Relationship between the* t- *or ABCD parameters.*

$$0 = t_{11}V_T - t_{12}I'_2 \tag{8.2}$$
$$-I'_1 = t_{21}V_T - t_{22}I'_2 \tag{8.3}$$

From equation (8.2) we see that

$$I'_2 = t_{11}V_T/t_{12}$$

and from equation (8.3)

$$-I'_1 = t_{21}V_T - t_{22}t_{11}V_T/t_{12}$$
$$= (t_{21} - t_{22}t_{11}/t_{12})V_T$$

The reciprocity theorem states that if the network does not contain voltage or current sources, then $I'_1 = I_{2SC}$. That is

$$-I_{2SC} = (t_{21} - t_{22}t_{11}/t_{12})V_T$$

but, from equation (8.1) $V_T = t_{12}I_{2SC}$, or

$$-I_{2SC} = (t_{21} - t_{22}t_{11}/t_{12})t_{12}I_{2SC}$$

or

$$-1 = t_{12}t_{21} - t_{11}t_{22}$$

that is

$$t_{11}t_{22} - t_{12}t_{21} = 1$$

In terms of the *ABCD* parameters this becomes

$$AD - BC = 1$$

As a simple example, we will calculate the *t*-parameters of the passive 2-port in figure 8.2.

The parameter t_{11} is calculated by applying $1\angle 0°$ V to the input and open-circuiting the output ($I_2 = 0$), the ratio V_1/V_2 giving the value of t_{11} as follows

$$V_2 = 1 \times 5/(20 + 5) = 1/5 \text{ V}$$

hence

$$t_{11} = V_1/V_2 = 1/(1/5) = 5$$

We calculate t_{12} by short-circuiting the output ($V_2 = 0$) and applying $1\angle 0°$ V to the input. The current in the short-circuit (which *leaves* the top output terminal) is

$$I_2 = V_1/20 = 1/20 = 0.05 \text{ A}$$

that is

$$t_{12} = V_1/I_2 = 1/0.05 = 20 \ \Omega$$

To determine t_{21} we open-circuit the output ($I_2 = 0$) and apply a current of $1\angle 0°$ A at the input. By current division, the current in the 5 Ω resistor is

$$1 \times 10/(10 + (20 + 5)) = 0.286 \text{ A}$$

and

$$V_2 = 5 \times 0.286 = 1.43 \text{ V, giving}$$
$$t_{21} = I_1/V_2 = 1/1.43 = 0.7$$

Finally, we evaluate t_{22} by short-circuiting the output ($V_2 = 0$) and connect a $1\angle 0°$ A current source to the input. The current in the short-circuit (*leaving* the output) is

$$I_2 = 1 \times 10/(10 + 20) = 0.333 \text{ A}$$

hence

$$t_{22} = I_1/I_2 = 1/0.333 = 3$$

also

$$t_{11}\, t_{22} - t_{12}\, t_{21} = (5 \times 3) - (20 \times 0.7) = 1$$

8.7 Relationships between the *y*-, *z*-, *h*- and *t*-parameters

It would be very tedious to deduce the relationship between each set of parameters, so we will merely indicate how this is done by relating the *y*- and *z*-parameters. The complete relationship between the *z*-, *y*-, *h*- and *t*-parameters is listed in table 8.1.

The *y*-parameter equations for a 2-port network are

$$I_1 = y_{11}V_1 + y_{12}V_2$$

$$I_2 = y_{21}V_1 + y_{22}V_2$$

Table 8.1 Transformation between *y*-, *z*-, *h*- and *t*-parameters*

To \ From	y		z		h		t	
y	y_{11}	y_{12}	$\frac{z_{22}}{\Delta_z}$	$\frac{-z_{12}}{\Delta_z}$	$\frac{1}{h_{11}}$	$\frac{-h_{12}}{h_{11}}$	$\frac{t_{22}}{t_{12}}$	$\frac{-\Delta_t}{t_{12}}$
	y_{21}	y_{22}	$\frac{-z_{21}}{\Delta_z}$	$\frac{z_{11}}{\Delta_z}$	$\frac{h_{21}}{h_{11}}$	$\frac{\Delta_h}{h_{11}}$	$\frac{-1}{t_{12}}$	$\frac{t_{11}}{t_{12}}$
z	$\frac{y_{22}}{\Delta_y}$	$\frac{-y_{12}}{\Delta_y}$	z_{11}	z_{12}	$\frac{\Delta_h}{h_{22}}$	$\frac{h_{12}}{h_{22}}$	$\frac{t_{11}}{t_{21}}$	$\frac{\Delta_t}{t_{21}}$
	$\frac{-y_{21}}{\Delta_y}$	$\frac{y_{11}}{\Delta_y}$	z_{21}	z_{22}	$\frac{-h_{21}}{h_{22}}$	$\frac{1}{h_{22}}$	$\frac{1}{t_{21}}$	$\frac{t_{22}}{t_{21}}$
h	$\frac{1}{y_{11}}$	$\frac{-y_{12}}{y_{11}}$	$\frac{\Delta_z}{z_{22}}$	$\frac{z_{12}}{z_{22}}$	h_{11}	h_{12}	$\frac{t_{12}}{t_{22}}$	$\frac{\Delta_t}{t_{22}}$
	$\frac{y_{21}}{y_{11}}$	$\frac{\Delta_y}{y_{11}}$	$\frac{-z_{21}}{z_{22}}$	$\frac{1}{z_{22}}$	h_{21}	h_{22}	$\frac{-1}{t_{22}}$	$\frac{t_{21}}{t_{22}}$
t	$\frac{-y_{22}}{y_{21}}$	$\frac{-1}{y_{21}}$	$\frac{z_{11}}{z_{21}}$	$\frac{\Delta_z}{z_{21}}$	$\frac{-\Delta_h}{h_{21}}$	$\frac{-h_{11}}{h_{21}}$	t_{11}	t_{12}
	$\frac{-\Delta_y}{y_{21}}$	$\frac{-y_{11}}{y_{21}}$	$\frac{1}{z_{21}}$	$\frac{z_{22}}{z_{21}}$	$\frac{-h_{22}}{h_{21}}$	$\frac{-1}{h_{21}}$	t_{21}	t_{22}

* For all parameter sets, $\Delta_p = p_{11}p_{22} - p_{12}p_{21}$

Solving for $\boldsymbol{V}_1$ gives

$$\boldsymbol{V}_1 = \frac{y_{22}\boldsymbol{I}_1 - y_{12}\boldsymbol{I}_2}{\Delta_y}$$

where $\Delta_y = y_{11}y_{22} - y_{12}y_{21}$
That is, the equation for $\boldsymbol{V}_1$ is

$$\boldsymbol{V}_1 = \frac{y_{22}}{\Delta_y}\boldsymbol{I}_1 - \frac{y_{12}}{\Delta_y}\boldsymbol{I}_2$$

When we look below at the equivalent z-parameter equation we see that

$$\boldsymbol{V}_1 = z_{11}\boldsymbol{I}_1 + z_{12}\boldsymbol{I}_2$$

hence

$$z_{11} = \frac{y_{22}}{\Delta_y} \quad \text{and} \quad z_{12} = \frac{-y_{12}}{\Delta_y}$$

similarly

$$z_{12} = \frac{-y_{21}}{\Delta_y} \quad \text{and} \quad z_{22} = \frac{y_{11}}{\Delta_y}$$

8.8 Interconnection between two-port networks

As mentioned at the outset of this chapter, each parameter set has its own advantages for a given application. One of these advantages is that each set of parameters allows other 2-port networks using the same parameter set to be interconnected in a particular way, thereby simplifying the mathematics involved. We look at the interconnections here.

y-*parameters*

These are particularly useful when ports are connected in parallel both at the input and at the output, as shown in figure 8.12. For network L

$$[I_L] = [y_L][V_L]$$

where

$$[I_L] = \begin{bmatrix} I_{L1} \\ I_{L2} \end{bmatrix} \quad [V_L] = \begin{bmatrix} V_{L1} \\ V_{L2} \end{bmatrix}$$

and $[y_L]$ are the y-parameters for network L. Similarly for network M

$$[I_M] = [y_M][V_M]$$

Since the networks are in parallel with one another

$$[V_L] = [V_M] = [V]$$

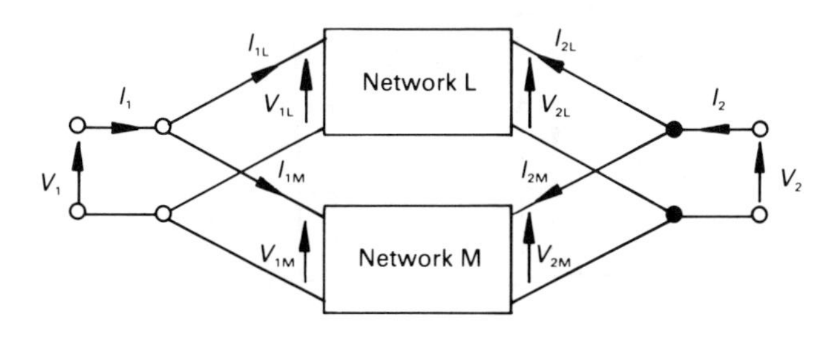

Figure 8.12 *Parallel connection of two 2-port networks.*

and

$$[I] = [I_L + I_M]$$

That is

$$[I] = ([y_I] + [y_M])\,[V] = [y][V]$$

Consequently, each parameter of the resulting parallel network is the sum of the corresponding y-parameters of the individual networks, that is

$$y_{11} = y_{11L} + y_{11M}, \text{ etc.}$$

This can be extended to any number of 2-port networks connected in parallel at the input and at the output.

z-parameters

The z-parameters are useful where the 2-port networks are connected in series both at the input and at the output, as shown in figure 8.13.

In this case

$$[I] = [I_L] = [I_M]$$

and

$$\begin{aligned}[V] &= [V_L + V_M] = [z_L][I_L] + [z_M][I_M] \\ &= ([z_L] + [z_M])[I] = [z][I]\end{aligned}$$

so that $z_{11} = z_{11L} + z_{11M}$, etc.

h-*parameters*

The primary advantage of h-parameters is the ease with which they may be measured in transistors. An analysis of 2-port networks similar to that undertaken above shows that the h-parameters of two 2-ports may be

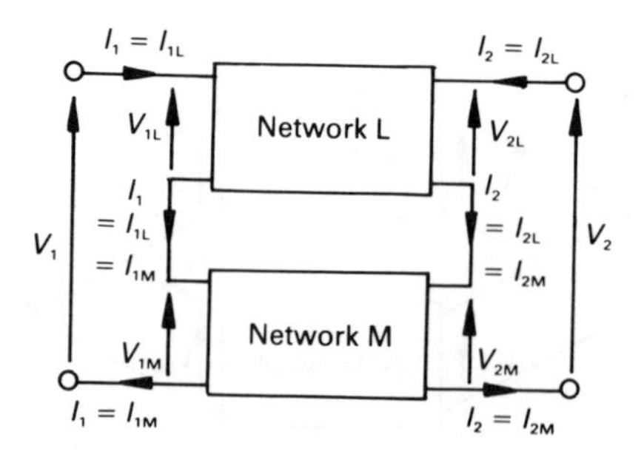

Figure 8.13 *Series connection of two 2-port networks at the input and at the output.*

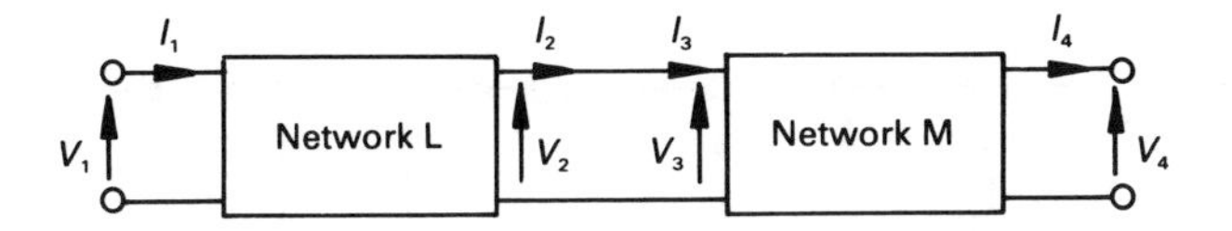

Figure 8.14 *Cascaded 2-port networks.*

added together if the two networks are connected in series at the input (see figure 8.13) and in parallel at the output (see figure 8.12). This is an unusual configuration, and is little used.

t-*parameters*

Transmission parameters are useful where 2-port networks are cascaded as shown in figure 8.14, such as transmission line sections. In this case, the current leaving the output port of one network is the input current of the next network. That is

$$V_2 = V_3 \quad \text{and} \quad I_2 = I_3$$

For network L

$$\begin{bmatrix} V_1 \\ I_1 \end{bmatrix} = [t_L] \begin{bmatrix} V_2 \\ I_2 \end{bmatrix} = [t_L] \begin{bmatrix} V_3 \\ I_3 \end{bmatrix}$$

and for network M

$$\begin{bmatrix} V_3 \\ I_3 \end{bmatrix} = [t_M] \begin{bmatrix} V_4 \\ I_4 \end{bmatrix}$$

That is

$$\begin{bmatrix} V_1 \\ I_1 \end{bmatrix} = [t_L][t_M] \begin{bmatrix} V_4 \\ I_4 \end{bmatrix} = [t] \begin{bmatrix} V_4 \\ I_4 \end{bmatrix}$$

Consequently the t-parameters of the cascaded network is given by the matrix product

$$[t] = [t_L][t_M]$$

This product *is not* the simple mathematical product of the corresponding elements in the matrix, but is *the matrix product*, as described in chapter 15.

As a simple example of the above, we will determine the t-parameters of the half-T network in figure 8.15(a). This can be considered to consist of the series impedance in diagram (b) cascaded with the shunt admittance in diagram (c). By inspection, the equations for figure 8.15(b) are

$$\begin{aligned} V_1 &= V_2 + ZI_2 \\ I_1 &= \qquad I_2 \end{aligned}$$

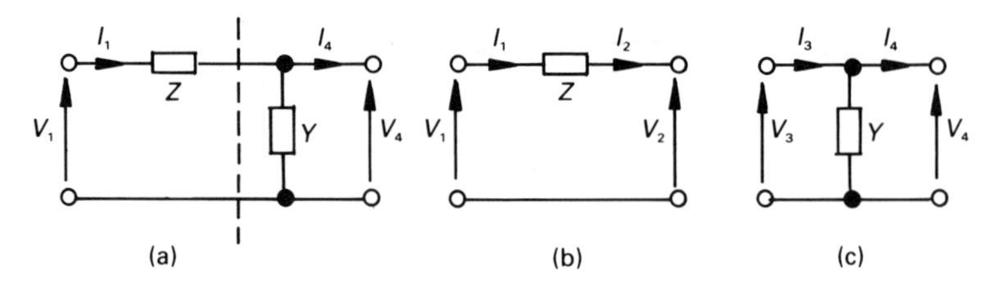

Figure 8.15 *The t-parameters of a half-T network.*

or

$$\begin{bmatrix} V_1 \\ I_1 \end{bmatrix} = \begin{bmatrix} 1 & Z \\ 0 & 1 \end{bmatrix} \begin{bmatrix} V_2 \\ I_2 \end{bmatrix}$$

and the equations for figure 8.15(c) are

$$\begin{aligned} V_3 &= V_4 \\ I_3 &= YV_4 + I_4 \end{aligned}$$

or

$$\begin{bmatrix} V_3 \\ I_3 \end{bmatrix} = \begin{bmatrix} 1 & 0 \\ Y & 1 \end{bmatrix} \begin{bmatrix} V_4 \\ I_4 \end{bmatrix}$$

The reader will observe that, for both circuits

$$t_{11}t_{22} - t_{12}t_{21} = 1$$

When the two are cascaded, as shown in figure 8.15(a), we may write

$$\begin{bmatrix} V_1 \\ I_1 \end{bmatrix} = \begin{bmatrix} 1 & Z \\ 0 & 1 \end{bmatrix} \begin{bmatrix} 1 & 0 \\ Y & 1 \end{bmatrix} \begin{bmatrix} V_2 \\ I_2 \end{bmatrix}$$

$$= \begin{bmatrix} 1 + YZ & Z \\ Y & 1 \end{bmatrix} \begin{bmatrix} V_2 \\ I_2 \end{bmatrix}$$

That is $t_{11} = 1 + YZ$, $t_{12} = Z$, $t_{21} = Y$ and $t_{22} = 1$; also $t_{11}t_{22} - t_{12}t_{21} = 1$.

Unworked problems

8.1. Calculate the input impedance of the non-port network in figure 8.16.
[9.1 Ω]

8.2. Evaluate R_{in} for the circuit in figure 8.17. Assume the op-amp to be ideal.
[2 kΩ]

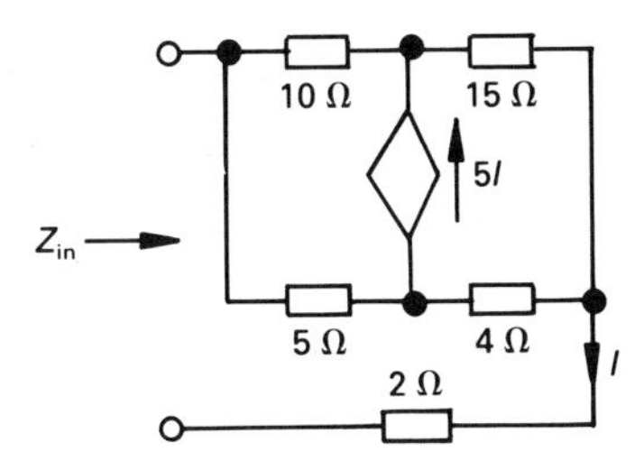

Figure 8.16

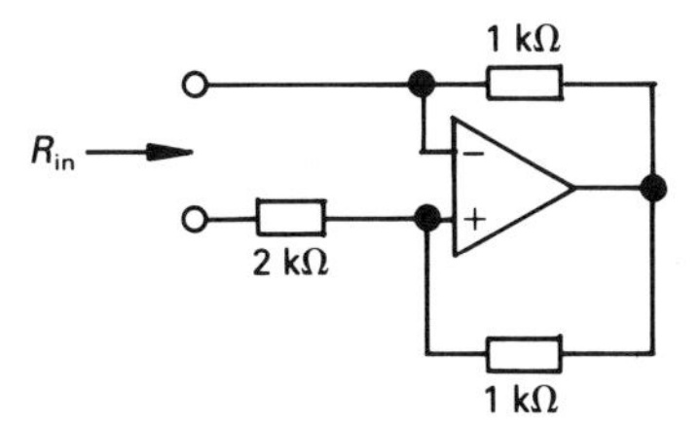

Figure 8.17

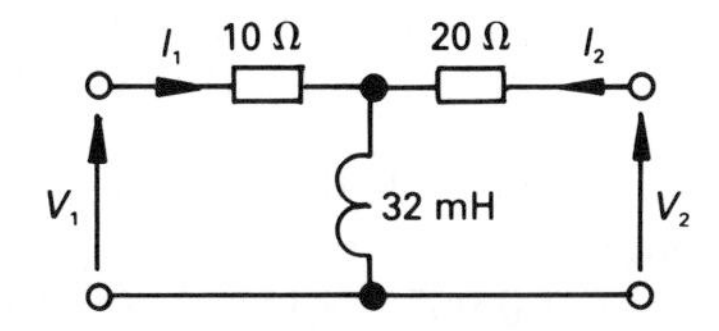

Figure 8.18

8.3. If all the resistance values in figure 8.16 are doubled, calculate the input admittance of the circuit.
[0.057 S]

8.4. Determine the y-parameters of the circuit in figure 8.18 at a frequency of 50 Hz.
[$\boldsymbol{y}_{11} = 0.062\angle{-29.8°}$ S; $\boldsymbol{y}_{12} = 0.028\angle{-146.4°}$ S; $\boldsymbol{y}_{21} = 0.028\angle{-146.4°}$ S; $\boldsymbol{y}_{22} = 0.039\angle{-11.3°}$ S]

8.5. If, in problem 8.4, $\boldsymbol{V}_1 = 10$ V at 50 Hz, calculate (a) the voltage gain, (b) the current gain, (c) $\boldsymbol{Z}_{\text{in}}$, and (d) the power gain of the network if a capacitor of reactance 20 Ω is connected between the output terminals.
[(a) $0.486\angle{-14.2°}$; (b) $0.44\angle{-63.5°}$; (c) $18.5\angle{40.7°}$ Ω; (d) 0]

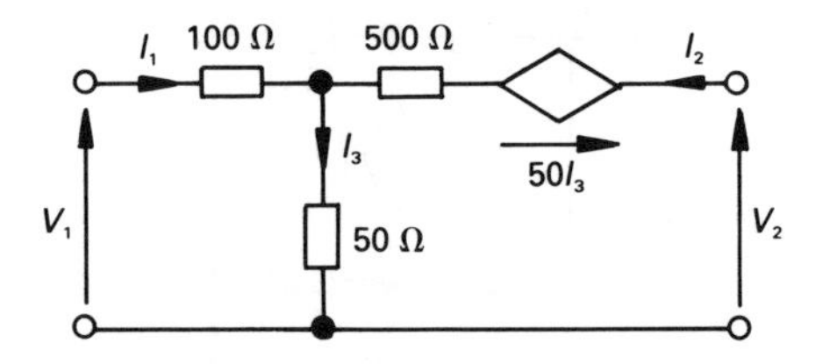

Figure 8.19

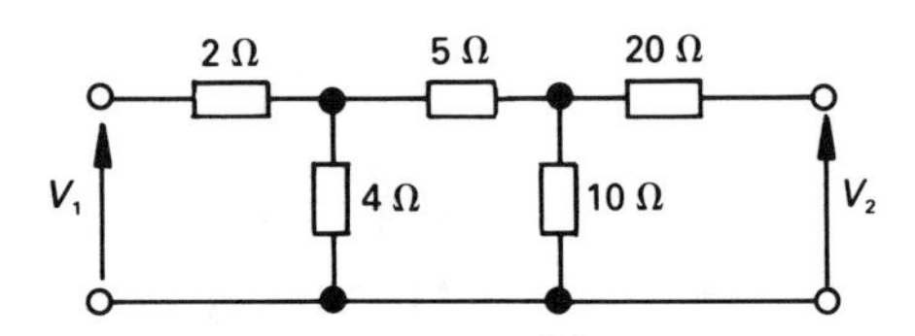

Figure 8.20

8.6. Evaluate the y-parameters of the 2-terminal network in figure 8.19.
$[y_{11} = 9.3 \times 10^{-3}$ S; $y_{12} = -2 \times 10^{-3}$ S;
$y_{21} = -2.7 \times 10^{-3}$ S; $y_{22} = 2 \times 10^{-3}$ S]

8.7. Determine the z-parameters of the network in figure 8.20.
$[z_{11} = 5.16\ \Omega$; $z_{12} = 2.1\ \Omega$; $z_{21} = 2.1\ \Omega$; $z_{22} = 24.7\ \Omega]$

8.8. Repeat problem 8.5, but for the circuit in figure 8.20 and the associated z-parameters.
[(a) $0.262\angle -50.1°$; (b) $0.067\angle -141.1°$; (c) $5.05\angle -1°$; (d) 0]

8.9. Calculate the z-parameters for the circuit in figure 8.21 at a frequency of 50 Hz.
$[z_{11} = 5.93\angle -90°\ \Omega$; $z_{12} = 5.07\angle 57.3°\ \Omega$; $z_{21} = 4.08\angle 90°\ \Omega$;
$z_{22} = 13.63\angle 17.4°\ \Omega]$

8.10. Repeat problem 8.5 but for the circuit in figure 8.21 and the associated z-parameters.
[(a) $0.69\angle 131°$; (b) $0.2\angle -39.2°$; (c) $5.75\angle -80.2°$; (d) 0]

8.11. If the value of each resistor in figure 8.20 is changed to 10 Ω, calculate the **h**-parameters for the circuit.
$[h_{11} = 16\ \Omega$; $h_{12} = 0.2$; $h_{21} = -0.2$; $h_{22} = 0.06$ S]

8.12. A transistor has the following h-parameters: $h_{11} = 1200\ \Omega$, $h_{12} = 5 \times 10^{-4}$, $h_{21} = 50$, $h_{22} = 20\ \mu$S. If the load connected to its output terminals is 10 kΩ, calculate the voltage gain and the current gain of the transistor amplifier.
$[G_V = -420.2$; $G_I = 41.69]$

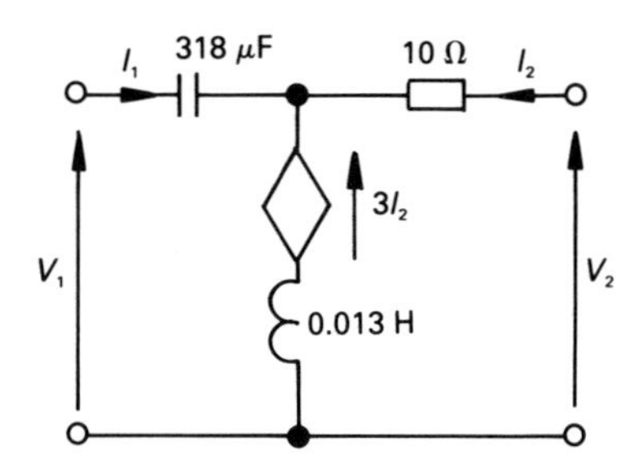

Figure 8.21

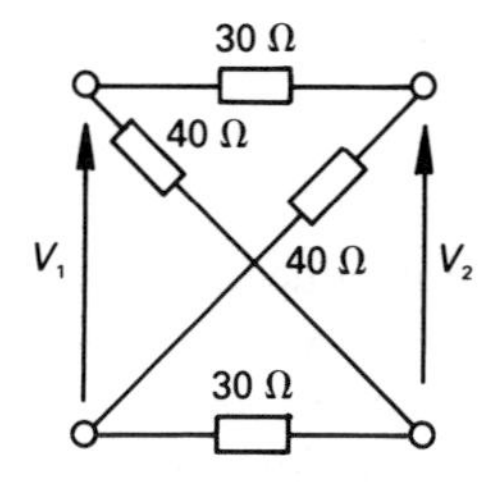

Figure 8.22

8.13. Determine the z-parameters of the lattice network in figure 8.22.
[z_{11} = 35 Ω; z_{12} = 5 Ω; z_{21} = 5 Ω; z_{22} = 35 Ω]

8.14. A two-port network has the parameters t_{11} = 0.02; t_{12} = 2 Ω; t_{21} = 2 mS; t_{22} = 0.1. If the input source has an internal resistance of 10 Ω, and the resistor connected to the output terminals has a resistance of 20 Ω, calculate (a) the voltage gain and (b) the current gain.
[(a) 14.29; (b) 7.143]

8.15. A communications line is represented by a symmetrical-π network, in which the series element is 120∠60° Ω, and each shunt admittance is 2.5×10^{-3}∠90° S. Calculate the value of the t-parameters of the line.
[$t_{11} = t_{22}$ = 0.755∠11.4°; t_{12} = 120∠60° Ω; $t_{21} = 4.4 \times 10^{-3}$∠95° S]

9

The Transformer

9.1 Introduction

The transformer was briefly introduced in chapter 4, when mutual inductance was discussed. The majority of transformers have two electrical circuits, one – known as the *primary circuit* – contains a power source, and the second – known as the *secondary circuit* – is connected to the load. There are two broad classes of transformer, namely the ideal transformer and the linear transformer.

An *ideal transformer* is one in which the coils are wound on an iron core, and the magnetic coupling between the windings is near-perfect; that is, the magnetic coupling coefficient is practically unity. This class of transformer includes the power transformer.

A *linear transformer* is one whose magnetic coupling coefficient is less than unity, in fact its value can be very low indeed! This class of transformer includes the radio-frequency transformer.

Certain types of transformer with an iron core, such as those used in the output circuit of an electronic amplifier, are rather less than electrically 'perfect'. The reason is that their winding resistance is generally high, and the magnetic circuit design is such that the magnetic coupling coefficient is less than unity.

Since the ideal (power) transformer is the type most frequently encountered, we begin our studies by looking at this type. With this sort of transformer, we are not particularly interested in its magnetic coupling coefficient, since we can assume its value to be unity.

9.2 The ideal transformer

An 'ideal' transformer is one in which the coils are wound on an iron core and the windings have a very high inductance (ideally infinite), and the magnetic coupling coefficient is unity. In practice, such an ideal cannot be achieved, but is closely approached in a well-designed iron-cored transformer.

Assuming that the magnetic flux in the core varies sinusoidally, the instantaneous flux is given by

$$\phi = \Phi_{\mathrm{m}} \sin \omega t$$

where Φ_{m} is the maximum value of flux in the core, and the instantaneous value of the induced e.m.f. in a coil of N turns is

$$e = -N\,\mathrm{d}\phi/\mathrm{d}t = -N\omega\Phi_{\mathrm{m}} \cos \omega t$$

Since the form of the equation for the induced e.m.f. in each winding is the same, it follows that the e.m.f.s induced in the windings are in phase with one another. However, this is a philosophical argument because, unless the windings are electrically connected to one another at a common point we cannot, at this stage, be absolutely certain what the phase relationship is. Let us assume, therefore, that the induced voltages are in phase with one another. Moreover, since the winding resistance of a power transformer is very small, we can say that the voltage across the winding is

$$v \approx -e = N\omega\Phi_{\mathrm{m}} \cos \omega t$$

The maximum value of the voltage across the coil is

$$V_{\mathrm{m}} = N\omega\Phi_{\mathrm{m}}$$

and the r.m.s. voltage in the coil is

$$\sqrt{2} \times N \times 2\pi f\Phi_{\mathrm{m}} = 4.44Nf\Phi_{\mathrm{m}}$$

Since we are dealing with an ideal transformer whose magnetic coupling coefficient is unity, then we may say that this voltage is induced in *every coil* on the core. Provided that the winding resistance is low (which is usually the case in a power transformer) then, to a first approximation, the voltage across the coil is equal to the e.m.f. induced in the coil. That is, in coil 1

$$V_1 = |E_1|$$

and in coil 2

$$V_2 = |E_2| \text{ etc.}$$

and we may also say

$$V_1 = 4.44N_1 f\Phi_{\mathrm{m}}$$

$$V_2 = 4.44N_2 f\Phi_{\mathrm{m}}, \text{ etc.}$$

Hence

$$\frac{V_2}{V_1} = \frac{N_2}{N_1} = a \tag{9.1}$$

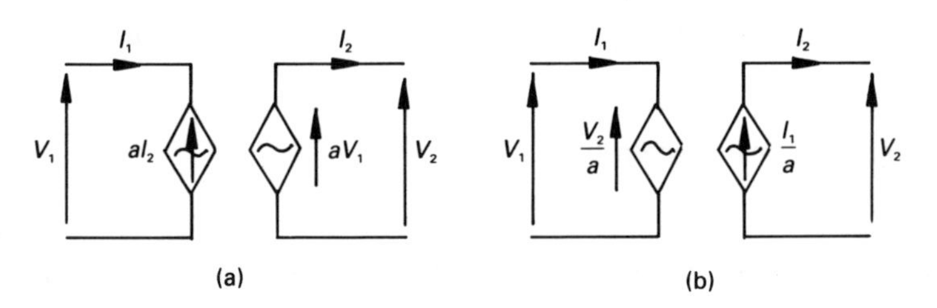

Figure 9.1 *Simple equivalent circuits of the ideal transformer.*

where a is known as the *turns ratio* of the transformer. An equivalent circuit corresponding to equation (9.1) is shown in figure 9.1(a). The equation may be re-written as follows

$$\frac{V_1}{N_1} = \frac{V_2}{N_2}$$

That is, *each winding in an ideal transformer supports the same number of volts per turn*. This applies even if the transformer has many separate windings on its magnetic core.

Furthermore, since the transformer is ideal, the windings have little resistance, and the power loss in them is very low. Assuming that we are dealing with a two-winding transformer, we may say, to a first approximation

$$V_1 I_1 \cos \phi = V_2 I_2 \cos \phi$$

where ϕ is the phase angle between the voltage and current. Hence *each winding in an ideal transformer supports the same number of volt-amperes*, that is

$$\frac{V_2}{V_1} = \frac{I_1}{I_2} = a \tag{9.2}$$

An equivalent circuit corresponding to equation (9.2) is shown in figure 9.1(b). Combining equations (9.1) and (9.2) gives the following relationship for an ideal transformer

$$\frac{V_2}{V_1} = \frac{I_1}{I_2} = \frac{N_2}{N_1} = a$$

From this we see that

$$I_1 N_1 = I_2 N_2$$

That is, *each winding in an ideal transformer supports the same number of ampere-turns*.

9.3 Phasor diagram for an ideal transformer on no-load

It was shown in section 9.2 that the voltage across each winding on the transformer were all in phase with one another and, with a sinusoidal flux waveform, the voltage has a cosine waveform. That is, *the voltage across each winding leads the flux waveform by 90°*. Consequently, the phasor diagram for the primary and secondary winding for a transformer on no-load is as shown in figure 9.2.

The phasor diagrams for the primary and secondary windings are shown separately in figure 9.2 because the two windings are electrically isolated from one another.

To maintain the magnetic flux in the core, the primary winding carries a *no-load current*, I_0. This current consists of two components:

(1) The *magnetising current component*, I_{mag}, which lags behind the current by 90° and produces the flux Φ.
(2) The *core loss component*, I_C, which is in phase with V_1 and supplies the power loss dissipated in the magnetic core (known as the *core loss* or *iron loss*).

The phase angle of I_0 is ϕ_0 which, in an ideal transformer, approaches 90°, that is

$$\boldsymbol{I_0} = I_C + jI_{mag}$$

and the core loss is

$$P_0 = V_1 I_C$$

When an inductive load is connected to the secondary of the transformer, the phasor diagrams are shown in figure 9.3. The secondary current, $\boldsymbol{I_2}$, lags behind the secondary terminal voltage, $\boldsymbol{V_2}$, by ϕ_2. Since ampere–turn

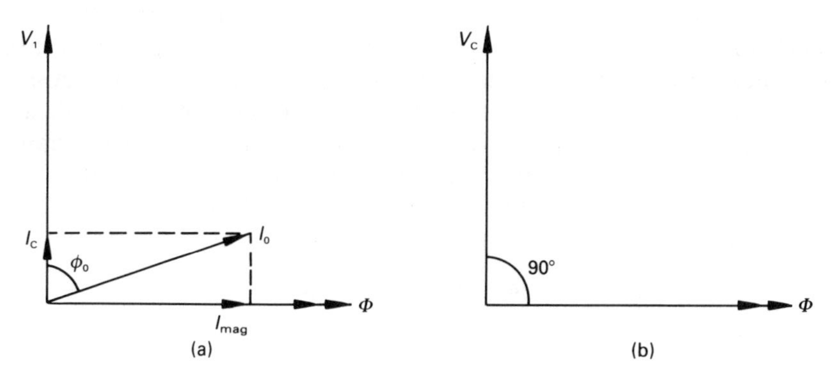

Figure 9.2 *No-load phasor diagram for an ideal transformer (a) primary winding, and (b) secondary winding.*

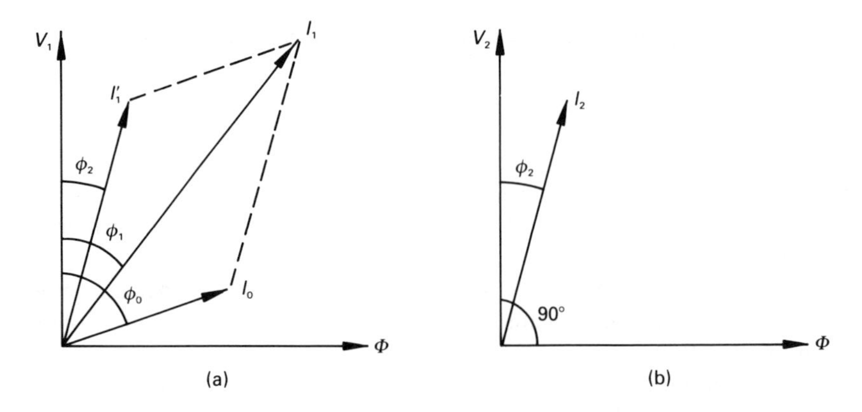

Figure 9.3 *Phasor diagram for a transformer with an inductive load (a) primary winding, (b) secondary winding.*

balance is maintained on the two windings, a corresponding current $\boldsymbol{I}_1'$ flows in the primary winding where

$$\boldsymbol{I}_1' N_1 = \boldsymbol{I}_2 N_2$$

or

$$\boldsymbol{I}_1' = \boldsymbol{I}_2 N_2 / N_1$$

Accordingly, $\boldsymbol{I}_1'$ is shown lagging $\boldsymbol{V}_1$ by ϕ_2. The current in the primary winding is the phasor sum of $\boldsymbol{I}_0$ and $\boldsymbol{I}_1'$ as follows

$$\boldsymbol{I}_1 = \boldsymbol{I}_0 + \boldsymbol{I}_1'$$

Figure 9.4 shows a simplified equivalent circuit of a power transformer which represents the phasor diagram in figure 9.3. The resistor R_C is the path through which the core loss current flows, and X_m is a pure inductive reactance through which the magnetising current flows (no magnetising current flows in the ideal transformer itself). The value of R_C and X_m are determined by means of a *no-load test* or *open-circuit test* on the transformer, in which the primary winding is excited at its nominal voltage and frequency, at which time the no-load current and power loss, P_0, are measured. From these values we can say

$$\phi_0 = \cos^{-1}(P_0 / V_1 I_0)$$

$$I_C = I_0 \cos \phi_0$$

$$I_{mag} = I_0 \sin \phi_0$$

$$R_C = V_1 / I_C$$

$$X_m = V_1 / I_{mag}$$

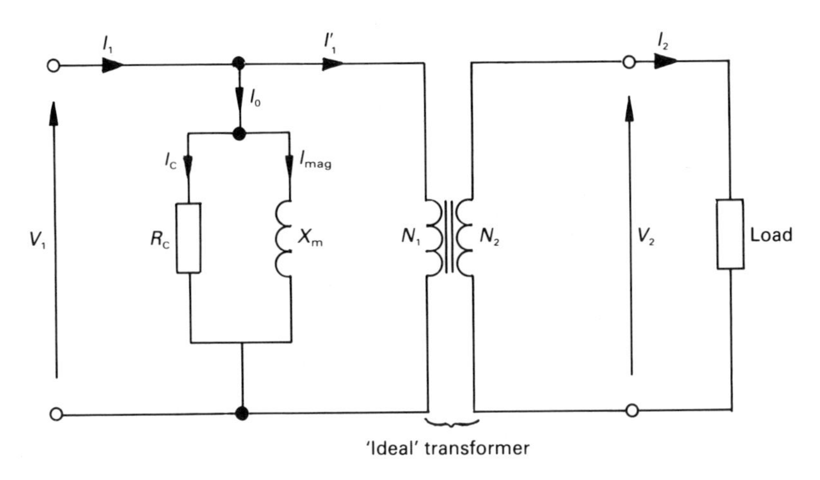

Figure 9.4 *Simplified equivalent circuit of a power transformer taking account of the no-load current.*

9.4 Leakage flux in a transformer

When a load is connected to the secondary of a transformer, as shown in figure 9.5, a current $\boldsymbol{I}_2$ flows in the secondary.

Although the coils are wound on an iron core, the coupling coefficient in a practical power transformer is not quite unity, so that some of the flux produced by *each winding* links only with that particular coil, and with no other coil. Consequently, this flux induces an e.m.f. only in the originating coil. Moreover, one-half of the magnetic path for this flux lies in the air surrounding the coil, and the reluctance of this flux path is very high indeed. This flux, which is produced by each current-carrying winding, is known as the *leakage flux* associated with that winding; because of the high reluctance of the magnetic path, the leakage flux is proportional to the current in the coil. In figure 9.5, the primary winding leakage flux is Φ_{L1}, and the secondary winding leakage flux is Φ_{L2}.

Since each leakage flux is linked with one winding only, it induces a 'back' e.m.f. in that winding, which opposes the current flow in the winding. The greater the leakage flux, the greater the voltage drop due to this cause. Good transformer design aims to reduce the leakage flux to a low level.

Unfortunately, the simple winding construction in figure 9.5 gives a large leakage flux, and is not used in practice. The leakage flux is kept to a minimum either by winding both coils on the same limb (the *shell construction*), or by winding one-half of each winding on each of the vertical limbs (the *core construction*).

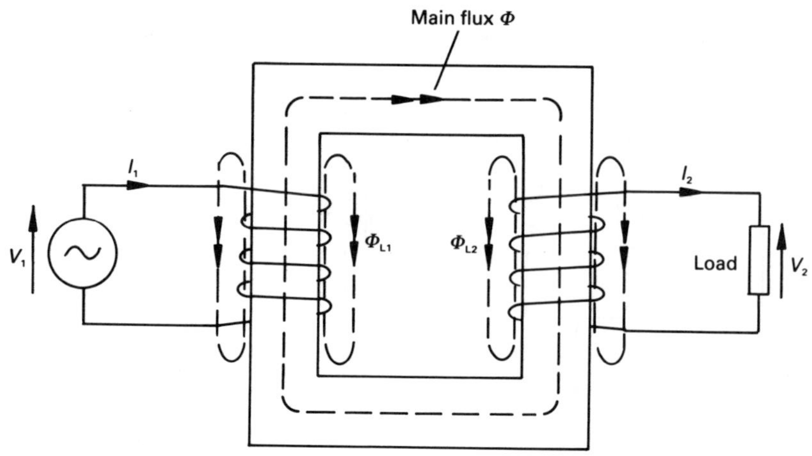

Figure 9.5 *Leakage flux.*

As mentioned above, the leakage flux has the effect of inducing a back e.m.f. in the winding which produced it. We regard this voltage drop as being due to a *leakage reactance*, which can be thought of as being in series with each winding of the transformer. That is, in figure 9.6, we have a *primary winding leakage reactance* X_1, and a *secondary winding leakage reactance* X_2.

A 'practical' transformer can be thought of as an ideal transformer together with circuit elements which allow for no-load current, winding resistance, leakage reactance, etc. The sole function of the ideal transformer is to carry the main flux and to provide a 'turns ratio'; it neither stores nor dissipates energy, so that it does not draw magnetising current, has no hysteresis loss, and has no I^2R loss in its windings.

A fairly complete form of equivalent circuit for the transformer discussed so far is shown in figure 9.6, in which X_1 and X_2 are the respective primary and secondary winding leakage reactances, R_1 and R_2 are the primary and secondary winding resistance, and R_C and X_m are as described earlier.

9.5 Impedance matching with an ideal transformer

An iron-cored transformer is widely used as an interface device between the output of an electronic amplifier and a loudspeaker; both of these usually have widely differing values of resistance. The reader will recall that, for maximum power transmission, the resistance of the load should have the same resistance as the output resistance of the source (or the amplifier in this case).

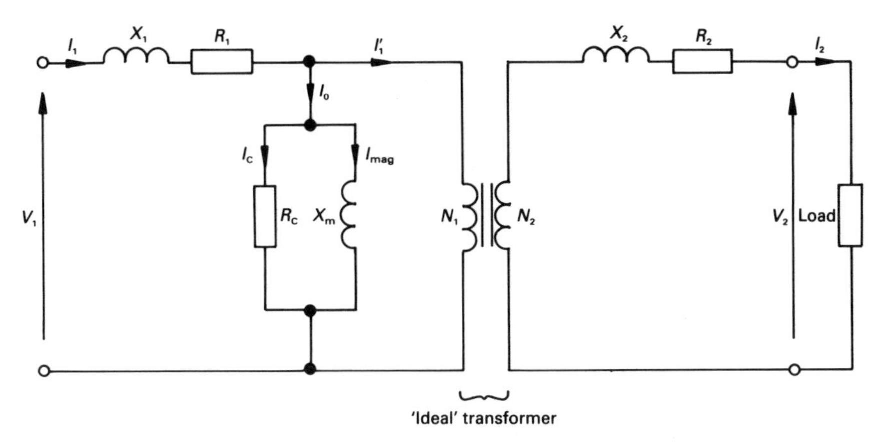

Figure 9.6 *A complete equivalent circuit of a single-phase transformer.*

In electronic circuits the load is frequently regarded as a pure resistor, and the secondary current is

$$I_2 = V_2/R_L \quad \text{or} \quad R_L = V_2/I_2$$

For an ideal transformer

$$I_1 = aI_2 \quad \text{and} \quad V_1 = V_2/a$$

The apparent resistance, R_1, 'seen' by the primary supply source is

$$R_1 = \frac{V_1}{I_1} = \frac{V_2}{a}\frac{1}{aI_2} = \frac{1}{a^2}\frac{V_2}{I_2} = \frac{R_L}{a^2}$$

Suppose that a loudspeaker of resistance 8 Ω is to be matched to an electronic amplifier of output resistance 100 Ω. To obtain maximum power transfer from the amplifier to the load, we need to satisfy the condition

$$100 = R_L/a^2 = 8/a^2$$

where a is the turns ratio of the interface transformer. That is

$$a = N_2/N_1 = 1/\sqrt{(100/8)} = 1/3.54$$

that is $N_1/N_2 = 3.54$.

An audio-frequency transformer is generally less 'perfect' than a power transformer, and it will be found that impedance matching is a little less perfect than suggested here!

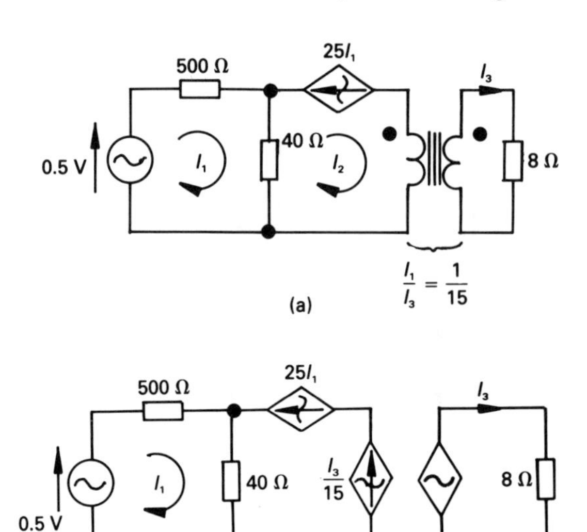

Figure 9.7 *(a) A class A power amplifier and (b) an equivalent circuit.*

Worked example 9.5.1

An equivalent circuit of a 'class A' electronic power amplifier is shown in figure 9.7(a). Calculate the current in the 8 Ω load if the r.m.s. input voltage is 0.5 V. Also determine the power gain of the amplifier.

Solution

The transformer which couples the amplifier to the load has a voltage step-down ratio of 15:1 (a = 1/15), and is replaced in figure 9.7(b) by an equivalent circuit (see also figure 9.1(a)). The equation for the mesh containing $\boldsymbol{I}_1$ is

$$0.5 = 500\boldsymbol{I}_1 + 40(1 + 25)\boldsymbol{I}_1 = 1540\boldsymbol{I}_1$$

or

$$\boldsymbol{I}_1 = 3.25 \times 10^{-4} \text{ A}$$

and the input power to the circuit is

$$0.5 \times 3.25 \times 10^{-4} = 1.625 \times 10^{-4} \text{ W}$$

In the mesh in which $\boldsymbol{I}_2$ flows we have

$$\boldsymbol{I}_3/15 = 25\boldsymbol{I}_1$$

or $$I_3 = 375 I_1 = 0.122 \text{ A}$$

with a corresponding output power of

$$0.122^2 \times 8 = 0.119 \text{ W}$$

Hence $$G_p = 0.119/1.625 \times 10^{-4} = 732.3$$

9.6 The ideal transformer as a two-port network

The ideal transformer can conveniently be described as a 2-port network using the t-parameters as follows (see chapter 8 for details).

Since $a = \dfrac{N_2}{N_1} = \dfrac{V_2}{V_1} = \dfrac{I_1}{I_2}$, the V–I relations for the transformer are $V_1 = V_2/a$ and $I_1 = aI_2$, so that the t-parameter equations for the ideal transformer are as follows

$$\begin{bmatrix} V_1 \\ I_1 \end{bmatrix} = \begin{bmatrix} 1/a & 0 \\ 0 & a \end{bmatrix} \begin{bmatrix} V_2 \\ I_2 \end{bmatrix}$$

that is $t_{11} = 1/a$, $t_{12} = t_{21} = 0$ and $t_{22} = a$.

If the secondary winding of the transformer has a load connected to it, as shown in figure 9.8, the t-parameters of the load (see also chapter 8) are $t_{11} = 1$, $t_{12} = 0$, $t_{21} = Y_L$ and $t_{22} = 1$. The matrix equation for the transformer and load in figure 9.8 is therefore

$$\begin{bmatrix} V_1 \\ I_1 \end{bmatrix} = \begin{bmatrix} 1/a & 0 \\ 0 & a \end{bmatrix} \begin{bmatrix} 1 & 0 \\ Y_L & a \end{bmatrix} \begin{bmatrix} V_4 \\ I_4 \end{bmatrix}$$

$$= \begin{bmatrix} 1/a & 0 \\ aY_L & a \end{bmatrix} \begin{bmatrix} V_4 \\ I_4 \end{bmatrix}$$

that is

$$V_1 = V_4/a$$

$$I_1 = Y_L a V_4 + a I_4$$

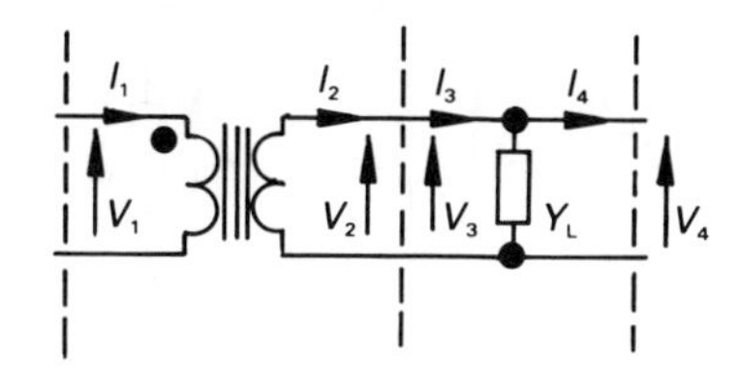

Figure 9.8 *A loaded transformer.*

Since no other load than $\boldsymbol{Y}_L$ is connected to the secondary, then $\boldsymbol{I}_4 = 0$ and

$$\boldsymbol{Z}_{in} = \frac{\boldsymbol{V}_1}{\boldsymbol{I}_1} = \frac{\boldsymbol{V}_4/a}{\boldsymbol{Y}_L a \boldsymbol{V}_4} = \frac{1}{\boldsymbol{Y}_L a^2} = \frac{\boldsymbol{Z}_L}{a^2}$$

where $\boldsymbol{Z}_L = 1/\boldsymbol{Y}_L$. This theme is capable of development, and a loaded transformer can be analysed by means of a chain of *t*-parameters or *ABCD* parameters.

9.7 Thévenin's equivalent circuit of an ideal transformer and voltage regulation

We can simplify the solution of many problems involving transformer circuits by using equations developed earlier, together with Thévenin's theorem.

If we consider the circuit in figure 9.9(a), in which we have an ideal transformer with an impedance $\boldsymbol{Z}_1$ in the primary circuit (which may, for example, represent the winding resistance and leakage reactance of the transformer), and a load $\boldsymbol{Z}_L$ in the secondary circuit, we may simplify the complete circuit by 'reflecting' the load impedance into the primary circuit, as shown in figure 9.9(b).

The 'reflected' load impedance is $\boldsymbol{Z}_L/a^2$ (see section 9.6), which is the input impedance of an ideal transformer with a load $\boldsymbol{Z}_L$ connected to its secondary terminals. Now $\boldsymbol{I}_1 = \boldsymbol{V}_S/(\boldsymbol{Z}_1 + \boldsymbol{Z}_L/a^2)$ and $\boldsymbol{I}_z = \boldsymbol{I}_1/a$, also $\boldsymbol{V}_2' = \boldsymbol{I}_1\boldsymbol{Z}_L/a^2$, hence

$$\boldsymbol{V}_2 = a\boldsymbol{V}_2' = \boldsymbol{I}_1\boldsymbol{Z}_L/a$$

Similarly, we can 'reflect' the primary circuit values into the secondary circuit as follows. From figure 9.9(b)

$$\boldsymbol{I}_1 = \boldsymbol{V}_S/(\boldsymbol{Z}_1 + \boldsymbol{Z}_L/a^2)$$

and since $\boldsymbol{I}_2 = \boldsymbol{I}_1/a$, then

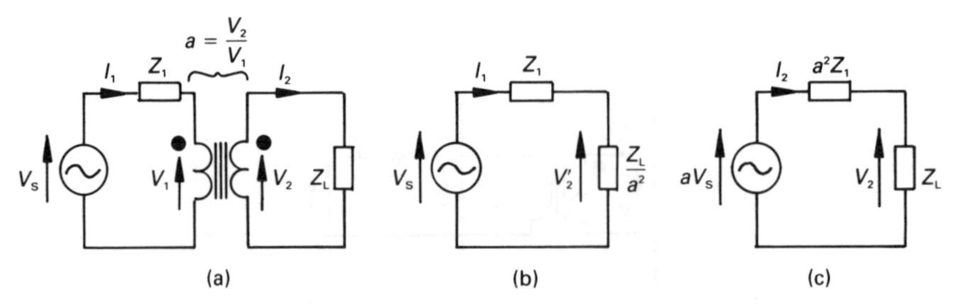

Figure 9.9 *An ideal transformer with an impedance in the primary and in the secondary circuit.*

$$\mathbf{I}_1 = \frac{1}{a}\frac{\mathbf{V}_S}{(\mathbf{Z}_1 + \mathbf{Z}_L/a^2)} = \frac{\mathbf{V}_S}{a\mathbf{Z}_1 + \mathbf{Z}_L/a}$$

$$= \frac{a\mathbf{V}_S}{a^2\mathbf{Z}_1 + \mathbf{Z}_L}$$

The corresponding equivalent circuit for values reflected into the secondary circuit is shown in figure 9.9(c).

Power engineers are concerned with maintaining the electrical voltage at the consumers terminals. The variation in voltage at the terminals of the consumer is often quoted in terms of the *voltage regulation* as follows

$$\text{Per unit voltage regulation} = \frac{\text{modulus of the no-load voltage} - \text{modulus of the full-load voltage}}{\text{modulus of the no-load voltage}}$$

$$= \frac{V_{OC} - V_L}{V_{OC}}$$

If $\mathbf{Z}_L$ represents an impedance which draws full-load current from the transformer, the equivalent circuits in figure 9.9 can be used to calculate the voltage regulation of the transformer (in many cases the voltage regulation is given as a per cent figure).

Referring to figure 9.9, the equation for voltage regulation is
for diagram (b)

$$\text{Voltage regulation} = (V_S - V_2')/V_S$$

for diagram (c)

$$\text{Voltage regulation} = (aV_S - V_2)/aV_S$$

The reader should note that if the dot notation in figure 9.9 is reversed on one of the windings, there is no change in the magnitude of the voltages and currents involved, and no change in the voltage regulation. However, the secondary voltage and current are phase shifted by 180°.

Worked example 9.7.1

A 100 kVA, 3300/250 V single-phase transformer has a total equivalent resistance and leakage reactance of 3 Ω and 12 Ω, respectively, both referred to the primary winding. What is the full-load current at (a) 0.8 power factor lagging, (b) 0.8 power factor leading? Calculate the voltage regulation in each case.

Solution

The modulus of the full-load secondary current is

$$I_2 = 100\ 000/250 = 400 \text{ A}$$

Hence
(a) for a lagging load of 0.8 power factor

$$\boldsymbol{I}_2 = 400\angle -36.9^\circ \text{ A}$$

(b) for a leading load of 0.8 power factor

$$\boldsymbol{I}_2 = 400\angle 36.9^\circ \text{ A}$$

The voltage regulation is calculated as follows. The magnitude of the impedance connected to the secondary winding to produce full-load current is

$$Z_L = 250/400 = 0.625\ \Omega$$

At this point we have the choice of calculating the voltage regulation based on values referred to the primary winding (see figure 9.9(b)), or to values referred to the secondary winding (see figure 9.9(c)). Selecting the former, we need to calculate the load impedance referred to the primary winding, which is

$$Z'_L = Z_L/a^2 = 0.625/(250/3300)^2 = 108.9\ \Omega$$

(a) For a 0.8 power factor lagging load

$$\boldsymbol{Z}'_L = 108.9\angle 36.9^\circ = 87.1 + j65.39\ \Omega$$

By potential division in figure 9.9(b)

$$\begin{aligned} \boldsymbol{V}'_2 &= \boldsymbol{V}_S \frac{\boldsymbol{Z}'_L}{\boldsymbol{Z}_1 + \boldsymbol{Z}'_L} \\ &= 3300 \frac{108.9\angle 36.9^\circ}{(3 + j12) + (87.1 + j65.39)} \\ &= 3026\angle -3.76^\circ \text{ V} \end{aligned}$$

Hence

$$\begin{aligned} \text{Voltage regulation} &= \frac{3300 - 3026}{3300} \\ &= 0.083 \text{ per unit (p.u.) or } 8.3 \text{ per cent} \end{aligned}$$

That is, the secondary voltage is reduced by 8.3 per cent when full load (lagging) is connected.

(b) For a 0.8 power factor leading load

$$Z'_L = 108.9\angle -36.9° = 87.1 - j65.39\ \Omega$$

By potential division in figure 9.7(b)

$$V'_2 = V_S \frac{Z'_L}{Z_1 + Z'_L}$$
$$= 3300 \frac{108.9\angle -36.9°}{(3 + j12) + (87.1 - j65.39)}$$
$$= 3431\angle -6.25°\ V$$

Hence

$$\text{Voltage regulation} = \frac{3300 - 3431}{3300} = -0.04 \text{ or } -4 \text{ per cent}$$

That is, the secondary voltage rises by 4 per cent when the leading load is connected.

9.8 The linear transformer

The windings on a linear transformer are wound on an air core, and the magnetic coupling coefficient between the windings usually has a low value. The linear transformer is widely used in electronic and communications circuits, possibly with both primary and secondary circuits being tuned or resonant.

Figure 9.10(a) shows a *two-winding* linear transformer, having primary and secondary winding resistance together with a load impedance. Using the dot notation, outlined in chapter 4, the equivalent circuit of diagram (a) is shown in diagram (b). The mesh equations are

$$V_1 = (R_1 + j\omega L_1)I_1 - j\omega M I_2$$
$$0 = -j\omega M I_1 + (R_2 + j\omega L_2 + Z_L)I_2$$

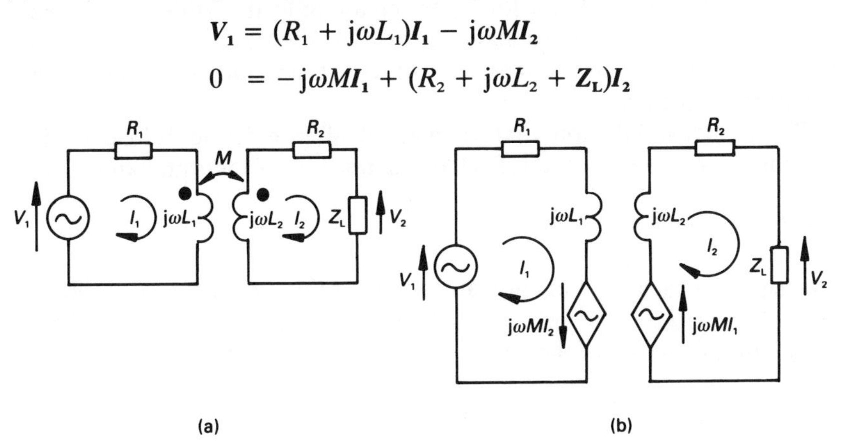

Figure 9.10 *(a) A linear transformer, (b) an electrical equivalent circuit.*

Using the notation adopted for mesh analysis, we may say that

$$\mathbf{Z}_{11} = R_1 + j\omega L_1 \quad \text{and} \quad \mathbf{Z}_{22} = R_2 + j\omega L_2 + \mathbf{Z}_L$$

The mesh equations can therefore be re-written in the form

$$\mathbf{V}_1 = \mathbf{Z}_{11}\mathbf{I}_1 - j\omega M\mathbf{I}_2$$

$$0 = -j\omega M\mathbf{I}_1 + \mathbf{Z}_{22}\mathbf{I}_2$$

hence

$$\mathbf{I}_2 = -j\omega M\mathbf{I}_1/\mathbf{Z}_{22}, \quad \text{or}$$

$$\mathbf{V}_1 = \mathbf{Z}_{11}\mathbf{I}_1 + \omega^2 M^2\mathbf{I}_1/\mathbf{Z}_{22}$$

That is to say, the *input impedance* of a linear transformer is

$$Z_{\text{in}} = \frac{\mathbf{V}_1}{\mathbf{I}_1} = \mathbf{Z}_{11} + \frac{\omega^2 M^2}{\mathbf{Z}_{22}}$$

The input impedance clearly consists of the primary winding self-impedance, $\mathbf{Z}_{11}$, in addition to an impedance $\omega^2 M^2/\mathbf{Z}_{22}$, known as the *reflected impedance*, which is inversely related to the total impedance of the secondary circuit. Writing

$$\mathbf{Z}_{22} = R_{22} + jX_{22}$$

then we may rationalise the denominator (see chapter 15 for details) of the reflected impedance term as follows

$$\frac{\omega^2 M^2}{\mathbf{Z}_{22}} = \frac{\omega^2 M^2}{R_{22} + jX_{22}} = \frac{\omega^2 M^2 R_{22}}{R_{22}^2 + X_{22}^2} - j\frac{\omega^2 M^2 X_{22}}{R_{22}^2 + X_{22}^2}$$

That is, if the secondary circuit has a net inductive reactance, it *reflects* a capacitive reactance into the primary circuit. However, $\mathbf{Z}_{11}$ has a sufficiently large inductive reactance for $\mathbf{Z}_{\text{in}}$ to remain inductive (this is illustated in worked example 9.8.1). Similarly, it may be shown that if the secondary circuit has a net capacitive reactance, it reflects an inductive reactance into the primary winding.

The reader should note that when an 'ideal' transformer is loaded with a reactance of any kind, the reflected reactance is of the same kind.

Worked example 9.8.1

The circuit values for a linear transformer of the type in figure 9.10(a) are $L_1 = 75$ mH, $L_2 = 150$ mH, $M = 80$ mH and $R_1 = R_2 = 10\ \Omega$. Calculate the input impedance of the circuit at a frequency of 1000 rad/s for a load of (a) 0.4 H inductance, (b) a 2.5 μF capacitance.

Solution

The self-impedance of the primary circuit is

$$\mathbf{Z}_{11} = R_1 + j\omega L_1 = 10 + j(1000 \times 7.5 \times 10^{-3})$$
$$= 10 + j75\ \Omega$$

Also $$\omega^2 M^2 = 1000^2 \times (80 \times 10^{-3})^2 = 6400\ \Omega^2$$

(a) *Inductive load*
The load impedance is

$$\mathbf{Z}_L = j\omega L = j1000 \times 0.4 = j400\ \Omega$$

and

$$\mathbf{Z}_{22} = R_2 + j\omega L_2 + \mathbf{Z}_L$$
$$= 10 + j(1000 \times 150 \times 10^{-3}) + j400$$
$$= 550.1\angle 88.96^\circ\ \Omega$$

hence

$$\mathbf{Z}_{in} = \mathbf{Z}_{11} + (\omega M)^2/\mathbf{Z}_{22}$$
$$= (10 + j75) + 6400/550.1\angle 88.96^\circ$$
$$= 10.21 + j63.38\ \Omega$$

(b) *Capacitive load*
The load impedance is

$$\mathbf{Z}_C = 1/j\omega C = 1/j(1000 \times 2.5 \times 10^{-6}) = -j400\ \Omega$$

hence

$$\mathbf{Z}_{22} = (10 + j150) - j400 = 10 - j250$$
$$= 250.2\angle -87.7^\circ\ \Omega$$

giving

$$\mathbf{Z}_{in} = \mathbf{Z}_{11} + (\omega M)^2/\mathbf{Z}_{22}$$
$$= (10 + j75) + 6400/250.2\angle -87.7^\circ$$
$$= 11.03 + j100.6\ \Omega$$

The reader will note that, in both cases, the resistance of the secondary circuit is reflected into the primary circuit as an increase in the resistance element of the input impedance, that is, the input resistance is higher than the 10 Ω resistance in the primary circuit.

In case (a), the effect of the inductive load is to reduce the input inductive reactance below the j75 Ω reactance of the primary circuit alone, while in case (b) the capacitive load has the opposite effect.

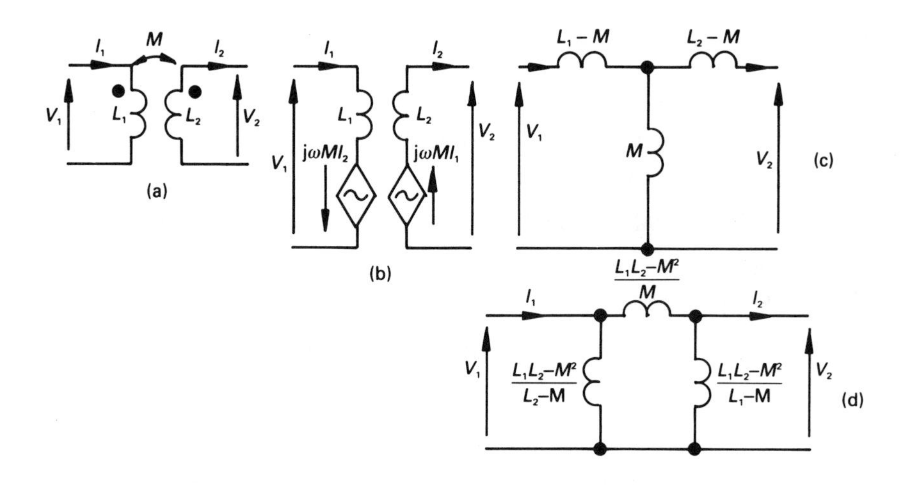

Figure 9.11 *(a) Idealised linear transformer, (b) an equivalent circuit, (c) the electrically connected T-equivalent circuit and (d) the π-equivalent circuit.*

9.9 T- and π-equivalent circuit of a linear transformer

Although a linear transformer has two magnetically coupled, but electrically isolated windings, it is sometimes convenient to view them as though they are electrically coupled.

If our coupled circuit has the dot notation in figure 9.11(a), its equivalent circuit is shown in figure 9.11(b), whose mesh equations are

$$\boldsymbol{V}_1 = \mathrm{j}\omega L_1\boldsymbol{I}_1 - \mathrm{j}\omega M\boldsymbol{I}_2 = \mathrm{j}\omega([L_1 - M] + M)\boldsymbol{I}_1 - \mathrm{j}\omega M\boldsymbol{I}_2$$

$$\boldsymbol{V}_2 = -\mathrm{j}\omega M\boldsymbol{I}_1 + \mathrm{j}\omega L_2\boldsymbol{I}_2 = -\mathrm{j}\omega M\boldsymbol{I}_1 + \mathrm{j}\omega([L_2 - M] + M)\boldsymbol{I}_2$$

If we look at these equations as though they refer to electrically coupled elements, we conclude that the circuit is the one shown in figure 9.11(c). The corresponding π-equivalent circuit is shown in figure 9.11(d).

If the position of the dot on either of the windings in figure 9.11(a) is reversed, that is, the connections to any one winding are reversed, the T-equivalent circuit would have an inductance of $-M$ in the common vertical branch, an inductance of $(L_1 + M)$ in the top left-hand branch, and an inductance of $(L_2 + M)$ in the top right-hand branch.

Unworked problems

9.1. Calculate $\boldsymbol{I}_2$ in figure 9.12. The supply frequency is 100 rad/s. [8.19∠35° mA]

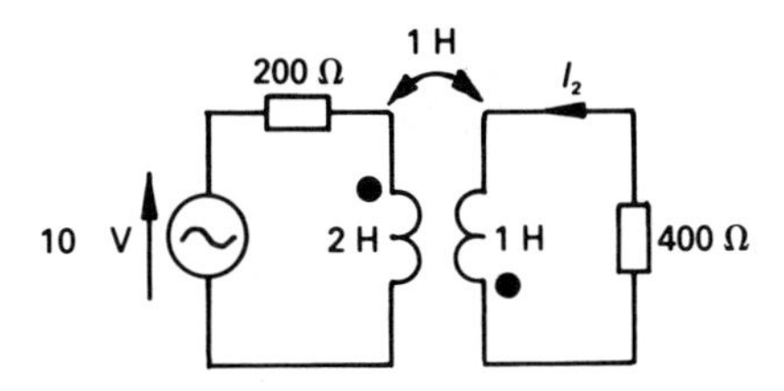

Figure 9.12

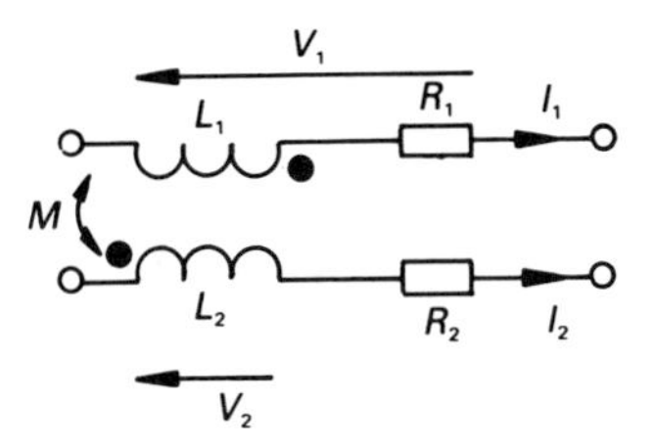

Figure 9.13

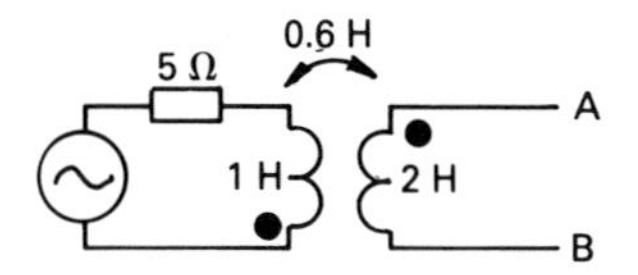

Figure 9.14

9.2. Calculate the input impedance of the circuit in figure 9.12.
[$296\angle 41°\ \Omega$]

9.3. Deduce an expression for $\boldsymbol{V}_1$ and $\boldsymbol{V}_2$ in figure 9.13.
[$(R_1 + j\omega L_1)\boldsymbol{I}_1 - j\omega M\boldsymbol{I}_2$; $j\omega(M - L_2)\boldsymbol{I}_2$]

9.4. If the source in figure 9.14 is (a) a $5\angle 0°$ V voltage source (positive terminal at the top), (b) a $5\angle 0°$ A current source (current flowing upwards), calculate V_{AB} if the frequency is 1 rad/s.
[(a) $0.588\angle -101.3°$ V; (b) $3\angle -90°$ V]

9.5. If a load of 20 Ω is connected between A and B in figure 9.14, calculate V_{AB} for both sources.
[(a) $0.584\angle -107°$ V; (b) $2.986\angle -95.7°$ V]

9.6. Calculate the total power consumed in figure 9.15 if f = 100 Hz.
[0.41 W]

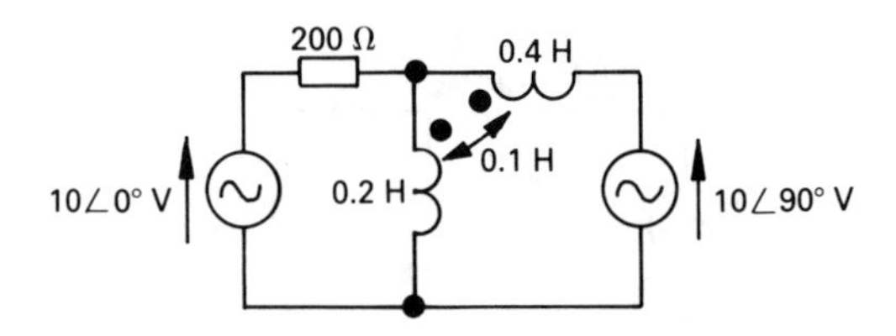

Figure 9.15

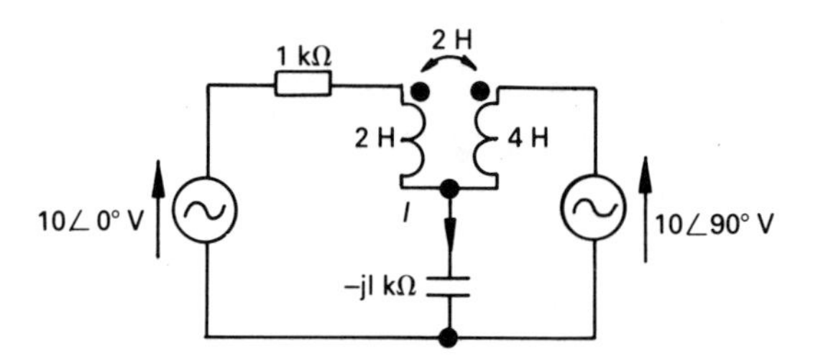

Figure 9.16

9.7. Calculate the current I in figure 9.16 if the supply frequency is 1000 rad/s.
[0.03∠0° A]

9.8. The primary winding of a coupled circuit has a resistance of 20 Ω and a self-inductance of 6 mH. The corresponding values for the secondary circuit are 10 Ω and 4 mH, respectively. If the mutual inductance between the two windings is 2 mH, and a pure resistive load of 40 Ω is connected to the terminals of the secondary winding, calculate the input impedance if the supply frequency is 5000 rad/s.
[36.49∠53.5° Ω]

9.9. Calculate the load current in problem 9.8 if (a) a voltage source of 10 ∠0° V, (b) a current source of 1∠0° A is connected to the primary winding.
[(a) 5.09∠14.74° mA; (b) 0.186∠68.2° A]

9.10. The primary inductance of a coupled circuit is 0.5 H, that of the secondary is 0.1 H, and the mutual inductance is 60 mH; the primary winding resistance is 12 Ω and that of the secondary is 20 Ω. If a capacitor of capacitance 1 μF is connected in series with the primary winding, and a capacitor of 0.2 μF is connected between the terminals of the secondary winding, calculate the input impedance at (a) 5000 rad/s, (b) 800 rad/s.
[(a) 2478∠89.2° Ω; (b) 850∠−89.2° Ω]

9.11. A 50 kVA, 3300/250 V single-phase transformer has a primary winding resistance of 5 Ω and a secondary winding resistance of 0.03 Ω. If the total leakage reactance referred to the secondary winding is 0.13 Ω, calculate the voltage regulation of the transformer for full load at a power factor of (a) 0.8 lagging, (b) unity power factor.

[(a) 10 per cent; (b) 4.7 per cent]

10

Transient Solution of Electrical Circuits

10.1 Introduction

In electrical circuits containing one or more energy storage elements, a *transient state* exists whenever the circuit conditions change. A circuit containing one storage element is described as a *first-order circuit* or a *single-energy circuit*, and one containing a capacitor and an inductor is known as a *second-order circuit* or *double-energy circuit*, which is characterised by a linear differential equation containing a second-order derivative.

Transient analysis of circuits is probably one of the most difficult areas of work covered in a course and, by its nature, involves a higher mathematical content than many other topics. In many cases, transient analysis is first approached via classical mathematical methods and, later, by the D operator method. However, using these methods, many solutions are obtained via a 'back door approach', and one needs to know many special techniques (or should they be called 'tricks of the trade'?) to obtain the correct answer.

Moreover, with some forcing functions, solutions cannot be obtained using classical methods. Also, if initial conditions exist in the circuit, such as a charge on a capacitor or an initial current in an inductor, obtaining a solution using classical methods can be very tedious.

If only to highlight some of the difficulties which may be encountered, we will take a brief look in sections 10.2 and 10.3 at the solution of first- and second-order circuits, respectively, using classical methods. Following this we turn our attention to solution by the Laplace transform method, since this has a number of advantages from an engineering viewpoint.

However, it should be pointed out that there is a small 'drawback' of the Laplace transform method, and this is that one has to traverse a basic mathematical 'fog' before one enters the calmer waters of electrical circuits. The reader is encouraged to understand the basic mathematics of Laplace transforms; in any event, he can refer back to the mathematics when studying the worked examples. Generally speaking the mathematics involves integral and differential calculus, but no more than is found in a first year subject covering these topics.

10.2 Classical solution of first-order systems

The circuits in figures 10.1 and 10.2 contain a single energy storage element (see chapter 4), in which v_S is the supply voltage and i is the circuit current at time t seconds after closing the switch. At this time voltages v_R, v_L and v_C are across the components.

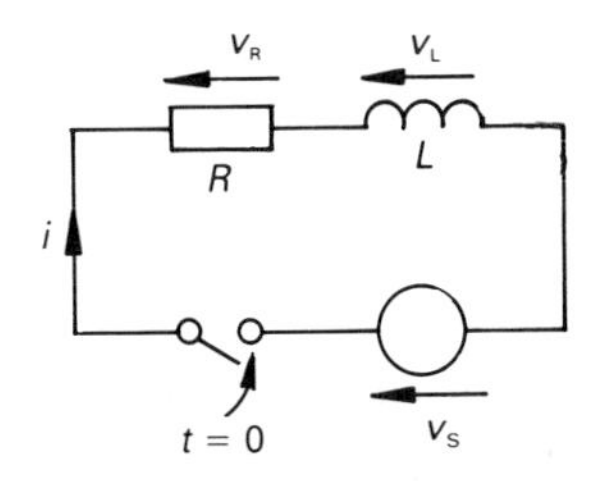

Figure 10.1 *Simple first-order* R–L *system.*

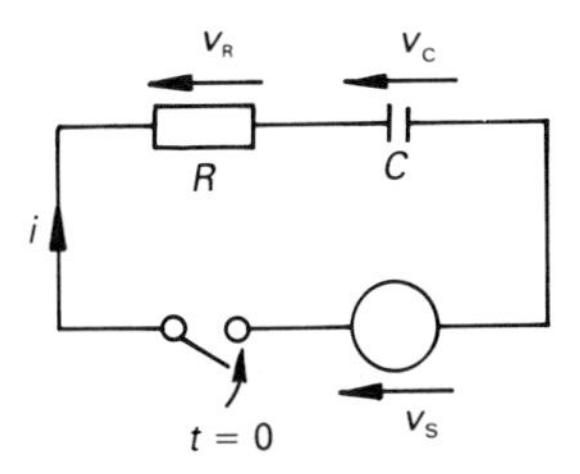

Figure 10.2 *Simple first-order* R–C *system.*

Inductive circuit

For Figure 10.1, by KVL

$$v_S = v_R + v_L$$

But

$$v_L = L\frac{di}{dt}$$

and therefore

$$v_S = iR + L\frac{di}{dt}$$

or

$$\frac{v_S}{R} = i + \frac{L}{R}\frac{di}{dt} \qquad (10.1)$$

Capacitive circuit

For Figure 10.2, by KVL

$$v_S = v_R + v_C$$
$$= iR + v_C$$

Now, since

$$i = C\frac{dv_C}{dt}$$

then

$$v_S = RC\frac{dv_C}{dt} + v_C \qquad (10.2)$$

In general

$$F = x + \tau\frac{dx}{dt} \qquad (10.3)$$

where F = driving function = steady-state value of x, in which x = circuit variable, τ = circuit time constant and dx/dt is the first derivative.

Using the D operator, where $D = d/dt$, then $F = x + \tau Dx$.

The solution of linear, first-order equations of this type, where the variables can be separated on the two sides of the equation consists of two parts

$$x = x_t + x_{ss}$$

where x_t is the *transient solution* obtained by making $F = 0$, and x_{ss} is the *steady-state solution*, that is, the value of x as $t \rightarrow \infty$.

For the transient solution of the equation,

$$x_t + \tau\frac{dx_t}{dt} = 0$$

x_t is always of the form $Ae^{-t/\tau}$, where A is a constant to be determined from the circuit's initial conditions, that is, the value of x at $t = 0$.

In general, the complete solution is of the form

$$x = Ae^{-t/\tau} + x_{ss} \qquad (10.4)$$

Constant voltage source,* v_S = E *volts, and the switch is closed at* t = *0

For the inductive circuit shown in figure 10.1, from equation (10.1)

$$i + \frac{L}{R}\frac{\mathrm{d}i}{\mathrm{d}t} = \frac{E}{R}$$

or $$i + \tau \mathrm{D} i = \text{final current} \quad \text{where } \tau = L/R$$

Since the inductor has no voltage across it when the current is steady,

$$i_{\mathrm{SS}} = \frac{E}{R}$$

hence the complete solution for the current in the circuit in figure 10.1 is

$$i = A\mathrm{e}^{-t/\tau} + \frac{E}{R} \tag{10.5}$$

For the capacitive circuit shown in figure 13.2, from equation (10.2)

$$v_{\mathrm{C}} + RC\frac{\mathrm{d}v_{\mathrm{C}}}{\mathrm{d}t} = E$$

so

$$v_{\mathrm{C}} + \tau \mathrm{D} v_{\mathrm{C}} = \text{final voltage}$$

Since the capacitor is fully charged in the steady state, no current flows in it and $v_{\mathrm{SS}} = E$.

The complete solution for the capacitor voltage in the circuit in figure 10.2 is

$$v_{\mathrm{C}} = \mathrm{A}e^{-t/\tau} + E \tag{10.6}$$

Initial conditions

Substitute the values of *i* and *v* at the instant the switches are closed in the two circuits. Let these be zero and V_1 volts respectively.

Since the current through the inductor and the voltage across a capacitor cannot be changed instantly, then, at $t = 0$, in equation 10.5 for the inductive circuit in figure 10.1

$$0 = A + \frac{E}{R}$$

that is

$$A = -\frac{E}{R}$$

Hence

$$i = \frac{E}{R}(1 - e^{-Rt/L}) \tag{10.7}$$

In equation 10.6 for the capacitive circuit in figure 10.2

$$V_1 = A + E$$

then

$$A = V_1 - E$$

Hence

$$v = E - (E - V_1)e^{-t/RC} \tag{10.8}$$

In both cases the solution is of the form:

$$\text{instantaneous value} = \text{final value} - \text{step size} \times e^{-t/T}$$

which is illustrated graphically for two cases in figure 10.3. In diagram (a), final value > initial value, while in diagram (b), final value < initial value.

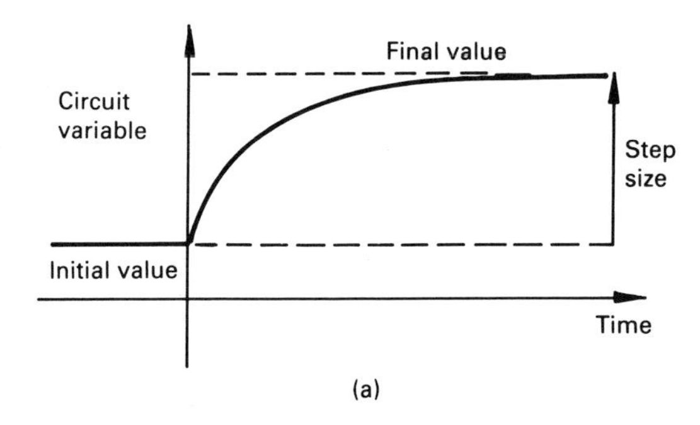

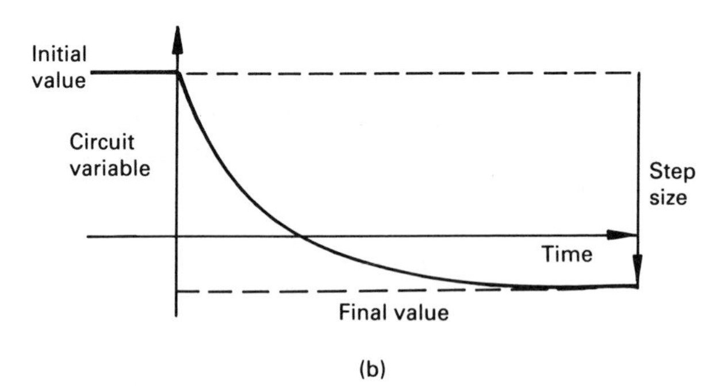

Figure 10.3 *Transients in a first-order system with an initial condition.*

While these solutions are relatively straightforward, circuit analysis becomes increasingly difficult if the network is excited by, say, a sinewave or a complex function, particularly when initial conditions are involved.

10.2.1 The time constant τ of a circuit

In many first-order circuits, a quantity known as the *time constant*, τ, gives an indication of the transient performance of the circuit, particularly when it is excited by a d.c. source.

Where the response of a circuit is described by the expression

$$f(t) = Ae^{-t/\tau}$$

the value of $f(t)$ falls from a value equal to A at $t = 0$, to zero when t approaches infinity. More interestingly, $f(t)$ reaches a value less than 0.01 A when $t = 5\tau$. The period 5τ is often described as the *settling time* of the transient and, beyond this time, we can assume that steady-state conditions have been reached. Similarly, if the response is described by

$$f(t) = A(1 - e^{-t/\tau})$$

the value of $f(t)$ rises from zero when $t = 0$ to a final value of A when t approaches infinity. In fact, it can be shown to rise to a value greater than 0.99 A when $t = 5\tau$. Once again, we can think of 5τ as being the settling time of the transient.

It is also useful to think of a time of 0.7τ as the 'half-life' of the transient since, in both the above cases, $f(t)$ changes by 50 per cent for each 0.7τ time change. Also, $f(t)$ changes by 63.2 per cent (either a reduction or an increase) when $t = \tau$.

In the case of a simple RC circuit

$$\tau = 1/RC \text{ s} \quad (R \text{ in } \Omega,\ C \text{ in F})$$

and in an RL circuit

$$\tau = L/R \text{ s} \quad (R \text{ in } \Omega,\ L \text{ in H})$$

10.2.2 Circuits reducible to first-order format

Circuits which have a single energy storage element may be solved by first obtaining Thévenin's equivalent circuit. The circuit time constants then become L/R_{Th} or CR_{Th}.

10.3 Classical solution of second-order systems

The voltage equation for the series RLC circuit in figure 10.4 is given by KVL

$$v_R + v_L + v_C = v_S$$

that is

$$iR + L\frac{di}{dt} + \frac{1}{C}\int i dt = v_S$$

Differentiating this equation with respect to time to give

$$R\frac{di}{dt} + L\frac{d^2i}{dt^2} + \frac{i}{C} = \frac{dv_S}{dt}$$

or

$$\frac{d^2i}{dt^2} + \frac{R}{L}\frac{di}{dt} + \frac{i}{LC} = \frac{1}{L}\frac{dv_S}{dt}$$

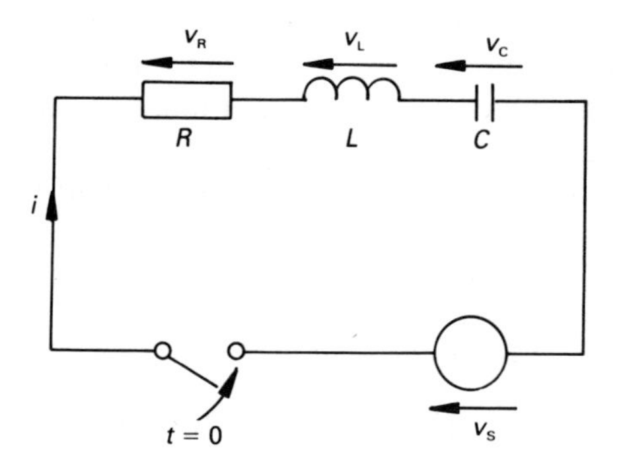

Figure 10.4 *Second-order series* RLC *circuit.*

The highest derivative in the output variable, i, is second order and the solution is again made up of two parts: namely

$$i = i_t + i_{SS}$$

The *steady-state solution* i_{SS} is determined by single-phase theory. For example, when the source is a constant voltage, i_{SS} is zero because of the capacitor.

The *transient solution*, i_t is obtained by equating the left-hand side of the equation to zero. For convenience in interpretation, the coefficient of the highest derivative is reduced to unity. Hence, using the D operator for clarity

$$\mathrm{D}^2 i_t + \frac{R}{L}\mathrm{D}i_t + \frac{1}{LC} i_t = 0 \qquad (10.9)$$

The 'standard form' of this linear second-order differential equation is

$$(\mathrm{D}^2 + 2\zeta\omega_n \mathrm{D} + \omega_n^2) i_t = 0 \qquad (10.10)$$

where ω_n is the *undamped natural frequency* of the oscillations and ζ is the *damping factor*.

Comparing equations 10.9 and 10.10, shows for the series circuit that

$$\omega_n = \frac{1}{\sqrt{LC}} \quad \text{and } \zeta = \frac{R}{2\omega_n L} = \frac{R}{2}\sqrt{\frac{C}{L}}$$

The condition $\zeta = 0$ implies zero circuit resistance.

The solutions to equation 10.10 are determined by the value of ζ as follows:

1. $\zeta < 1$, damped oscillations

$$i_t = A\mathrm{e}^{-\zeta\omega_n t} \sin(\omega_o t + \theta)$$

2. $\zeta = 1$, critical damping

$$i_t = (A + Bt)\mathrm{e}^{-\omega_n t}$$

3. $\zeta > 1$, overdamped response

$$i_t = (A\mathrm{e}^{\omega_o t} + B\mathrm{e}^{-\omega_o t})\mathrm{e}^{-\zeta\omega_n t}$$

4. $\zeta = 0$, no damping, continuous oscillation

$$i_t = A \sin(\omega_n t + \theta)$$

where A, B, and θ are constants determined by the initial conditions of the circuit, and ω_o is the frequency of the damped oscillations. This is given by

$$\omega_o = \omega_n \sqrt{|(1 - \zeta^2)|}$$

Initial conditions

With constant voltage e volts applied to the circuit in figure 10.4, by KVL

$$Ri + L\mathrm{D}i + v_C = E$$

Let the voltage across the capacitor be V_1 and the circuit current be zero before closing the switch, hence at $t = 0$

$$L\mathrm{D}i + V_1 = E$$

$$\mathrm{D}i = \frac{(E - V_1)}{L}$$

Steady-state solution

$i_{ss} = 0$, since the capacitor is fully charged to E.

Hence the complete solution is given by $i = i_t$. Substituting these values into 1, 2, 3 and 4 above gives:

(1) $\zeta < 1$

$$i = \frac{(E - V_1)}{\omega_o L} e^{-\zeta\omega_n t} \sin \omega_o t$$

(2) $\zeta = 1$

$$i = \frac{(E - V_1)}{L} t e^{-\omega_n t}$$

(3) $\zeta > 1$

$$i = \frac{(E - V_1)}{2\omega_o L} e^{-\zeta\omega_n t} (e^{\omega_o t} - e^{-\omega_o t})$$

$$= \frac{(E - V_1)}{\omega_o L} e^{-\zeta\omega_n t} \sinh \omega_o t$$

(4) $\zeta = 0$

$$i = \frac{(E - V_1)}{\omega_n L} \sin \omega_n t$$

As with first-order systems, the solution can be very involved if the circuit is excited by a sinusoidal signal or complex wave, particularly when initial conditions are involved.

10.4 The Laplace transform

The *Laplace transform*, $\boldsymbol{F}(s)$, of a function of time $f(t)$ is defined as

$$\boldsymbol{F}(s) = \mathscr{L}[f(t)] = \int_0^\infty f(t)e^{-st}dt$$

The Laplace transform is a one-sided transform, which assumes that $f(t)$ only exists for $t \geqslant 0$. The variable $s = \sigma + j\omega$ (see also chapter 11) is also complex, and a restriction on its value is that its real part must be large enough to make the integral convergent.

The lower limit on the integral is $t = 0$ but, because of possible discontinuities at $t = 0$ we should, strictly speaking, use a value of t which is just less than zero. We generally denote this as $t = 0^-$. The integral then becomes

$$\boldsymbol{F}(s) = \int_{0^-}^{\infty} f(t)\mathrm{e}^{-st}\mathrm{d}t$$

Thus the operator $\mathscr{L}[f(t)]$ transforms a function $f(t)$ in the *time domain* into a function $\boldsymbol{F}(s)$ in the s-*domain* or *complex frequency domain*. Also, given $\boldsymbol{F}(s)$, we can obtain $f(t)$ by the inverse process as follows

$$f(t) = \mathscr{L}^{-1}\,[f(t)]$$

However, because the direct use of this method needs results from complex-variable theory, we shall (at almost any cost!) avoid employing it. Instead, we will develop a table of Laplace transforms together with the functions from which they are derived, and use it whenever we need to find the inverse of a transform. Trust an engineer to find an easier solution!

After investigating a number of special functions associated with electrical circuits, we will devote some time to the derivation of a few basic Laplace transforms before we move on to circuit analysis.

10.5 Step, impulse and ramp functions

Any one of many forms of signal may be applied to an electrical circuit, and we take a look here at three important forms, namely the *step*, the *impulse* and the *ramp*.

The unit step signal, u(t)

The simplest step signal is the *unit step* ($u(t)$), which is illustrated in figure 10.5(a), and is mathematically described as follows

$$u(t) = \begin{cases} 0 \text{ for } -\infty < t < 0 \\ 1 \text{ for } \quad 0 \leqslant t < \infty \end{cases}$$

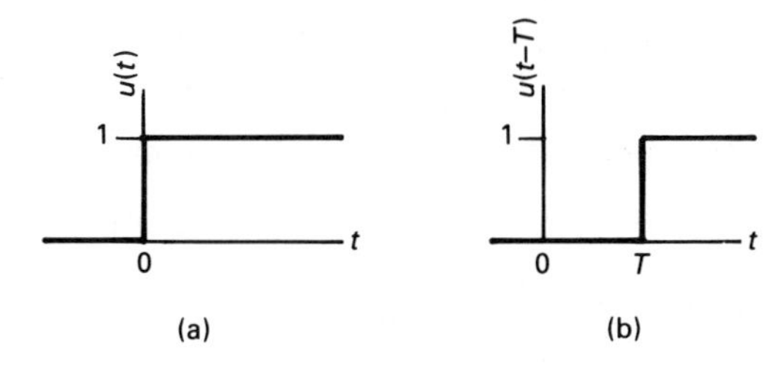

Figure 10.5 *(a) The unit step function, (b) a shifted unit step.*

The function is not continuous at $t = 0$ and, in rigorous mathematical terms, the derivative of $u(t)$ does not exist for $t = 0$.

We can time-shift the unit step function to cause the sudden transition from zero to unity to occur at some time T as follows. We define the *shifted unit step* as follows

$$u(t - T) = \begin{cases} 0 \text{ for } -\infty < t < T \\ 1 \text{ for } \quad T \leqslant t < \infty \end{cases}$$

This is illustrated in figure 10.5(b) and, in this case, the function is not continuous at $t = T$.

The unit step function can also be amplitude-scaled to represent a step of any magnitude merely by multiplying the unit step by a factor K, that is $Ku(t)$ or $Ku(t - t_1)$.

A forcing function frequently encountered in electrical circuits is the unit voltage pulse; this can be represented by two shifted unit steps as follows. Consider the shifted unit step functions $u(t - t_1)$ and $-u(t - t_2)$; these are illustrated in figure 10.6(a), and their sum (the *unit pulse*) $(u(t - t_1) - u(t - t_2))$ is shown in figure 10.6(b). The equivalent electrical circuit which produces the unit pulse is illustrated in figure 10.6(c).

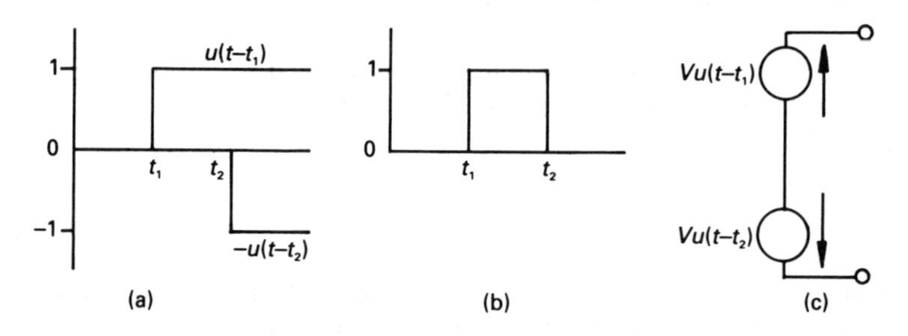

Figure 10.6 *(a) The shifted unit step functions* u(t – t$_1$) *and* u(t – t$_2$) *together generate the unit pulse forcing function in (b). The corresponding equivalent electrical circuit is shown in (c).*

The unit pulse is often used in electrical circuits as a sampling pulse. For example, if we wish to apply a burst of high-frequency oscillations to an electronic system for, say 0.2 μs, we can represent the resulting wave as

$$v(t) = [u(t - t_1) - u(t - t_2)]V_m \sin \omega t \quad \text{V}$$

where $t_2 = t_1 + 0.2 \times 10^{-6}$. A typical resulting waveform may be as shown in figure 10.7.

We can also use the delayed step function to describe the operation of a switch which closes at time $t = t_1$ as follows. If the voltage source KV_1 shown in figure 10.8(a) is connected to an external circuit when $t = t_1$, the

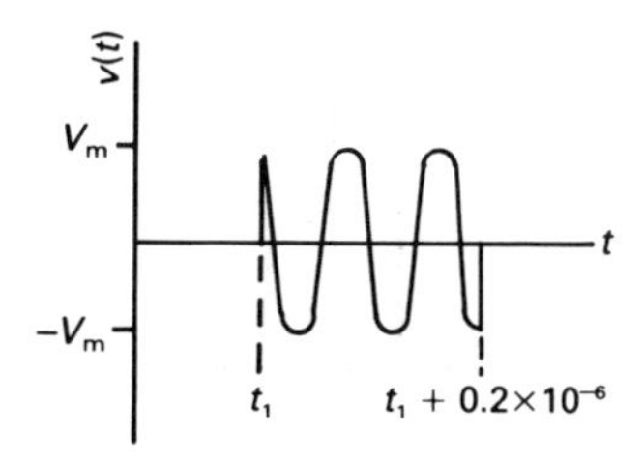

Figure 10.7 *A sinusoidal high-frequency burst.*

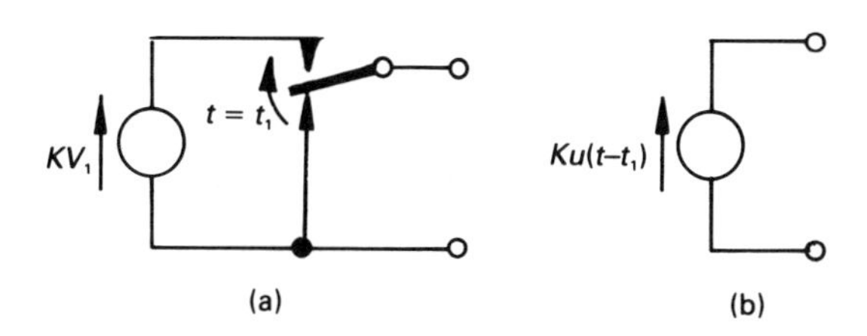

Figure 10.8 *(a) Actual switching circuit, (b) its equivalent circuit.*

complete circuit can be replaced by its mathematical equivalent in diagram (b).

The unit impulse function or delta function, δ(t)

If we differentiate the unit step function, $u(t)$, with respect to time, its value is zero for all time except at $t = 0$. At the latter time $u(t)$ is mathematically indeterminate, and its differential is infinity, that is, the differential is a pulse of infinite value and of zero width. However, all is not lost, for we can look more closely at the unit step function.

A practical unit step cannot change its value from zero to unity in zero time, so let us assume that its change in value occurs over the period $-\Delta/2$ to $\Delta/2$ (see figure 10.9(a)). The derivative of the practical unit step function is shown in figure 10.9(b), and we see that its amplitude is $1/\Delta$. Clearly, as Δ approaches zero, the amplitude of the pulse approaches infinity, thereby approximating to the ideal impulse. Moreover, the area of the impulse in diagram (b) is unity, hence mathematically

$$\int_{-\infty}^{\infty} \delta(t)\,\mathrm{d}t = \int_{0^-}^{0^+} \delta(t)\,\mathrm{d}t = 1$$

and

$$\frac{d}{\mathrm{d}t}\,[u(t)] = \delta(t)$$

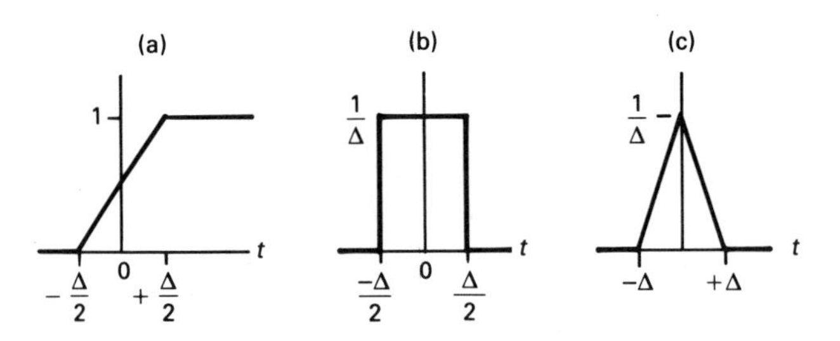

Figure 10.9 *(a) A practical unit step function, (b) the derivative of the practical unit step, (c) a triangular pulse with similar characteristics to pulse (b).*

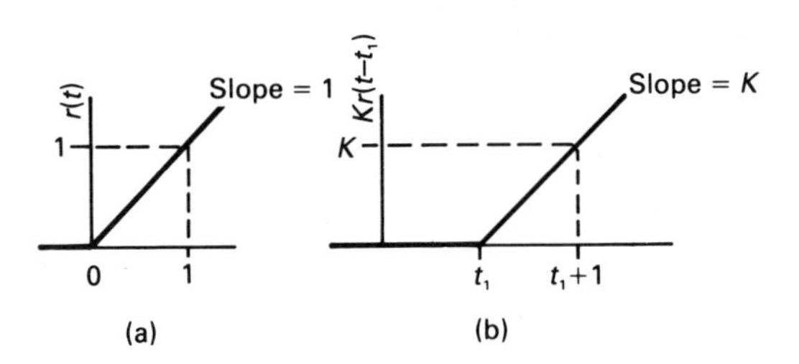

Figure 10.10 *(a) The unit ramp function* r(t), *(b) the magnitude-scaled and time-shifted ramp function* $\mathrm{Kr}(\mathrm{t} - \mathrm{t}_1)$.

The reader should note that, in many texts, it is assumed that the slope of the practical unit step function commences at $t = 0$, and completes the slope at $t = \Delta$. The overall result is the same as that outlined above, other than the unit step function is time shifted by $\Delta/2$.

As with the unit step, we can scale the unit impulse to any magnitude simply by multiplying by the factor K, that is, it becomes $K\delta(t)$.

The unit ramp function, r(t)

The unit ramp function $r(t)$ in figure 10.10(a) is defined methamatically by

$$r(t) = \begin{cases} 0 \text{ for } -\infty < t < 0 \\ t \text{ for } \quad 0 \leqslant t < \infty \end{cases}$$

As with the other 'unit' functions, we can both magnitude-scale and time-shift the ramp function, as shown in figure 10.10(b).

Relationship between the unit step, unit impulse and unit ramp functions

The mathematical relationship between the three functions is

$$\frac{d}{dt}[r(t)] = u(t) \qquad r(t) = \int_{-0}^{\infty} u(t)\ dt$$

$$\frac{d}{dt}[u(t)] = \delta(t) \qquad u(t) = \int_{-0}^{\infty} \delta(t)\ dt$$

The lower limit associated with the integral may be, in fact, any value less than zero.

Worked example 10.5.1

The voltage

$$v(t) = -2u(t) + \tfrac{1}{2}\, r(t) - \tfrac{1}{2}\, r(t-8) - 2u(t-8) \quad \text{V}$$

is applied to a capacitor of 1.5 F capacitance. Determine an expression for the current in the circuit, and sketch its waveshape.

Solution

At this stage it is helpful to sketch the voltage waveform in order to appreciate what type of function we are dealing with. The step function $-2u(t)$ corresponds to a step of -2 V which occurs at $t = 0$, as shown in figure 10.11(a). Next, the function $\frac{1}{2}r(t)$ is a ramp of slope 0.5, commencing at $t = 0$ (see figure 10.11(b)). The ramp function $-\frac{1}{2}r(t-8)$ is a ramp of slope -0.5 (that is, negative-going), commencing at $t = 8$ s (see figure 10.11(c)). Finally, the step function $-2u(t-8)$ is a step of -2 V, commencing at $t = 8$ s (figure 10.11(d)).

The forcing function is the sum of these components, and is shown in figure 10.11(e). The capacitor current is

$$i = C\frac{dv}{dt} = 1.5\frac{dv}{dt} \quad \text{A}$$

hence

$$i = 1.5\left[\frac{d}{dt}[-2u(t)] + \frac{d}{dt}[\tfrac{1}{2}r(t)] + \frac{d}{dt}[-\tfrac{1}{2}r(t-8)] + \frac{d}{dt}[-2u(t-8)]\right]$$

and, using the relationships deduced earlier

$$\begin{aligned} i &= 1.5\,[-2\delta(t) + \tfrac{1}{2}u(t) - \tfrac{1}{2}u(t-8) - 2\delta(t-8)] \\ &= -3\delta(t) + 0.75u(t) - 0.75u(t-8) - 3\delta(t-8) \end{aligned}$$

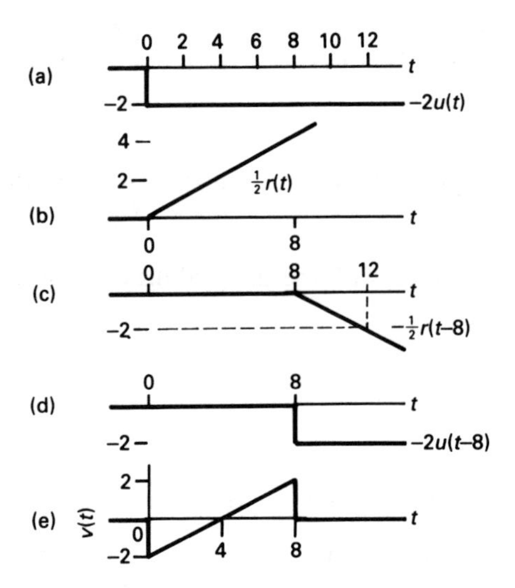

Figure 10.11 *The build-up of* v(t).

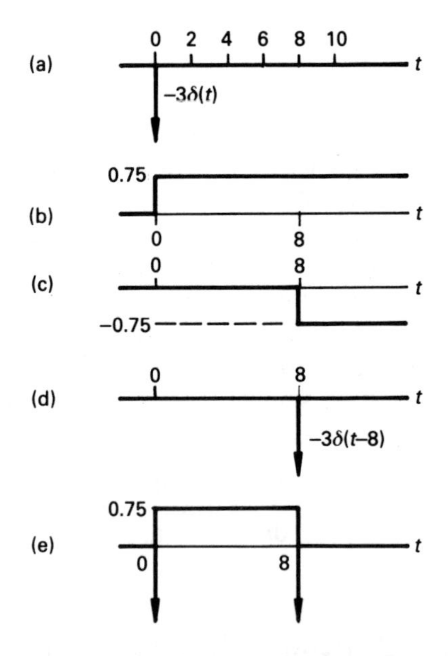

Figure 10.12 *Construction of the current waveform through the 1.5 F capacitor.*

The corresponding current waveform is constructed in figure 10.12. The sudden negative-going transition of current at the beginning and end of figure 10.12(e) would (theoretically at least) result in a current pulse of amplitude $-\infty$! In practice, of course, this could never happen because the output resistance of the source would limit its magnitude.

10.6 Laplace transforms of some useful functions

In the following we develop Laplace transforms of a number of the more useful functions in electrical engineering. These and other Laplace transform pairs are given in table 10.1.

Table 10.1 Short table of Laplace transforms

$f(t) = \mathscr{L}^{-1}[F(s)]$	$F(s) = \mathscr{L}[f(t)]$
1 or $u(t)$ (unit step)	$1/s$
A	A/s
$e^{-\alpha t}$	$1/(s + \alpha)$
$\delta(t)$ (unit impulse)	1
$t\,u(t)$ (ramp)	$1/s^2$
$u(t - T)$ (delayed step)	e^{-sT}/s
Rectangular pulse	$(1 - e^{-sT})/s$
$\int_0^t f(t)\,dt$	$F(s)/s$
$\dfrac{df(t)}{dt}$	$sF(s) - f(0)$
$\dfrac{d^2f(t)}{dt^2}$	$s^2F(s) - sf'(0) - f(0)$
$\sin \omega t$	$\omega/(s^2 + \omega^2)$
$\cos \omega t$	$s/(s^2 + \omega^2)$
$e^{-\alpha t}\left(\cos \omega t - \dfrac{\alpha}{\omega} \sin \omega t\right)$	$s/((s + \alpha)^2 + \omega^2)$
$\sin (\omega t \pm \theta)$	$(\omega \cos \theta \pm s \sin \theta)/(s^2 + \omega^2)$
$\cos (\omega t \pm \theta)$	$(s \cos \theta \pm \omega \sin \theta)/(s^2 + \omega^2)$
t	$1/s^2$
t^n (n is a positive integer)	$n!/s^{n+1}$
$t \sin \omega t$	$2\omega s/(s^2 + \omega^2)^2$
$t \cos \omega t$	$(s^2 - \omega^2)/(s^2 + \omega^2)^2$
$e^{-\alpha t}t^n$	$n!/(s + \alpha)^{n+1}$
$\sinh \alpha t$	$\alpha/(s^2 - \alpha^2)$
$\cosh \alpha t$	$s/(s^2 - \alpha^2)$
$e^{-\alpha t} \cosh \beta t$	$(s + \alpha)/((s + \alpha)^2 - \beta^2)$
$e^{-\alpha t} \sinh \beta t$	$\beta/((s + \alpha)^2 - \beta^2)$
$e^{-\alpha t} \cos \omega t$	$(s + \alpha)/((s + \alpha)^2 + \omega^2)$
$e^{-\alpha t} \sin \omega t$	$\omega/((s + \alpha)^2 + \omega^2)$

Worked example 10.6.1

Derive the Laplace transform of the unit step function $u(t)$.

Solution

The step function is defined as follows

$$u(t) = \begin{cases} 0 \text{ for } -\infty \leqslant t \leqslant 0 \\ 1 \text{ for } \quad 0 \leqslant t \leqslant \infty \end{cases}$$

hence

$$\mathscr{L}[u(t)] = \int_0^\infty u(t)\mathrm{e}^{-st}\,\mathrm{d}t = \int_0^\infty \mathrm{e}^{-st}\,\mathrm{d}t$$

$$= -\frac{1}{s}\left[\mathrm{e}^{-st}\right]_0^\infty = -\frac{1}{s}\left[0 - 1\right] = \frac{1}{s}$$

If the step has amplitude A, then

$$\mathscr{L}[Au(t)] = \frac{A}{s}$$

Worked example 10.6.2

Deduce the Laplace transform of $f(t) = t^n$.

Solution

This expression can be integrated by parts ($\int u\,\mathrm{d}v = uv - \int v\,\mathrm{d}u$), so that if $u = t^n$ and $\mathrm{d}v = \mathrm{e}^{-st}\mathrm{d}t$, then $v = -\mathrm{e}^{-st}/s$ and $\mathrm{d}u = nt^{n-1}\mathrm{d}t$ as follows

$$\mathscr{L}[t^n] = \int_0^\infty t^n\mathrm{e}^{-st}\mathrm{d}t = -\frac{t^n}{s}\,\mathrm{e}^{-st} - \int_0^\infty \left(-\frac{nt^{n-1}}{s}\right)\mathrm{e}^{-st}\,\mathrm{d}t$$

$$= 0 + \left[\frac{nt^{n-1}}{s^2}\,\mathrm{e}^{-st}\right]_0^\infty - \int_0^\infty \frac{n(n-1)t^{n-2}}{-s^2}\,\mathrm{e}^{-st}\,\mathrm{d}t$$

$$= 0 + 0 + \ldots - \int_0^\infty \frac{n!}{-s^n}\,\mathrm{e}^{-st}\,\mathrm{d}t$$

$$= \frac{n!}{s^{n+1}}$$

Note: n is a positive integer. It also follows that

$$\mathscr{L}\left[\frac{t^{n-1}}{(n-1)!}\right] = \frac{1}{s^n}$$

and

$$\mathscr{L}[t] = 1/s^2$$

The above example allows us to write down the Laplace transform of the ramp voltage $v(t) = kt$. Its Laplace transform is

$$\mathscr{L}[kt] = k/s^2$$

Worked example 10.6.3

Evaluate the Laplace transform of $f(t) = e^{\alpha t}$.

Solution

$$\mathscr{L}[e^{\alpha t}] = \int_0^\infty e^{\alpha t}\, e^{-st}\, dt = \int_0^\infty e^{-(s-\alpha)t}\, dt$$

$$= \left[-\frac{1}{(s-\alpha)} e^{-(s-\alpha)t}\right]_0^\infty = \frac{1}{s-\alpha}$$

This assumes, of course that the real part of s is greater than α. It also follows from the above that

$$\mathscr{L}[e^{-\alpha t}] = 1/(s+\alpha)$$

The function $e^{-\alpha t}$ occurs frequently in electrical circuits where $1/\alpha$ is the time constant, τ, of the circuit. The function is a decaying exponential which falls to slightly less than 1 per cent of its initial value when $t = 5\tau = 5/\alpha$. Similarly, the function $(1 - e^{-\alpha t})$ is the equation of a rising exponential curve, which rises to slightly more than 99 per cent of its final value when $t = 5\tau = 5/\alpha$.

For exponential curves of this kind, the period 5τ is known as the *settling time*, which is the time taken for the transient period to have practically settled out.

Worked example 10.6.4

Determine $\mathscr{L}[\sin \omega t]$ and $\mathscr{L}[\cos \omega t]$.

Solution

Writing $\sin \omega t$ as $\frac{1}{2j}(e^{j\omega t} - e^{-j\omega t})$ we get

$$\mathscr{L}[\sin \omega t] = \frac{1}{2j}\int_0^\infty (e^{j\omega t} - e^{-j\omega t})\, e^{-st}\, dt$$

$$= \frac{1}{2j}\left(\frac{1}{s - j\omega} - \frac{1}{s + j\omega}\right) = \frac{\omega}{s^2 + \omega^2}$$

and since $\cos \omega t$ is $\frac{1}{2}(e^{j\omega t} + e^{-j\omega t})$ then

$$\mathscr{L}[\cos \omega t] = s/(s^2 + \omega^2)$$

Worked example 10.6.5

Evaluate $\mathscr{L}[t \sin \omega t]$.

Solution

If we differentiate $e^{-st} \sin \omega t$ with respect to s, we get

$$\frac{d}{ds}(\sin \omega t\, e^{-st}) = -t \sin \omega t\, e^{-st}$$

and, using the result from worked example 10.6.4

$$\int_0^\infty \frac{d}{ds}(\sin \omega t\, e^{-st})\, dt = \frac{d}{ds}\frac{\omega}{(s^2 + \omega^2)}$$

or

$$-\int_0^\infty t \sin \omega t\, e^{-st}\, dt = \frac{-2\omega s}{(s^2 + \omega^2)^2}$$

hence

$$\mathscr{L}[t \sin \omega t] = 2\omega s/(s^2 + \omega^2)^2$$

Worked example 10.6.6

Determine the Laplace transform of $f(t) = \sinh \alpha t$.

Solution

Since $\sinh \alpha t = \frac{1}{2}(e^{\alpha t} - e^{-\alpha t})$, then

$$\mathscr{L}[t \sinh \alpha t] = \frac{1}{2}\int_0^\infty (e^{\alpha t} - e^{-\alpha t})\, e^{-st}\, dt$$

$$= \frac{1}{2}\left(\frac{1}{s - \alpha} - \frac{1}{s + \alpha}\right) = \frac{\alpha}{s^2 - \alpha^2}$$

Worked example 10.6.7

Determine $\mathscr{L}[\sin(\omega t + \theta)]$.

Solution

Expanding $\sin(\omega t + \theta)$ as $\sin\theta.\cos\omega t + \cos\theta.\sin\omega t$, then

$$\mathscr{L}[\sin(\omega t + \theta)] = \mathscr{L}[\sin\theta.\cos\omega t + \cos\theta.\sin\omega t]$$

$$= \frac{s\sin\theta}{s^2 + \omega^2} + \frac{\omega\cos\theta}{s^2 + \omega^2}$$

$$= \frac{s\sin\theta + \omega\cos\theta}{s^2 + \omega^2}$$

10.7 Properties of the Laplace transform

The Laplace transform has several properties of particular interest to electrical and electronics engineers, and we will look at a number of them.

Scalar multiplication

$$\mathscr{L}[k\,f(t)] = kF(s)$$

For example

$$\mathscr{L}[10\cos\omega t] = 10\,\mathscr{L}[\cos\omega t] = \frac{10s}{s^2 + \omega^2}$$

Addition

$$\mathscr{L}[f_1(t) \pm f_2(t)] = F_1(s) \pm F_2(s)$$

For example

$$\mathscr{L}[10t + 3e^{-2t} - 5\sin 4t]$$

$$= 10\mathscr{L}[t] + 3\mathscr{L}[e^{-2t}] - 5\mathscr{L}[\sin 4t]$$

$$= \frac{10}{s^2} + \frac{3}{s + 2} - \frac{20}{s^2 + 16}$$

Time differentiation

$$\mathscr{L}\left[\frac{\mathrm{d}f(t)}{\mathrm{d}t}\right] = sF(s) - f(0)$$

where $f(0)$ is the initial condition which exists in the circuit.

Time integration

$$\mathscr{L}\left[\int_0^t f(t)\,\mathrm{d}t\right] = \frac{1}{s}\,F(s)$$

Time shift

The *time-shift theorem* states that if a time function is delayed by time a in the time domain, then the result in the frequency domain is multiplied by e^{-as}.

$$\mathscr{L}\,[f(t-a)u(t-a)] = \mathrm{e}^{-as}F(s)$$

For example, the rectangular voltage pulse in figure 10.13(a) can be described by the following expression.

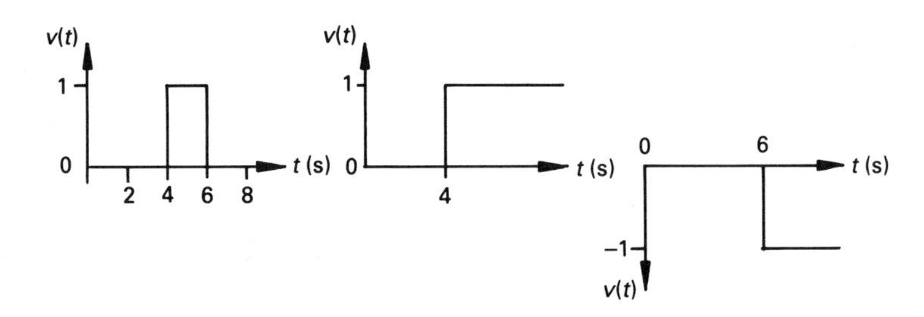

Figure 10.13 *(a) The rectangular pulse shown can be considered to be the sum of the unit step functions in (b) and (c).*

$$v(t) = u(t-4) - u(t-6)$$

Since the Laplace transform of $u(t)$ is $1/s$ then, since $u(t-4)$ is $u(t)$ delayed by 4 s, the transform of the first delayed function is e^{-4s}/s. Similarly, the transform of the second delayed step function is e^{-6s}/s, hence the overall Laplace transform is

$$\mathscr{L}\,[v(t)] = V(s) = \frac{\mathrm{e}^{-4s} - \mathrm{e}^{-6s}}{s}$$

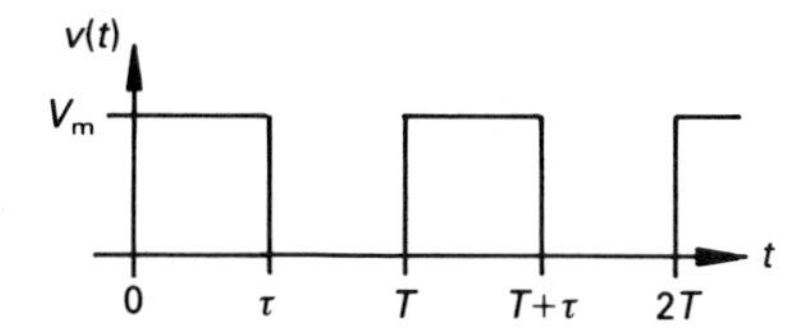

Figure 10.14 *The Laplace transform of a train of rectangular pulses.*

This theorem can also be used to determine the Laplace transform of a periodic time function as follows. If $f(t)$ is a function with a period of T for positive values of T, then

$$F(s) = F_1(s)/(1 - e^{-sT})$$

where $F_1(s) = \int_0^T f(t)e^{-sT}\, dt$, is the transform of the period of the time function. The $(1 - e^{-sT})$ in the denominator accounts for the periodicity of the function.

Consider the rectangular pulse train in figure 10.14. The Laplace transform of the first period of the function is

$$\mathscr{L}[V_m(u(t) - u(t - \tau))] = \frac{V_m}{s}[1 - e^{-s\tau}]$$

That is, the transform for the pulse train is

$$V(s) = \frac{V_m}{s}\frac{(1 - e^{-s\tau})}{(1 - e^{-sT})}$$

Frequency shift

$$\mathscr{L}[e^{-\alpha t} f(t)] = \mathbf{F}(s + \alpha)$$

This theorem states that replacing s by $(s + \alpha)$ in the frequency-domain corresponds to multiplying $f(t)$ by $e^{-\alpha t}$ in the time domain. Thus, making reference to table 10.1, we have

$$\mathscr{L}[e^{-\alpha t}\cos \omega t] = \frac{s + \alpha}{(s - \alpha)^2 + \omega^2}$$

By inference we can also say that

$$\mathscr{L}[e^{\alpha t} f(t)] = \mathbf{F}(s - \alpha)$$

Frequency differentiation

$$\mathscr{L}[-tf(t)] = \frac{d\mathbf{F}(s)}{ds}$$

This theorem states that differentiation with respect to s in the frequency domain is equivalent to multiplication by $-t$ in the time domain. By implication, we may say that

$$\mathscr{L}[t^n f(t)] = (-1)^n \frac{d^n f(s)}{ds^n}$$

For example

$$\mathscr{L}[t \sin \omega t] = (-1)^1 \frac{d}{ds}\left(\frac{\omega}{s^2 + \omega^2}\right)$$

$$= \frac{2\omega s}{(s^2 + \omega^2)^2}$$

Frequency integration

$$\mathscr{L}\left[\frac{f(t)}{t}\right] = \int_s^\infty F(s)\, ds$$

For example

$$\mathscr{L}\left[\frac{\sin \omega t}{t}\right] = \int_s^\infty \frac{\omega\, ds}{s^2 + \omega^2} = \left[\tan^{-1}\frac{s}{\omega}\right]_s^\infty$$

$$= \frac{\pi}{2} - \tan^{-1}\frac{s}{\omega}$$

Frequency scaling

For $k \geqslant 0$

$$\mathscr{L}[f(ft)] = \frac{1}{k} F\left(\frac{s}{k}\right)$$

This theorem allows us to obtain the frequency-scaled version of a function of time whose Laplace transform exists. For example, since the transform of a 1 rad/s sinewave is

$$\mathscr{L}[\sin t] = 1/(s^2 + 1)$$

then the transform of a 2 kHz (4000π rad/s) sinewave is

$$\mathscr{L}[\sin 4000t] = \frac{1}{4000\pi} \frac{1}{\left(\dfrac{s}{4000\pi}\right)^2 + 1}$$

$$= \frac{4000\pi}{s^2 + (4000\pi)^2}$$

Initial value theorem

The initial value of a function is given by

$$f(0^+) = \lim_{s \to \infty} sF(s)$$

where $f(0^+)$ is the value of $f(t)$ when t is zero (or, strictly speaking, marginally greater than zero!). This theorem states that the initial value of the function $f(t)$ is obtained by putting $s = \infty$ in the equation for $sF(s)$.

We shall illustrate this by using a function for which the initial value of $f(t)$ is known. One example is the cosine wave, where $f(0^+) = 1$. In this case

$$\mathscr{L}[\cos \omega t] = s/(s^2 + \omega^2)$$

and its initial value is

$$f(0^+) = \lim_{s \to \infty} sF(s) = \lim_{s \to \infty} (s \times s/(s^2 + \omega^2)) = 1$$

Final value theorem

Provided that all the poles of $sF(s)$ are completely within the left-hand part of the complex frequency plane (see chapter 11 for details), then the final value of the function is given by

$$f(\infty) = \lim_{s \to 0} sF(s)$$

where $f(\infty)$ is the value of $f(t)$ when t approached infinity. That is, the final value of $f(t)$ is obtained by letting s be zero in $sF(s)$.

If $f(t)$ is a pure sinusoid, $F(s)$ has poles on the $\mathrm{j}\omega$ axis, and use of the final value theorem would lead us to the conclusion that the final value is zero! However, we know that the final value of a sinusoid is indeterminate, so *beware of poles on the* $\mathrm{j}\omega$ *axis*. The only exception is a simple pole at $s = 0$.

Once again, to illustrate this theorem, we will consider a function of time whose final value we know. Such a function is $f(t) = 1 - \mathrm{e}^{-\alpha t}$, whose final value is unity. In this case

$$sF(s) = s\left(\frac{1}{s} - \frac{1}{s + \alpha}\right) = \frac{\alpha}{s + \alpha}$$

hence

$$f(\infty) = \lim_{s \to 0} sF(s) = \lim_{s \to 0} \frac{\alpha}{s + \alpha} = 1$$

10.8 Representation of circuit elements in the s domain

So far, we only considered circuit elements which do not have initial conditions imposed on them, that is, an initial charge on a capacitor, or an initial current in an inductor. The Laplace transform allows us a fairly easy method of dealing with these conditions. However, we look at basic circuit elements before dealing with complete circuits.

Resistors

The v–i equation in the time domain for a resistor is

$$v(t) = Ri(t)$$

and the Laplace transform of this equation is

$$\mathscr{L}[v(t)] = \mathscr{L}[Ri(t)] = R\,\mathscr{L}[i(t)]$$

that is we may say

$$V(s) = R\boldsymbol{I}(s)$$

The transfer function $\boldsymbol{V}(s)/\boldsymbol{I}(s) = \boldsymbol{Z}(s) = R$ is the s-domain *impedance* of the element, and the ratio $\boldsymbol{I}(s)/\boldsymbol{V}(s) = \boldsymbol{Y}(s) = 1/R$ is the s-domain *admittance* of the element. Clearly, the s-domain impedance and admittance are identical to the time-domain values.

Inductance

The time-domain v–i relationship for the circuit in figure 10.15(a) is

$$v(t) = L\,\frac{\mathrm{d}i(t)}{\mathrm{d}t}$$

Hence

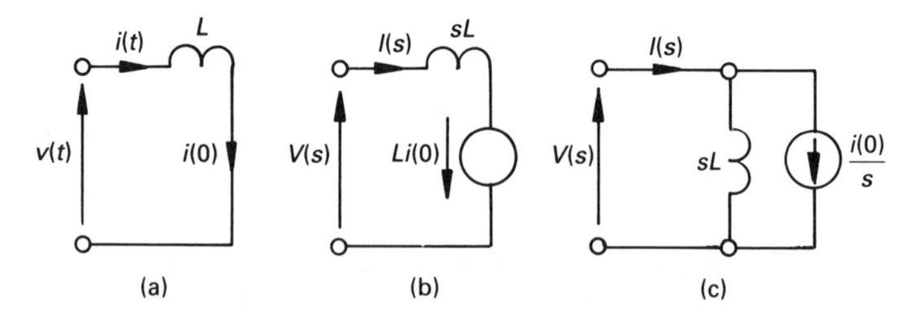

Figure 10.15 *(a) Inductor* L *with initial current* i*(0) in the time-domain, (b)* s*-domain Thévenin equivalent circuit, and (c) the Norton equivalent circuit.*

$$V(s) = \mathscr{L}\left[L\,\frac{\mathrm{d}i(t)}{\mathrm{d}t}\right] = sL\boldsymbol{I}(s) - Li(0)$$

The resulting equivalent circuit for the inductor is shown in figure 10.15(b). Re-writing the equation in the form

$$\boldsymbol{I}(s) = \frac{\boldsymbol{V}(s)}{sL} + \frac{i(0)}{s}$$

gives the s-domain equivalent circuit in figure 10.15(c).

If no initial current flows in the inductor then, in the case of figure 10.15(b), the initial condition generator is short-circuited and, in the case of figure 10.15(c), it is open-circuited.

Capacitance

The time-domain v–i equation for the circuit in figure 10.16(a) is

$$i(t) = C\,\frac{\mathrm{d}v(t)}{\mathrm{d}t}$$

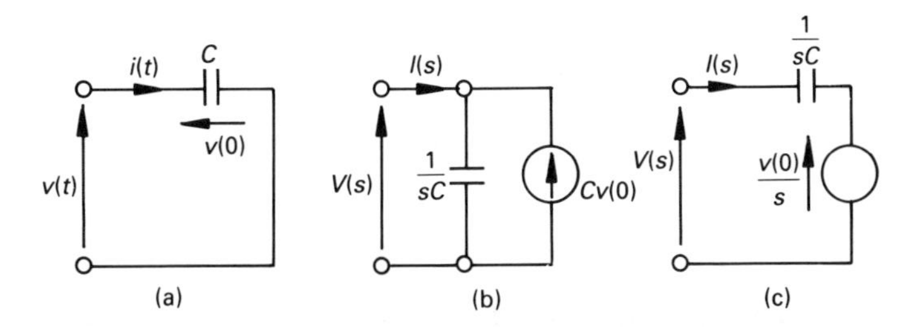

Figure 10.16 *(a) Time-domain circuit for a capacitor with an initial voltage* v*(0) between its terminals, (b) the Norton* s*-domain equivalent circuit and (c) the* s*-domain Thévenin circuit.*

that is

$$\boldsymbol{I}(s) = sC\boldsymbol{V}(s) - Cv(0)$$

from which the equivalent circuit in figure 10.16(b) is drawn. Re-writing the equation in the form

$$\boldsymbol{V}(s) = \frac{\boldsymbol{I}(s)}{sC} + \frac{v(0)}{s}$$

from which we may deduce the equivalent circuit in figure 10.16(c).

10.9 Introduction to analysis of first-order circuits using the Laplace transform

We will approach transient analysis of networks by considering systems containing a single energy storage element. The solutions provided are intended to give an indication of typical solutions, but are by no means exhaustive.

Worked example 10.9.1

Determine for figure 10.17(a) an expression for $i(t)$ and $v_L(t)$ if (a) $E_1 = 0$, (b) $E_1 = -20$ V. In both cases, calculate $i(t)$ when $t = 0.01$ s. The blade of the switch changes from A to B when $t = 0$; in both cases it can be assumed that the switch is ideal, and the changeover occurs instantaneously.

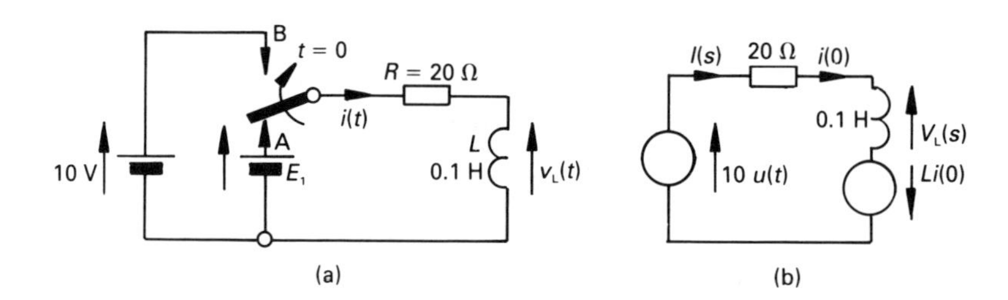

Figure 10.17 *(a) Circuit for worked example 10.9.1, (b) a s-domain equivalent circuit.*

Solution

This example will illustrate how easy it is with Laplace transforms to insert initial conditions into the circuit.

Assuming that steady-state conditions have been reached before the switch blade changes from A to B, the initial current in the circuit is $i(0) = E_1/R$. The circuit equation (see figure 10.17(b)), when $t \geqslant 0$ in the s-domain, is

$$10u(t) + Li(0) = (R + sL)\boldsymbol{I}(s)$$

or

$$10u(t) = (20 + 0.1s)\boldsymbol{I}(s) - 0.1i(0)$$

(a) In this case I(0) = 0/20 = 0 and the s-domain equation for the circuit is

$$\frac{10}{s} = (20 + 0.1s)\boldsymbol{I}(s)$$

hence

$$\boldsymbol{I}(s) = \frac{10}{s(20 + 0.1s)} = \frac{100}{s(200 + s)} = \frac{0.5}{s} - \frac{0.5}{200 + s}$$

$$= 0.5\left(\frac{1}{s} - \frac{1}{200 + s}\right)$$

The penultimate step in the above equation is obtained by the method of partial fractions (see chapter 15 for details), but any other method can be used.

Using the table of Laplace transforms we see that

$$i(t) = \mathscr{L}^{-1}\,[\boldsymbol{I}(s)] = 0.5(1 - \mathrm{e}^{-200t})\ \mathrm{A}$$

so that when $t = 0.01$ s, the current is

$$i_{0.01} = 0.5(1 - \mathrm{e}^{-200 \times 0.01}) = 0.432\ \mathrm{A}$$

We can evaluate the initial and final value of the current as follows

$$\text{initial value} = \lim_{s \to \infty} s\boldsymbol{I}(s) = \frac{100}{200 + s} = 0$$

$$\text{final value} = \lim_{s \to \infty} s\boldsymbol{I}(s) = \frac{100}{200 + s} = 0.5\ \mathrm{A}$$

Also $v_L(t) = L\,\mathrm{d}i/\mathrm{d}t$, then

$$\boldsymbol{V}_L(\mathbf{s}) = L(s\boldsymbol{I}(s) - f(0)) = L\,\frac{100}{200 + s} = \frac{10}{200 + s}$$

hence from the table of Laplace transforms

$$v_L(t) = 10\mathrm{e}^{-200t}\ \mathrm{V}$$

The corresponding waveforms for $i(t)$ and $v_L(t)$ are shown in figure 10.18(a) and (b), respectively.

(b) In this case, the initial current (in the direction shown in the figure) is $i(0) = -20/20 = -1$ A. The s-domain equation for the circuit is

$$\frac{10}{s} = (20 + 0.1s)\boldsymbol{I}(s) - 0.1(-1)$$

or

$$\boldsymbol{I}(\mathrm{s}) = \frac{\frac{10}{s} - 0.1}{20 + 0.1s} = \frac{10}{s(20 + 0.1s)} - \frac{0.1}{20 + 0.1s}$$

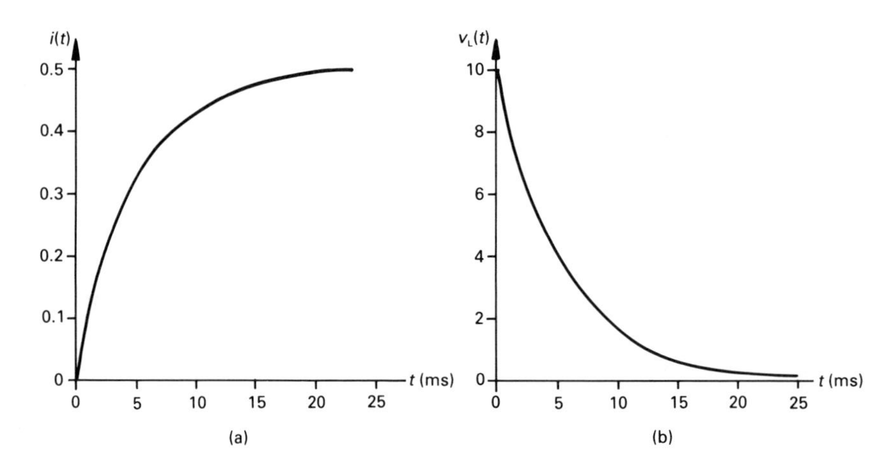

Figure 10.18 *(a)* i(t) *and (b)* v_L(t) *for part (a) of worked example 10.9.1.*

$$= \frac{100}{s(200 + s)} - \frac{1}{200 + s}$$

$$= \frac{0.5}{s} - \frac{0.5}{200 + s} - \frac{1}{200 + s} = \frac{0.5}{s} - \frac{1.5}{200 + s}$$

hence

$$i(t) = \mathscr{L}^{-1}\,[\boldsymbol{I}(s)] = 0.5 - 1.5\mathrm{e}^{-200t}\ \mathrm{A}$$

and the current when $t = 0.01$ s is

$$i_{0.01} = 0.5 - 1.5\mathrm{e}^{-2} = 0.297\ \mathrm{A}$$

Once again we can calculate the initial and final value of the current as follows.

$$\text{initial value} = \lim_{s \to \infty} s\boldsymbol{F}(s) = 0.5 - \frac{1.5s}{200 + s} = -1\ \mathrm{A}$$

$$\text{final value} = \lim_{s \to 0} s\boldsymbol{F}(s) = 0.5 - \frac{1.5s}{200 + s} = 0.5\ \mathrm{A}$$

The s-domain expression for the voltage across the inductor is

$$\boldsymbol{V}_{\mathrm{L}}(\mathrm{s}) = L(s\boldsymbol{F}(s) - f(0)) = 0.1\left(0.5 - \frac{1.5s}{200 + s} - (-1)\right)$$

$$= 0.1\left(1.5 - \frac{1.5s}{200 + s}\right) = 0.15\left(1 - \frac{s}{200 + s}\right)$$

$$= \frac{30}{200 + s}$$

hence

$$v_L(t) = \mathscr{L}^{-1}[V_L(s)] = 30e^{-200t} \text{ V}$$

A diagram showing $i(t)$ and $v_L(t)$ for the initial condition in the problem is given in figure 10.19(a) and (b), respectively.

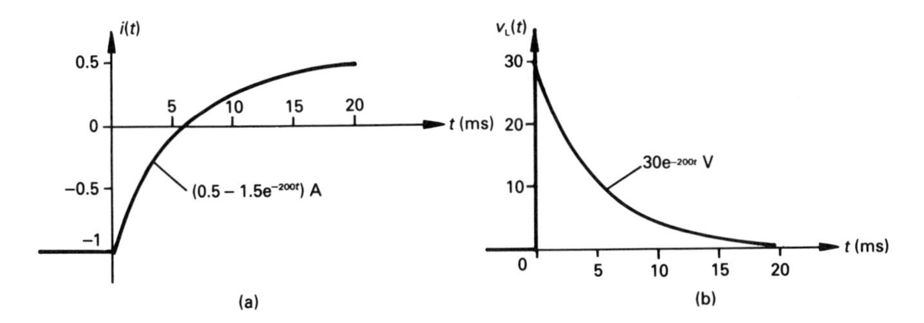

Figure 10.19 *(a)* i(t) *and (b)* v_L(t) *for part (b) of worked example 10.9.1.*

Worked example 10.9.2

Determine an expression for the voltage across the capacitor in figure 10.20(a) if the capacitor is initially charged to 40 V with the polarity shown. Also evaluate an expression for $i(t)$.

Solution

In this example we will illustrate the use of the parallel-connected initial condition source. The s-domain circuit is drawn in figure 10.20(b), in which

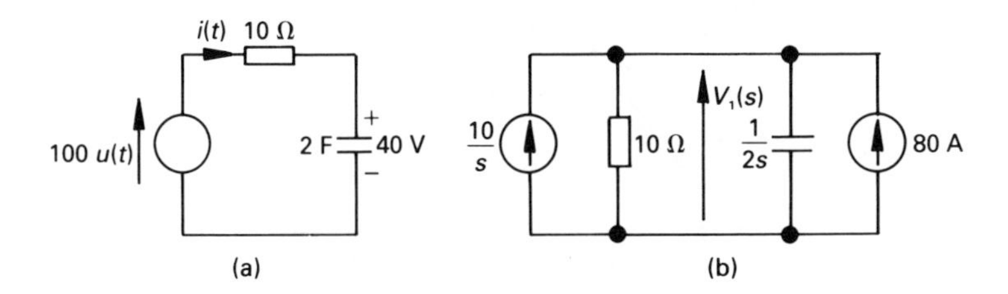

Figure 10.20 *Figure for worked example 10.9.2.*

the initial voltage across the capacitor is converted to its initial current source. The latter corresponds to a current source of $Cv(0) = 80$ A. The nodal equation for diagram (b) is

$$\mathscr{L}[10u(t)] + 80 = \left(\frac{1}{10} + 2s\right)V_1(s)$$

or

$$\frac{10}{s} + 80 = \frac{1 + 20s}{10} V_1(s)$$

therefore

$$V_1(s) = \frac{100}{s(1 + 20s)} + \frac{800}{1 + 20s}$$

$$= \frac{5}{s(0.05 + s)} + \frac{40}{0.05 + s}$$

$$= 100\left(\frac{1}{s} - \frac{1}{0.05 + s}\right) + \frac{40}{0.05 + s}$$

Hence the voltage across the capacitor is

$$v_1(t) = 100(1 - e^{-0.05t}) + 40e^{-0.05t} \text{ V}$$

$$= 100 - 60e^{-0.05t} \text{ V}$$

The resulting graph is shown in figure 10.21.

Since $i(t) = C\,dv_C(t)/dt$, the current in the capacitor can be calculated using the time-differentiation property

$$I(s) = C(sV_1(s) - f(0))$$

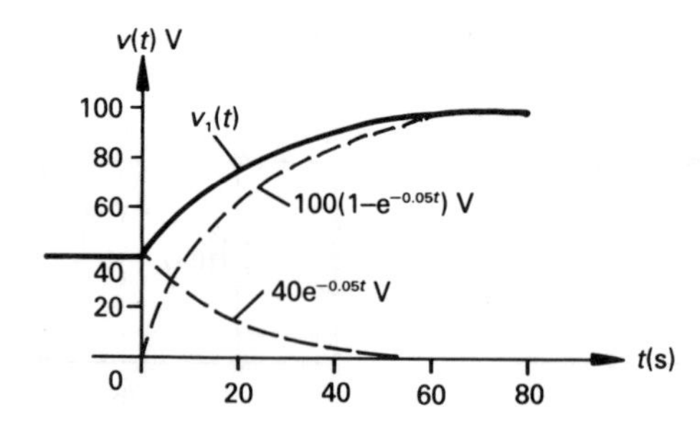

Figure 10.21 *Time variation of the voltage across* C *for* t > 0.

$$= 2\left(\frac{5}{0.05 + s} + \frac{40s}{0.05 + s} - 40\right)$$

$$= \frac{2}{0.05 + s}(5 + 40s - 40(0.05 + s))$$

$$= \frac{6}{0.05 + s}$$

that is

$$i(t) = 6e^{-0.05t}\ \text{A}$$

10.10 Sinusoidal excitation of first-order systems

Here we open our account with systems excited by an alternating current source. We illustrate this by means of an RC circuit in which the capacitor carries an initial charge.

Worked example 10.10.1

Determine an expression for $i(t)$ in the circuit of figure 10.22(a)

Solution

The equivalent s-domain circuit is drawn in figure 10.22(b) in which the alternating voltage source is expressed in the form

$$\mathscr{L}[20 \sin 2t] = 20 \times 2/(s^2 + 2^2) = 40/(s^2 + 4)$$

and the initial charge on the capacitor has the value

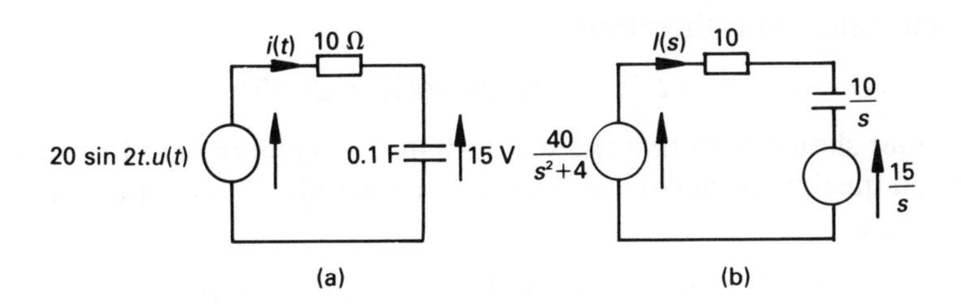

Figure 10.22 *(a) Electrical circuit, (b)* s*-domain circuit.*

$$v(0)/s = 15/s$$

The loop current equation for the circuit is

$$\mathrm{I}(s)\left(10 + \frac{10}{s}\right) = \frac{40}{s^2 + 4} - \frac{15}{s}$$

where $(10 + 10/s)$ is the s-domain impedance of the circuit. The first part of the right-hand side of the above equation is the result of the a.c. source, and the second part is produced by the initial charge on the capacitor. We will keep the two parts separate during the early section of the solution so that we can study the individual results.

$$I(s) = \frac{s}{(10s + 10)}\,\frac{40}{(s^2 + 4)} - \frac{s}{(10s + 10)}\,\frac{15}{s}$$

$$= \frac{4s}{(s + 1)(s^2 + 4)} - \frac{1.5}{s + 1}$$

$$= \left[-\frac{0.8}{s + 1} + \frac{0.8s + 3.2}{s^2 + 4}\right] - \frac{1.5}{s + 1}$$

The bracketed part of the solution was obtained by partial fractions (see chapter 15 for details), and relates to the effect of the a.c. source. From the table of Laplace transforms, the solution is

$$i(t) = (-0.8\mathrm{e}^{-t} + 0.8 \cos 2t + 1.6 \sin 2t) - 1.5\,\mathrm{e}^{-t}$$

$$= (-0.8\mathrm{e}^{-t} + 1.789 \sin(2t + 26.57°)) - 1.5\,\mathrm{e}^{-t}$$

At this point, the reader will observe that the action of connecting the a.c. source appears to produce a 'd.c.' transient current ($-0.8\mathrm{e}^{-t}$) with a time constant of 1 s, which will decay to zero in about $5\tau = 5$ s. The time constant is the natural time constant of the circuit ($1/RC$), and the initial 'd.c.' current of -0.8 A just compensates for the value of $1.789 \sin(2t + 26.57°)$ A when $t = 0$; the net result is an initial a.c. current of zero! The complete equation is therefore

$$i(t) = -2.3\mathrm{e}^{-t} + 1.789 \sin(2t + 26.57°)\ \mathrm{A}$$

The reader should note that the steady-state a.c. component of $i(t)$ can be predicted quickly using conventional theory as follows. The impedance of the circuit is

$$\mathrm{Z}(\mathrm{j}\omega) = R - \mathrm{j}/\omega C = 10 - \mathrm{j}/(2 \times 0.1) = 10 - \mathrm{j}5$$

$$= 11.18\angle{-26.57°}\ \Omega$$

and

$$I(j\omega) = V(j\omega)/Z(j\omega) = 20/11.18\angle{-26.57°}$$
$$= 1.789\angle 26.57° \text{ A}$$

The value of current calculated here is the peak value because we have used the peak value of the alternating voltage.

10.11 Solution of series second-order circuits using Laplace transforms

A second-order or double-energy circuit is one containing two different types of energy storage element, such as L and C. The series circuit in figure 10.23 is typical of this kind; initial conditions may, of course, exist in the circuit. The s-domain impedance of the circuit is

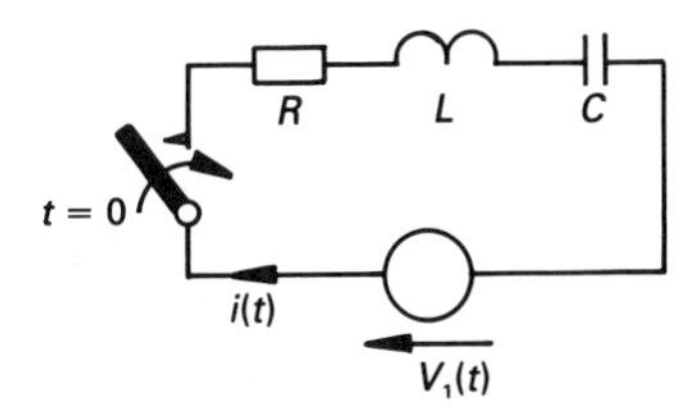

Figure 10.23 *Double-energy series circuit.*

$$Z(s) = R + sL + 1/sC$$

and the current in the circuit is determined from the expression

$$I(s) = \mathcal{L}[V_1(t)]/Z(s)$$

The s-domain expression for the current depends, of course, on the nature of the supply voltage. For the moment, we will assume that a step function of voltage is applied to the circuit, that is $\mathcal{L}[V_1(t)] = V_1/s$; in this case, the equation for $I(s)$ is

$$I(s) = \frac{V_1/s}{R + sL + \dfrac{1}{sC}} = \frac{V_1/L}{s^2 + \dfrac{R}{L}s + \dfrac{1}{LC}}$$

$$= \frac{V_1/L}{s^2 + 2\zeta\omega_n s + \omega_n^2} \qquad (10.11)$$

The denominator of the final expression represents the *characteristic equation* in what is known as the *standard form*, in which ω_n is the *undamped natural frequency*, and ζ is the *damping factor* of the circuit, which may be zero or any positive value. From the denominator of the above equations

$$\omega_n = 1/\sqrt{(LC)}$$

and $$2\zeta\omega_n = R/L$$

or $$\zeta = \frac{R}{2}\sqrt{\left[\frac{C}{L}\right]}$$

The response of the circuit is generally described in terms of the value of ζ namely

1. $\zeta = 0$ (*undamped response*)
2. $0 < \zeta < 1$ (*underdamped response*)
3. $\zeta = 1$ (*critically damped response*)
4. $\zeta > 1$ (*overdamped response*)

For a step voltage applied at $t = 0$, the response is as shown in figure 10.24. The reader will note that the undamped and underdamped response are oscillatory, implying that the characteristic equation of the system has complex poles in the left-hand half of the s-plane (see chapter 11).

The underdamped response is of particular interest to engineers because it corresponds to many practical systems. In this case $0 < \zeta < 1$, that is $\frac{R}{2}\sqrt{\left[\frac{C}{L}\right]} < 1$ or $R < 2\sqrt{(L/C)}$. In this case, the coefficient of s in the equation for $\boldsymbol{I}(s)$ is finite. Completing the square in the donominator of equation (10.11) gives

$$\boldsymbol{I}(s) = \frac{V_1/L}{(s + \zeta\omega_n)^2 + (\omega_n^2 - \zeta^2\omega_n^2)}$$

If $\omega_o^2 = \omega_n^2(1 - \zeta^2)$ or $\omega_o = \omega_n\sqrt{(1 - \zeta^2)}$ where ω_o is the *damped oscillatory frequency* or the *oscillatory frequency*, then

$$\boldsymbol{I}(s) = \frac{V_1/L}{(s + \zeta\omega_n)^2 + \omega_o^2} = \frac{\omega_o V_1/L}{\omega_o[(s + \zeta\omega_n)^2 + \omega_o^2]}$$

and, from the table of Laplace transforms, we see that

$$i(t) = \frac{V_1}{\omega_o L} e^{-\zeta\omega_n t} \sin \omega_o t$$

Also, from the equation for ω_o, we see that when $\zeta = 1$ then $\omega_o = 0$, that is, $\sin \omega_o t = 0$. That is, when the damping factor is unity, the output variable

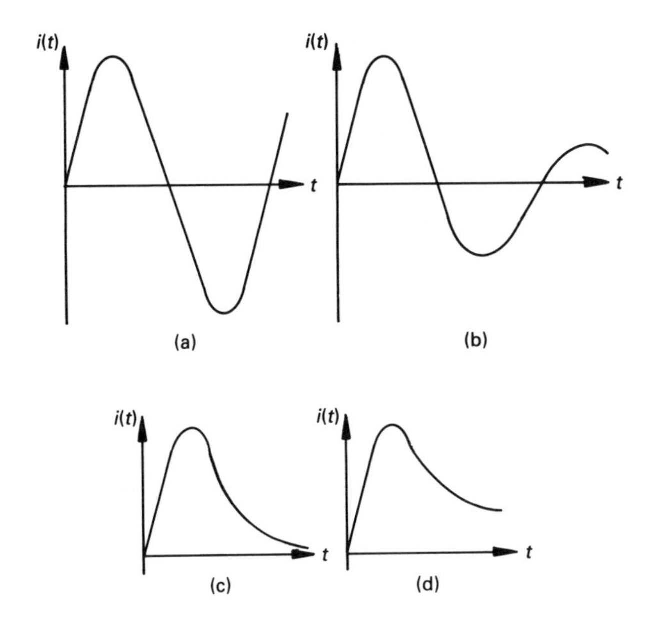

Figure 10.24 *Response of the circuit in figure 10.23 to a step change in applied voltage for (a) undamped response, (b) underdamped response, (c) critical response, (d) overdamped response.*

($i(t)$ in this case) just reaches its steady-state value without either overshoot or undershoot (see also figure 10.24(c)).

We will look at each of the four types of response in worked example 10.11.1.

Worked example 10.11.1

For a series circuit of the type in figure 10.23, in which $L = 2$ H, $C = 2$ F and $V_1(t) = 10$ V, determine expressions for $i(t)$ if the resistance R is (a) zero, (b) 1 Ω, (c) 2 Ω and (d) 3 Ω.

Solution

(a) R = *0 (undamped response)*
In this case

$$\zeta = \frac{R}{2}\sqrt{\left[\frac{C}{L}\right]} = 0 \quad \text{and} \quad \omega_n = 1/\surd(LC) = 0.5 \text{ rad/s}$$

hence

$$I(s) = \frac{V_1/L}{s^2 + \omega_n^2} = \frac{10/2}{s^2 + 0.25} = \frac{5}{0.5} \frac{\omega_n}{(s^2 + \omega_n^2)}$$

From the table of Laplace transforms we see that

$$i(t) = 10 \sin 0.5t \text{ A}$$

That is, the current has a peak value of 10 A, and it oscillates at a frequency of 0.5 rad/s.

(b) $R = 1\ \Omega$ *(underdamped response)*

$$\zeta = \frac{1}{2}\sqrt{\left[\frac{2}{2}\right]} = 0.5 \quad \text{and} \quad \omega_n = 0.5 \text{ rad/s}$$

consequently

$$I(s) = \frac{10/2}{s^2 + (2 \times 0.5 \times 0.5s) + 0.5^2} = \frac{5}{s^2 + 0.5s + 0.25}$$

Completing the square in the denominator gives

$$I(s) = \frac{5}{(s + 0.25)^2 + (0.25 - 0.25^2)}$$

$$= \frac{5}{(s + 0.25)^2 + 0.1875}$$

$$= \frac{5}{0.433} \frac{0.433}{[(s + 0.25)^2 + 0.433^2]}$$

$$= 11.55 \frac{\omega_o}{[(s + \alpha)^2 + \omega_o^2]}$$

where $\omega_o = 0.433$ and $\alpha = 0.25$. The table of transforms gives

$$i(t) = 11.55\ e^{-0.25t} \sin 0.433t \text{ A}$$

(c) $R = 2\ \Omega$ *(critically damped response)*

$$\zeta = \frac{2}{2}\sqrt{\left[\frac{2}{2}\right]} = 1,\ \omega_n = 0.5 \text{ rad/s and } \omega_o = 0, \text{ that is}$$

no oscillations. The s-domain expression for the current is

$$I(s) = \frac{V_1/L}{s^2 + 2\omega_n s + \omega_n^2} = \frac{V_1/L}{(s + \omega_n)^2} = \frac{10/2}{(s + 0.5)^2}$$

$$= 5 \frac{1!}{(s + \alpha)^{1+1}}$$

and from the table of Laplace transforms we see that

$$i(t) = 5te^{-0.5t}\ \text{A}$$

The general form of response is shown in figure 10.24(c). This is a case where we should be rather careful when using the concept that 'the transient has decayed in about 5 time constants'. In this case, apparently $\tau = 1/0.5 = 2$ s; if we use this length of time in the solution, we will find that the current is 0.34 A, which is not insignificant when compared with the peak current of about 3.7 A (which occurs after about 2 s). In fact, it takes about 16 s for the transient to decay to an insignificant value.

(d) $R = 3\ \Omega$ *(overdamped response)*
In this case

$$\zeta = \frac{3}{2}\sqrt{\left[\frac{2}{2}\right]} = 1.5 \quad \text{and} \quad \omega_n = 0.5\ \text{rad/s}$$

and the s-domain equation for the circuit current is

$$\boldsymbol{I}(s) = \frac{10/2}{s^2 + (2 \times 1.5 \times 0.5s) + 0.5^2} = \frac{5}{s^2 + 1.5s + 0.25}$$

Completing the square in the denominator gives

$$\boldsymbol{I}(s) = \frac{5}{(s + 0.75)^2 - 0.3125} = \frac{5}{0.559}\left[\frac{\beta}{(s+\alpha)^2 - \beta^2}\right]$$

and, from the table of Laplace transforms

$$i(t) = 8.945\ e^{-0.75t} \sinh^{0.559t}\ \text{A}$$

While the solution has some mathematical merit, it does not present an engineer with a mental picture of what is happening. More information can be obtained by converting the result into its exponential form as follows

$$\begin{aligned} i(t) &= 8.945\ e^{-0.75t}\,\frac{1}{2}\,[(e^{0.559t} - e^{-0.559t})] \\ &= 4.4725(e^{-0.191t} - e^{-1.309t})\ \text{A} \end{aligned}$$

The longest of the two time constants is $1/0.191 = 5.24$ s, so that it will take $5 \times 5.24 = 26.2$ s for the current to have fallen to zero (or nearly so).

The reader should note that it takes much longer for the overdamped circuit to settle than does the critically damped circuit.

Worked example 10.11.2

We will look in this worked example at the operation of an electromechanical system, namely a conventional contact-breaker ignition system of an automobile.

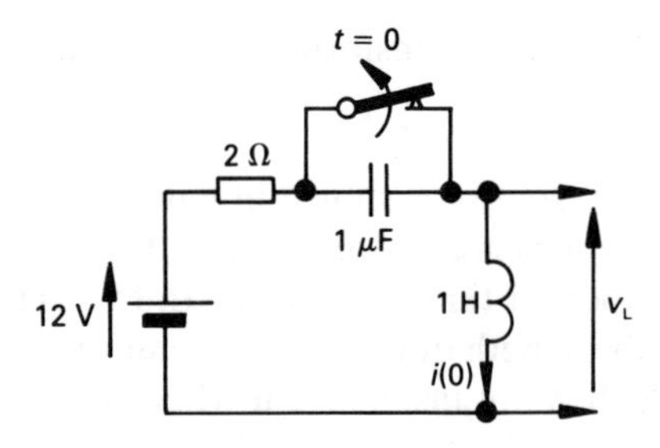

Figure 10.25 *Figure for worked example 10.11.2.*

The basic circuit is shown in figure 10.25, in which a resistor, a capacitor and an inductor are connected to the 12 V battery in the vehicle. The resistor is merely a current-limiting device; the capacitor is connected across a switch (known as the 'points'), and the voltage developed across the ignition coil is applied to the sparking plugs.

The points open and close periodically (depending on the speed of the engine) and, in this problem, we need to calculate the maximum voltage across the coil (assuming, that is, the sparking plugs have not broken down).

Solution

We can assume that the points are closed long enough for the current to have reached its steady-state value of 12 V/2 Ω = 6 A, before the points are opened.

The s-domain equivalent circuit is shown in figure 10.26, in which the initial condition source has a value of $Li(0) = 1 \times 6 = 6$ V (assisting the battery) and the corresponding mesh current equation is

$$\frac{12}{s} + 6 = \boldsymbol{I}(s)(2 + s + 10^6/s)$$

or

$$\boldsymbol{I}(s) = \frac{\dfrac{12}{s} + 6}{s + 2 + \dfrac{10^6}{s}} = \frac{12 + 6s}{s^2 + 2s + 10^6}$$

$$= 6\,\frac{s + 2}{(s + 1)^2 + [(10^3)^2 - 1]}$$

$$= 6\,\frac{s + 1}{(s + 1)^2 + (10^3)^2} - 6 \times 10^{-3}\,\frac{10^3}{(s + 1)^2 + (10^3)^2}$$

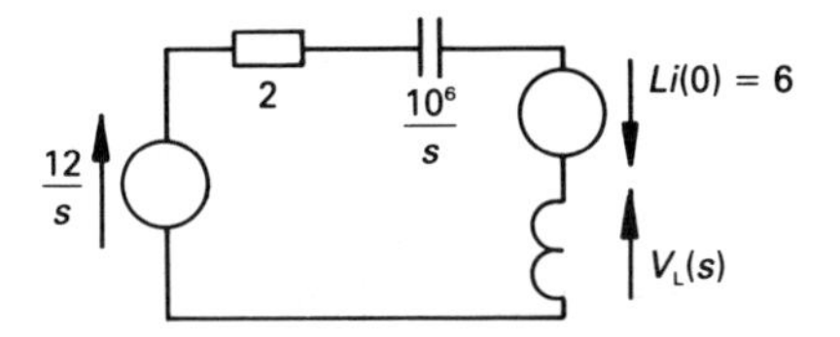

Figure 10.26 *Automobile ignition circuit in the s-domain.*

$$= 6\frac{s + \alpha}{(s + \alpha)^2 + \omega^2} - 6 \times 10^{-3}\frac{\omega}{(s + \alpha)^2 + \omega^2}$$

Taking the inverse transform gives

$$i(t) = 6e^{-t} \cos 1000t - 6 \times 10^{-3}e^{-t} \sin 1000t$$

For all practical purposes we may ignore the second term, hence

$$i(t) \approx 6e^{-t} \cos 1000t$$

The voltage across the ignition coil is, therefore

$$v_L(t) = L\frac{di(t)}{dt} = 6 \times (-1000) \times 6e^{-t} \sin 1000t$$

$$= -6000\ e^{-t} \sin 1000t\ \text{V}$$

The periodic time of the sinusoid is $T = 2\pi/\omega = 2\pi/1000$ s or 2π ms. The maximum voltage occurs when the sinusoid is maximum, or when $t = \pi/2$ ms, at which point $\sin 1000t = 1$. Hence

$$v_{L(max)} = -6000e^{-\pi/2000} = -5990\ \text{V}$$

10.12 *s*-Domain transfer functions

A transfer function is the ratio of the response of the circuit to the forcing function applied to the circuit. Having obtained the transfer function of the circuit or system, it is a matter of applying circuit analysis to obtain the output.

The technique for obtaining the *s*-domain transfer function follows the same general lines as in the frequency domain, and worked example 10.12.1 illustrates the general method.

Worked example 10.12.1

Determine the *s*-domain transfer function $\boldsymbol{H}(s) = \boldsymbol{V}_0(s)/\boldsymbol{V}_1(s)$ for the circuit in figure 10.27. Assume that the operational amplifier is ideal.

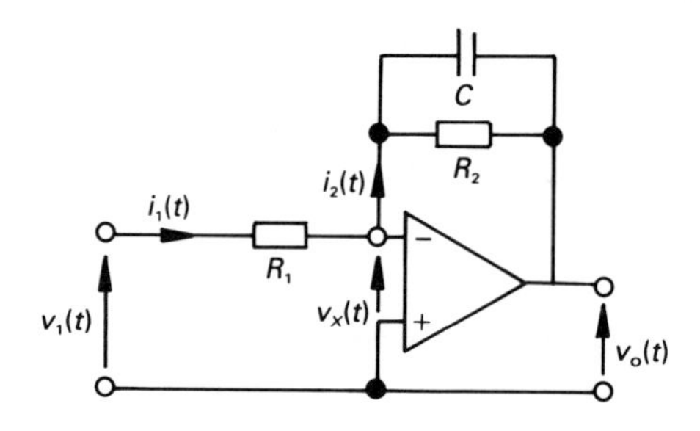

Figure 10.27 *Figure for worked example 10.12.1.*

Solution

Since the op-amp is ideal we may assume, for the purpose of analysis, that $v_X(t) = 0$. That is

$$i_1(t) = v_1(t)/R_1 \quad \text{or} \quad \boldsymbol{I}_1(s) = \boldsymbol{V}_1(s)/R_1$$

and

$$i_2(t) = i_1(t) \quad \text{or} \quad \boldsymbol{I}_2(s) = \boldsymbol{I}_1(s)$$

also

$$\boldsymbol{I}_2(s) = \frac{-\boldsymbol{V}_o(s)}{\dfrac{R_2}{sC} \Big/ (R_2 + 1/sC)} = -\boldsymbol{V}_o(s)\,\frac{1 + R_2Cs}{R_2}$$

Now, since $\boldsymbol{I}_2(s) = \boldsymbol{I}_1(s)$, then

$$\boldsymbol{H}(s) = \frac{\boldsymbol{V}_o(s)}{\boldsymbol{V}_1(s)} = \frac{-\boldsymbol{I}_2(s)R_2/(1 + R_2Cs)}{\boldsymbol{I}_1(s)R_2}$$

$$= \frac{-R_2}{R_1(1 + R_2Cs)}$$

10.13 Transients in magnetically coupled circuits

The time-domain equations for the coupled circuit in figure 10.28(a) are

$$v_1(t) = L_1\frac{\mathrm{d}i_1(t)}{\mathrm{d}t} + M\frac{\mathrm{d}i_2(t)}{\mathrm{d}t}$$

$$v_2(t) = M\frac{\mathrm{d}i_1(t)}{\mathrm{d}t} + L_2\frac{\mathrm{d}i_2(t)}{\mathrm{d}t}$$

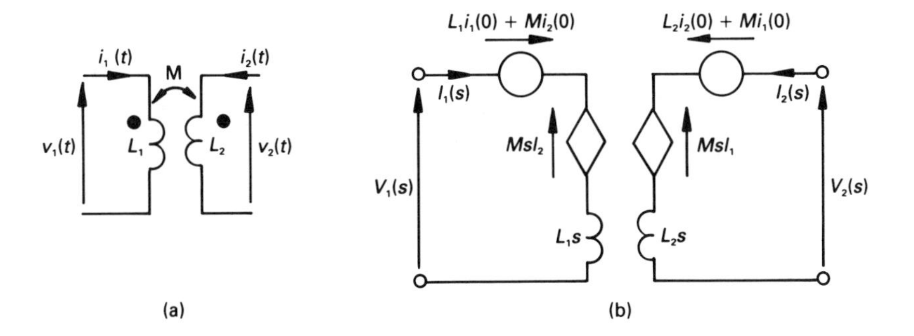

Figure 10.28 *(a) Basic two-winding magnetically coupled circuit, (b)* s*-domain equivalent circuit with isolated windings.*

Taking Laplace transforms of these equations gives

$$V_1(s) = L_1[s\boldsymbol{I}_1(s) - i_1(0)] + M[s\boldsymbol{I}_2(s) - i_2(0)]$$

$$V_2(s) = M[s\boldsymbol{I}_1(s) - i_1(0)] + L_2[s\boldsymbol{I}_2(s) - i_2(0)]$$

or

$$V_1(s) = L_1 s\boldsymbol{I}_1(s) + Ms\boldsymbol{I}_2(s) - [L_1 i_1(0) + Mi_2(0)]$$

$$V_2(s) = Ms\boldsymbol{I}_1(s) + L_2 s\boldsymbol{I}_2(s) - [L_2 i_2(0) + Mi_1(0)]$$

where i_1 (0) and i_2 (0) are initial currents in L_1 and L_2, respectively. The *s*-domain equivalent circuit in figure 10.28(b) is modelled on these equations. Worked example 10.13.1 illustrates how these equations can be used to solve for the transient solution in a magnetically coupled circuit.

Worked example 10.13.1

Derive expressions for the current in the primary and secondary winding of the transformer in figure 10.29(a). There are no initial conditions in the circuit.

Solution

We should note in figure 10.29(a) that (1) the primary winding is excited by a step function of direct voltage and (2) the current is assumed to leave the dotted terminal of the secondary winding. The second point means that the *s*-domain equivalent circuit appears as shown in figure 10.29(b). The mesh equation for the primary circuit is

$$\frac{10}{s} = (20 + 0.1s)\boldsymbol{I}_1(s) - 0.05\boldsymbol{I}_2(s)$$

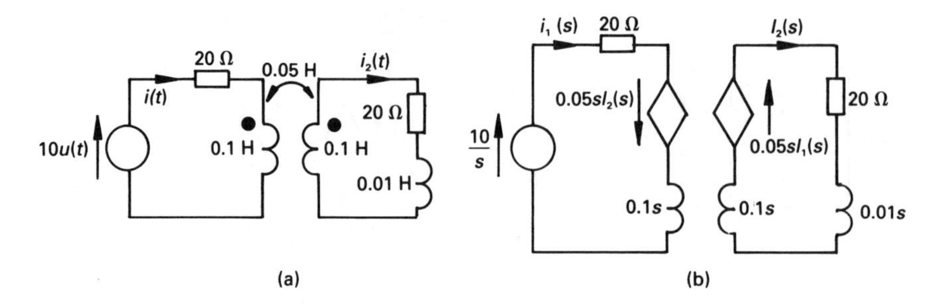

Figure 10.29 *Figure for worked example 10.13.1.*

and for the secondary is

$$0 = -0.05s\boldsymbol{I}_1(s) + (20 + [0.1 + 0.01]s)\boldsymbol{I}_2(s)$$
$$= -0.05s\boldsymbol{I}_1(s) + (20 + 0.11s)\boldsymbol{I}_2(s)$$

The equation for the secondary circuit shows that

$$\boldsymbol{I}_1(s) = \frac{20(20 + 0.11s)}{s}\,\boldsymbol{I}_2(s)$$

Substituting this expression into the primary circuit equation yields

$$\frac{10}{s} = \boldsymbol{I}_2(s)\left[\frac{(20 + 0.1s)20(20 + 0.11s)}{s} - 0.05s\right]$$

$$= \boldsymbol{I}_2(s)\left[\frac{0.17s^2 + 84s + 8000}{s}\right]$$

or

$$\boldsymbol{I}_2(s) = \frac{10}{0.17s^2 + 84s + 8000}$$

$$= \frac{58.82}{s^2 + 494s + 47\,059}$$

whence $\omega_n = \sqrt{47\,059} = 216.9$ rad/s, and $2\zeta\omega_n = 494$ or $\zeta = 1.14$. That is, $i_2(t)$ has an overdamped response. Completing the square in the denominator of the equation gives

$$\boldsymbol{I}_2(s) = \frac{58.82}{(s + 247)^2 - 118.2^2} = \frac{0.498 \times 118.2}{(s + 247)^2 - 118.2^2}$$

$$= 0.498 \frac{\beta}{(s + \alpha)^2 - \beta^2}$$

The table of transforms gives

$$\begin{aligned} i_2(t) &= 0.498\mathrm{e}^{-247t} \sinh 118.2t \\ &= 0.498\mathrm{e}^{-247t}(\mathrm{e}^{118.2t} - \mathrm{e}^{-118.2t})/2 \\ &= 0.249(\mathrm{e}^{-128.8t} - \mathrm{e}^{-365.2t}) \end{aligned}$$

The longest time constant is 1/128.8 = 7.76 ms, and the transient will have died away in 5 × 7.76 = 38.8 ms (say 40 ms).

The transform of the primary current is obtained by inserting the expression for $\boldsymbol{I}_2(s)$ into the mesh equation of the secondary as follows

$$\boldsymbol{I}_1(s) = \frac{20(20 + 0.11s)}{s} \boldsymbol{I}_2(s)$$

$$= \frac{20(20 + 0.11s)}{s} \frac{58.82}{s^2 + 494s + 216.9^2}$$

$$= 1176.4\left[\frac{20}{s(s^2 + 494s + 216.9^2)} + \frac{0.11}{s^2 + 494s + 216.9^2}\right]$$

The first expression inside the brackets is reduced into partial fraction form (see chapter 15 for details) to give the following

$$\boldsymbol{I}_1(s) = 1176.4\left[\frac{20}{216.9^2 s} - \frac{\dfrac{20s}{216.9^2} + \dfrac{494 \times 20}{216.9^2}}{s(s^2 + 494s + 216.9^2)} + \frac{0.11}{s^2 + 494s + 216.9^2}\right]$$

$$= \frac{0.5}{s} - 0.5 \frac{s + 494}{(s + 274)^2 - 118.2^2} + 1.095 \frac{118.2}{(s + 274)^2 - 118.2^2}$$

For simplicity, we will write $(s + 274)^2 - 118.2^2 = [\mathrm{D}]$, giving

$$\boldsymbol{I}_1(s) = \frac{0.5}{s} - 0.5 \frac{s + 247}{[\mathrm{D}]} - 0.5 \frac{247}{[\mathrm{D}]} + 1.095 \frac{118.2}{[\mathrm{D}]}$$

$$= \frac{0.5}{s} - 0.5 \frac{s + 247}{[\mathrm{D}]} + 0.05 \frac{118.2}{[\mathrm{D}]}$$

$$= \frac{0.5}{s} - 0.5 \frac{s + \alpha}{(s + \alpha)^2 - \beta^2} + 0.05 \frac{\beta}{(s + \alpha)^2 - \beta^2}$$

The inverse transformation gives

$$\begin{aligned} i_1(t) &= 0.5 - 0.5\mathrm{e}^{-247t} \cosh 118.2t + 0.05\mathrm{e}^{-247t} \sinh 118.2t \\ &= 0.5 - 0.225\mathrm{e}^{-128.8t} - 0.275\mathrm{e}^{-365.2t} \text{ A} \end{aligned}$$

The primary steady-state current is 0.5 A (as could be observed from the d.c. conditions in the circuit) and, once again, the longest time constant in the expression for $i_1(t)$ is 1/128.8 = 7.76 ms so that the transient part of $i_1(t)$ will have died away after 38.8 ms (say 40 ms).

Unworked Problems

10.1. A voltage given by

$$v(t) = u(t) - 4u(t-2) + 3u(t-4) \text{ V}$$

is applied to a 2 H inductor. Deduce an expression for the current in the inductor.
[$0.5r(t) - 2r(t-2) + 1.5r(t-4)$ A]

10.2. Deduce the Laplace transform of $(1 - e^{-\alpha t})$, where α is a constant.
[$\alpha/(s(s+\alpha))$]

10.3. Evaluate the Laplace transform of $f(t) = u(t) - u(t-a)$.
[$(1 - e^{-sa})/s$]

10.4. A cyclic sawtooth waveform whose first cycle is described by $f(t) = (t)$ for $0 \leqslant t \leqslant 1$, after which it instantly becomes zero again. Determine the Laplace transform of the wave.
[$(1 - (s+1)e^{-s})/((1 - e^{-s})s^2)$]

10.5. Determine the Laplace transform of the function
(a) $(t-a)e^{-\alpha(t-a)}u(t-a)$
(b) $\delta(t) + (a-b)e^{-bt}u(t)$.
[(a) $e^{-as}/(s+\alpha)^2$; (b) $(s+a)/(s+b)$]

10.6. A voltage E is applied at $t = 0$ to a series circuit containing a resistor R, an inductor L and a capacitor C, with no initial conditions in the circuit. Derive an expression for the current, $i(t)$, if the circuit is (a) underdamped, (b) critically damped, (c) overdamped.

$$\left[\begin{array}{l} \text{(a)}\ \dfrac{E}{\omega_0 L} e^{-\zeta\omega_n t} \sin \omega_0 t;\ \text{(b)}\ \dfrac{E}{L} t e^{-\omega_n t}; \\[2ex] \text{(c)}\ \dfrac{E}{aL} e^{-\zeta\omega_n t} \sinh at, \text{ where } a = \omega_n \sqrt{(\zeta^2 - 1)} \end{array}\right]$$

10.7. A capacitor is charged through a pure resistor by a battery of constant voltage E; derive an expression for the instantaneous charge on the capacitor if it is initially uncharged. If $C = 10\ \mu$F and $R = 1$ MΩ, calculate the time taken for the capacitor to receive 85 per cent of its final charge.
[$CE(1 - e^{-t/RC})$; 18.97 s]

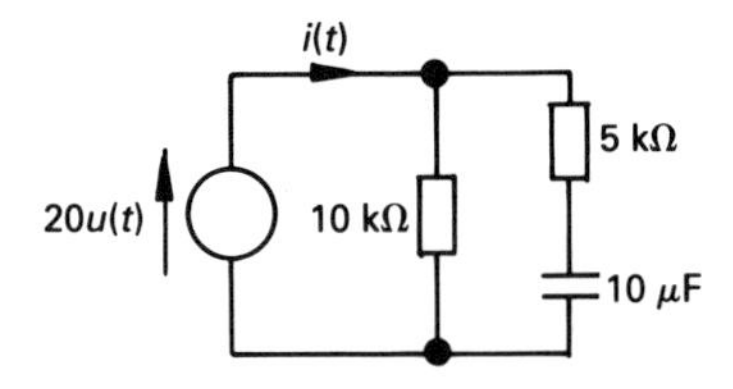

Figure 10.30

10.8. Calculate the time taken for $i(t)$ in figure 10.30 to fall to 4 mA.
[0.035 s]

10.9. A direct voltage, V_1, is applied to a coil of resistance 10 Ω and inductance 1 H, and is maintained constant until all transients have settled out. At this time the voltage is changed to V_2. (a) Derive an expression for the current in the coil after the voltage has changed to V_2. If (b) $V_1 = 200$ V and $V_2 = 100$ V, (c) $V_1 = 100$ V and $V_2 = 200$ V, calculate the current in the coil at (i) 0.1 s, (ii) 0.3 s after the voltage has changed.
[(a) $(V_2 + (V_1 - V_2)e^{-Rt/L})/R$; (b) (i) 16.32 A, (ii) 19.5 A;
(c) (i) 13.68 A, (ii) 10.5 A]

10.10. A surge generator used for testing high-voltage electrical apparatus produces its voltage by initially charging a number of parallel-connected capacitors to a very high voltage. To produce the voltage surge, the capacitors are connected in series giving, in this case, a total voltage across the series-connected capacitors of 300 kV.

The equivalent circuit of the surge generator is shown in figure 10.31; the spark gap merely acts as a switch and, for the purpose of analysis, can be regarded as a short-circuit. The apparatus under test, connected between terminals A and B, can be regarded as having an infinite impedance. Determine an expression for the voltage across the apparatus under test.
[$242(e^{-0.02t} - e^{-4.98t})$ kV (t in μs)]

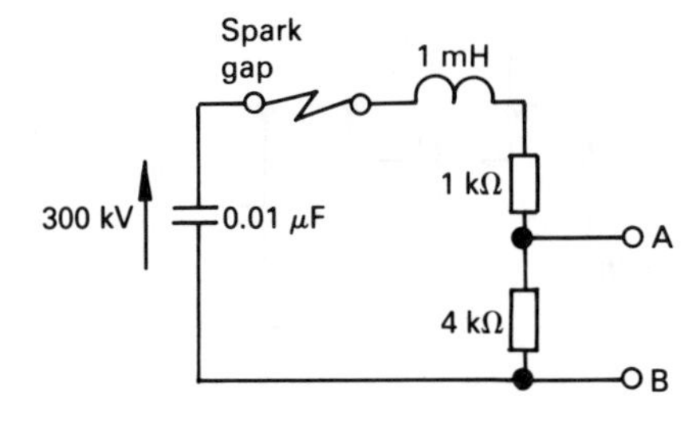

Figure 10.31

10.11. In problem 10.10, calculate the maximum voltage between A and B, and the time taken to reach the maximum value.
[235.7 kV; 1.11 μs]

10.12. A series circuit contains a 10 Ω resistor and a 0.1 H inductor. If a voltage of $10e^{-50t}u(t)$ is applied to the circuit, deduce an expression for the current in the circuit.
$[2(e^{-100t} - e^{-50t})]$

10.13. A circuit containing a resistance of 20 Ω in series with an inductance of 0.2 H is energised by a voltage $v(t) = 250\sin(300t + \phi)$ V. Deduce an expression for the current in the circuit if the supply is connected when (a) $\phi = 0°$, (b) $\phi = 90°$. There are no initial conditions in the circuit.
[(a) $3.95\sin(300t - 71.57°) + 3.75e^{-100t}$ A;
(b) $3.95\sin(300t + 18.43°) - 1.25e^{-100t}$ A]

10.14. A magnetically coupled circuit similar to that in figure 10.29 has the following values: d.c. supply = 20 V, resistance in the primary circuit = 10 Ω, L_1 = 0.2 H, L_2 = 0.4 H, M = 0.1 H. The load is a pure resistor of 10 Ω resistance, and there are no initial conditions in the circuit. Derive an expression for the current in the primary and in the secondary windings.
$[i_1(t) = 2 - 0.292e^{-22.66t} - 1.708e^{-63.06t}$ A; $i_2(t) = 0.707(e^{-22.66t} - e^{-63.06t})$ A]

10.15. Calculate, for problem 10.14, the maximum value of $i_2(t)$ and the time when it occurs.
[0.255 A; 0.025 s]

10.16. Derive the transfer function $v_2(s)/v_1(s)$ for the RC network in figure 10.32.

$$\left[\frac{(s + 1/(R_1C_1))(s + 1/(R_2C_2))}{s^2R_1R_2C_1C_2 + s(R_1C_1 + R_1C_2 + R_2C_2) + 1}\right]$$

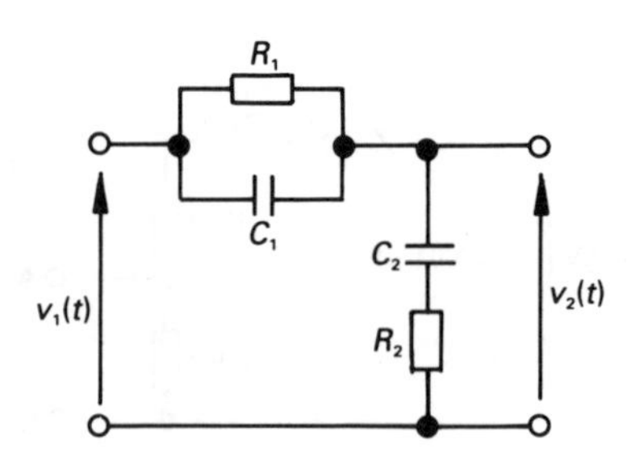

Figure 10.32

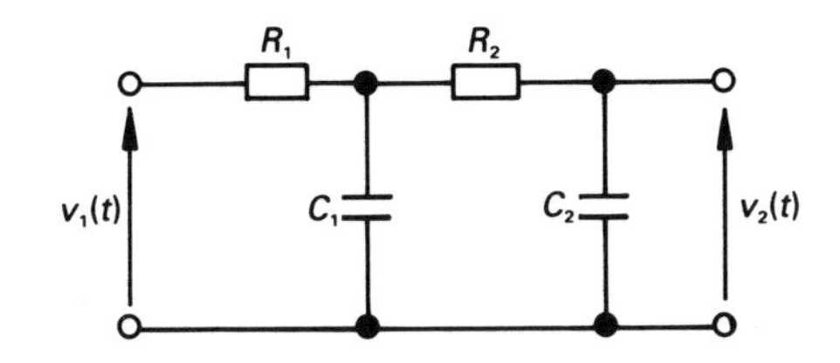

Figure 10.33

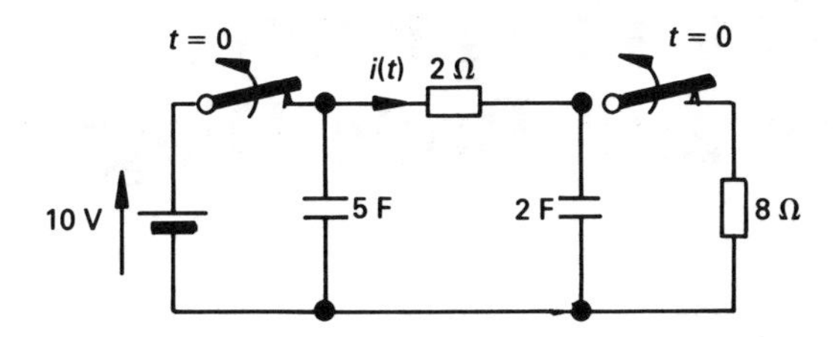

Figure 10.34

10.17. Derive $v_2(s)/v_1(s)$ for figure 10.33.
[$1/(s^2R_1R_2C_1C_2 + s(R_1C_1 + R_1C_2 + R_2C_2) + 1)$]

10.18. If both switches in figure 10.34 are closed when $t \leqslant 0$, and both capacitors are fully charged when $t = 0$, determine an expression for $i(t)$ for $t > 0$ if both switches are simultaneously opened when $t = 0$.
[$e^{-0.35t}$ A]

11

Complex Frequency, the s-Plane and Bode Diagrams

11.1 Introduction

This chapter introduces an important aspect of circuit analysis, namely complex frequency. This is a unifying concept which ties together many analytical techniques into a neat package.

Complex frequency is introduced by looking at an exponentially damped sinusoidal function, for example the voltage

$$v(t) = V_m e^{\sigma t} \cos (\omega t + \theta) \tag{11.1}$$

where σ is a real quantity, and usually has a negative value. While the reality of the equation (11.1) is not easy to grasp we have, in fact, looked at it in other forms on several occasions.

For example, in the case of a direct voltage, both σ and ω are zero, giving

$$v(t) = V_m \cos \theta = V_{dc}$$

In the general sinusoidal case, σ is zero, giving

$$v(t) = V_m \cos (\omega t + \theta)$$

If we allow ω to be zero, then we have the exponential case as follows

$$v(t) = V_m e^{\sigma t} \cos \theta = V_{dc} e^{\sigma t}$$

Thus the general expression in equation (11.1) includes the d.c. case, the general sinusoidal case and the exponential function.

Initially in this chapter we look at the mathematical aspects of complex frequency and the s-plane, and then we transform the impedance of elements into the s-domain. This is particularly important when we study topics such as the frequency response of circuits and Bode diagrams. Once we have mastered the concept of the s-domain 'impedance' of simple elements, it is a relatively simple step to determine the s-domain impedance of complete circuits.

11.2 The exponential form of a complex number

Expanding $e^{j\theta}$ as a Maclaurin series we get

$$\begin{aligned} e^{j\theta} &= 1 + j\theta + \frac{(j\theta)^2}{2!} + \frac{(j\theta)^3}{3!} + \ldots \\ &= \left(1 - \frac{\theta^2}{2!} + \frac{\theta^4}{4!} - \ldots\right) + j\left(\theta - \frac{\theta^3}{3!} + \frac{\theta^5}{5!} - \ldots\right) \\ &= \cos\theta + j\sin\theta \end{aligned}$$

It may also be shown that

$$e^{-j\theta} = \cos\theta - j\sin\theta$$

That is to say, $e^{j\theta}$ is a complex number, and can be represented by a point which is unit distance from the origin and subtending an angle θ with the real axis. Hence any point v, at distance V from the origin and subtending angle θ with the real axis can be represented in the form

$$v = Ve^{j\theta} = V(\cos\theta + j\sin\theta)$$

The reader should note that for any point v, there is an infinite number of other corresponding points, but differing from one another by an angle which is an integral number of 2π radians, that is their angle is $(\theta + 2\pi n)$, where n is an integer.

11.3 Complex frequency

If $\theta = \omega t$, we may write

$$e^{j\theta} = \cos\omega t + j\sin\omega t$$

or if $s = \sigma + j\omega$, where s is known as the *complex frequency*, then

$$\begin{aligned} e^{st} &= e^{(\sigma + j\omega)t} = e^{\sigma t}e^{j\omega t} \\ &= e^{\sigma t}(\cos\omega t + j\sin\omega t) \end{aligned}$$

If the magnitude of the expression is V, then

$$Ve^{st} = Ve^{\sigma t}(\cos\omega t + j\sin\omega t)$$

Let us consider the effect of various values of σ and ω in the above expression:

(1) $v(t) = 10$ $s = 0 + j0$
(2) $v(t) = 8e^{-5t}$ $s = -5 + j0$
(3) $v(t) = 4 \cos 20t$ $s_1 = 0 + j20$
$s_2 = s_1^* = 0 - j20$
(4) $v(t) = 7e^{-2t}\cos 40t$ $s_1 = -2 + j40$
$s_2 = s_1^* = -2 - j40$
(5) $v(t) = 6e^{7t} \cos (9t + 30°)$ $s_1 = 7 + j9$
$s_2 = s_1^* = 7 - j9$

Expression (1) corresponds to a 'd.c.' term in which there is no growth or decay of the function, nor is there any sinusoidal oscillation. Expression (2) is an exponentially decaying function in which $\sigma = -5$; the larger the negative value of σ, the more rapid the decay of the function. A positive value of σ corresponds to a function which grows exponentially.

The third expression is that of a sinusoid with no growth or decay in its magnitude. In this case, there is a pair of values of s, one being the conjugate of the other. Since each value of s possesses both real and imaginary parts, the sum of the two values identifies a real sinusoidal frequency; in this case the angular frequency is 20 rad/s.

In a similar manner there are, in expression (4), a pair of conjugate values; in this case there is an exponential decay associated with the $\sigma = -2$ component. Finally, expression (5) is related to a sinusoid which grows exponentially with time.

It is interesting to note that, in each conjugate pair of values, one has a negative value of ω! Such a value does, in fact, exist. For example, if we look at the expression for the compound angle

$$\begin{aligned}\cos(-40t + \phi) &= \cos(-40t).\cos\phi - \sin(-40t).\sin\phi \\ &= \cos 40t.\cos\phi + \sin 40t.\sin\phi \\ &= \cos(40t - \phi)\end{aligned}$$

This implies that it has the same frequency as the function $\cos(40t - \phi)$, but it has a different numerical value at time t.

Where there is a product term such as $9 \sin 8t.\cos 5t$, the roots contain the sum and difference of the frequencies. In this case the roots are $s_1 = j(8 + 5) = j13$, $s_2 = -j(8 + 5) = -j13$, $s_3 = j(8 - 5) = j3$, $s_4 = -j(8 - 5) = -j3$.

Generally speaking, the larger the value of σ, the more rapid the rate of growth or decay of the quantity; the greater the value of ω, the higher is the frequency.

11.4 The *s*-plane

We can map complex frequencies on the s-plane, as shown in figure 11.1, where σ is plotted in the horizontal or 'real' direction, and ω is plotted

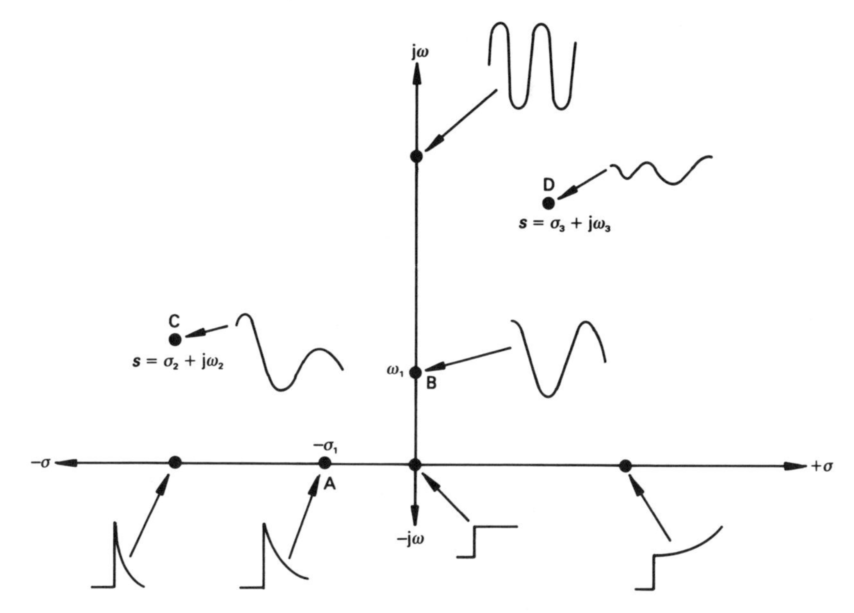

Figure 11.1 *Waveforms for complex frequencies in the s-plane.*

in the vertical or 'imaginary' direction. At the origin of the graph, where $s = 0 + j0$, we get a unit step or 'd.c.' condition. At point A, the function decays with time constant of $1/\sigma$; at point B, the function oscillates at a frequency of ω_1 rad/s (*note*: the conjugate point is not shown in the figure). If $|-\sigma| > |\omega|$, the rate of decay will is fairly small.

The reader is reminded that a pair of conjugate points is necessary to give a sinusoidal oscillation. A complex pair of points in the right-hand half of the s-plane (see point D – *note*: the conjugate point is not shown in the figure) gives rise to an oscillatory signal which grows exponentially to infinity.

In general, a conjugate pair of points in the right-hand half of the s-plane represents an unstable condition.

11.5 Transformation of impedance into the *s*-domain

Just as we have associated the sinusoid $I \cos(\omega t + \phi)$ with phasor $\boldsymbol{I} = Ie^{j\phi}$, we can relate an impedance with its s-domain equivalent as follows.

Resistance

The relation between the voltage applied to a resistor and the current through it is given by Ohm's law as follows

$$v = Ri$$

If $i = Ie^{\sigma t}e^{j(\omega t + \phi)} = Ie^{j\phi}e^{st}$, then the voltage across the resistor is

$$v = Ve^{\sigma t}e^{j(\omega t + \phi)} = Ve^{j\phi}e^{st}$$

That is

$$Ve^{j\phi}e^{st} = RIe^{j\phi}e^{st}$$

or

$$Ve^{j\phi} = RIe^{j\phi}$$

therefore

$$V\angle\phi = RI\angle\phi$$

hence

$$\mathbf{Z_R}(s) = \frac{V\angle\phi}{I\angle\phi} = R\angle 0°$$

and

$$\mathbf{Y_R}(s) = 1/R$$

Inductance

In a pure inductor $v_L = L(\mathrm{d}i/\mathrm{d}t)$; if the current is $i = Ie^{j\phi}e^{st}$ and $v_L = Ve^{j\theta}e^{st}$, the v–i relationship is

$$Ve^{j\theta}e^{st} = L\frac{\mathrm{d}}{\mathrm{d}t}(Ie^{j\phi}e^{st})$$

$$= LsIe^{j\theta}e^{st}$$

or

$$Ve^{j\theta} = LsIe^{j\theta}$$

hence

$$V\angle\theta = LsI\angle\theta$$

That is

$$\mathbf{V} = Ls\mathbf{I}$$

The s-domain impedance of a pure inductor therefore is

$$\mathbf{Z_L}(s) = \frac{\mathbf{V}}{\mathbf{I}} = Ls$$

and

$$\mathbf{Y_L}(s) = 1/Ls$$

Capacitance

Here the relationship is $i = C(\mathrm{d}v_C/\mathrm{d}t)$, and

$$Ie^{j\phi}e^{st} = C\frac{\mathrm{d}}{\mathrm{d}t}(Ve^{j\theta}e^{st})$$

$$= CsVe^{j\theta}e^{st}$$

or $$Ie^{j\phi} = CsVe^{j\theta}$$

hence $$I\angle\phi = CsV\angle\phi$$

that is $$\mathbf{I} = Cs\mathbf{V}$$

and the s-domain impedance of a pure capacitance is

$$\mathbf{Z}_C(s) = \frac{\mathbf{V}}{\mathbf{I}} = \frac{1}{Cs}$$

and $$\mathbf{Y}_C(s) = Cs$$

Series and parallel combination of similar elements

The reader should note that the usual rules apply to series and parallel combinations of similar elements. For example, if two resistors of resistance R are connected in series, then $\mathbf{Z}_R(s) = 2R$, and if connected in parallel then $\mathbf{Z}_R(s) = R/2$. If two inductors of inductance L are connected in series, then $\mathbf{Z}_L(s) = 2Ls$, and if connected in parallel then $\mathbf{Z}_L(s) = Ls/2$, etc.

11.6 Frequency response as a function of ω

We now consider the forced response of circuits as the supply frequency ω (or f) changes; during this section we assume that $\sigma = 0$, or $s = j\omega$. We can, therefore, either develop circuit equations in terms of s or in terms of $j\omega$. Generally speaking, the former is more useful since it is difficult, when terms have been multiplied, to convert an equation from the $j\omega$ form to the s form because it is not always possible to know how many 'j' terms were involved (*remember*: $j^2 = -1$, $j^3 = -j$, $j^4 = 1$, etc).

Initially we consider the frequency response of a fairly simple circuit – the RC parallel circuit in figure 11.2 – in order to define some basic terms. We look at the way in which both the magnitude and the phase shift vary with frequency. The impedance of the circuit is

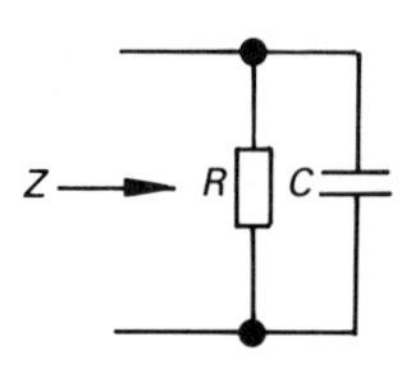

Figure 11.2 *A simple* RC *parallel circuit.*

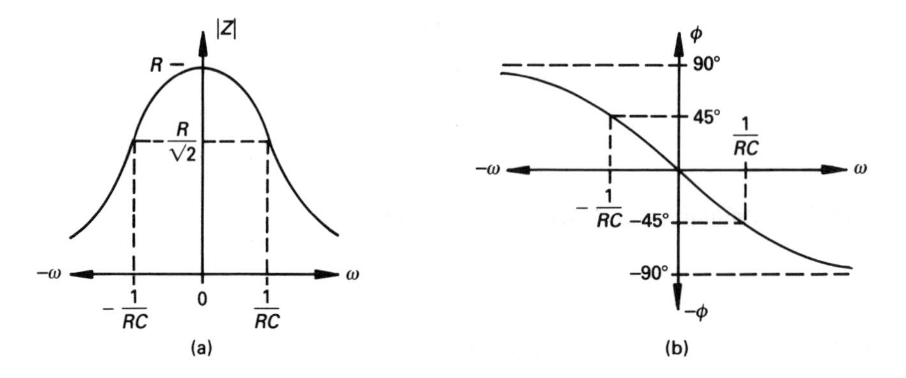

Figure 11.3 *(a) Magnitude–frequency plot for +ω and −ω of the impedance for figure 11.2, and (b) the phase–frequency plot.*

$$\mathbf{Z}(s) = \frac{\mathbf{Z}_R\mathbf{Z}_C}{\mathbf{Z}_R + \mathbf{Z}_C} = \frac{R(1/Cs)}{R + 1/Cs} = \frac{R}{1 + RCs}$$

Alternatively, we can evaluate $\mathbf{Z}(j\omega)$ simply by replacing s by $j\omega$ as follows:

$$\mathbf{Z}(j\omega) = \frac{R(1/j\omega C)}{R + 1/j\omega C} = \frac{R}{1 + j\omega RC}$$

$$= \frac{R}{\sqrt{(1 + [\omega RC]^2)}} \angle -\tan^{-1}(\omega RC) = |\mathbf{Z}| \angle \phi$$

where both $|\mathbf{Z}|$ and ϕ are functions of ω.

The curve showing how the magnitude of the impedance changes with frequency with both $+\omega$ and $-\omega$ is drawn in figure 11.3(a), and the corresponding phase–frequency plot is in figure 11.3(b). The two curves comprise the *frequency response* of the impedance $\mathbf{Z}$. It can be shown that the magnitude curve is symmetrical about the $\omega = 0$ axis; the phase angle has its sign reversed for $\omega < 0$. For these reasons, we generally need only interest ourselves in the frequency response for positive values of ω.

Consider now the power consumed by the circuit at the frequencies $\omega = 0$ and $\omega = 1/RC$; we will later refer to the latter frequency as ω_1. At $\omega = 0$, the capacitive reactance is $X_C = 1/\omega C = \infty$, and no current flows in it at this frequency. If the circuit is energised by a current source $\boldsymbol{I}$, then all the current flows through the resistor, and the power consumed at zero frequency is

$$P_0 = R|\boldsymbol{I}|^2$$

and is the maximum power consumed by the circuit. At frequency ω_1, the current in the resistor is

$$\boldsymbol{I} = \frac{1/\mathrm{j}\omega_1 C}{R + 1/\mathrm{j}\omega_1 C} = \frac{1}{1 + \mathrm{j}\omega_1 CR} = \frac{1}{1 + \mathrm{j}}$$

$$= \frac{1}{\sqrt{2}\angle 45^\circ} = \frac{1}{\sqrt{2}}\angle -45^\circ$$

and the power consumed at frequency ω_1 is

$$P_1 = \left|\frac{\boldsymbol{I}}{\sqrt{2}}\right|^2 R = \frac{|\boldsymbol{I}|^2 R}{2} = \frac{P_0}{2}$$

That is, the power consumed when $\omega_1 = 1/RC$ is one-half the maximum power. Consequently, we say that ω_1 is the *half-power frequency*, or the *half-power point*, or (for reasons given later) the *cut-off frequency*. In general, a frequency (and there may be more than one of these) which gives a power response which is one-half of the maximum response is known by one of the above names.

In the above case, the phase shift at the half-power frequency is $-45°$. In general, depending on the circuit, the phase shift at a half-power point is $\pm 45n°$, where n is an integer.

We will now look at two simple examples which are of particular interest when we study resonance (see chapter 12).

Worked example 11.6.1

Sketch the frequency response curve for the impedance of a two-branch parallel circuit which has a pure inductor, L, in one branch, and a pure capacitor, C, in the other.

Solution

Although this is an idealised situation (strictly speaking, there is no such thing as a 'pure' L and a 'pure' C), it will be very helpful in giving an overall view of the frequency response of this type of circuit. The complex frequency impedance is

$$\mathbf{Z}(s) = \frac{Ls(1/Cs)}{Ls + 1/Cs} = \frac{Ls}{1 + LCs^2}$$

or, replacing s by $\mathrm{j}\omega$

$$\mathbf{Z}(\mathrm{j}\omega) = \frac{\mathrm{j}\omega L}{1 + LC(\mathrm{j}\omega)^2} = \frac{\omega L\angle 90^\circ}{1 - \omega^2 LC}$$

when $\omega = 0$, then $\mathbf{Z}(\mathrm{j}\omega) = 0\angle 90°$. As ω increases in a positive direction, the magnitude of $\mathbf{Z}(\mathrm{j}\omega)$ increases and the phase angle remains constant at 90°. However, at some frequency ω_0, the denominator has zero value; this is when $\omega_0^2 LC = 1$, or

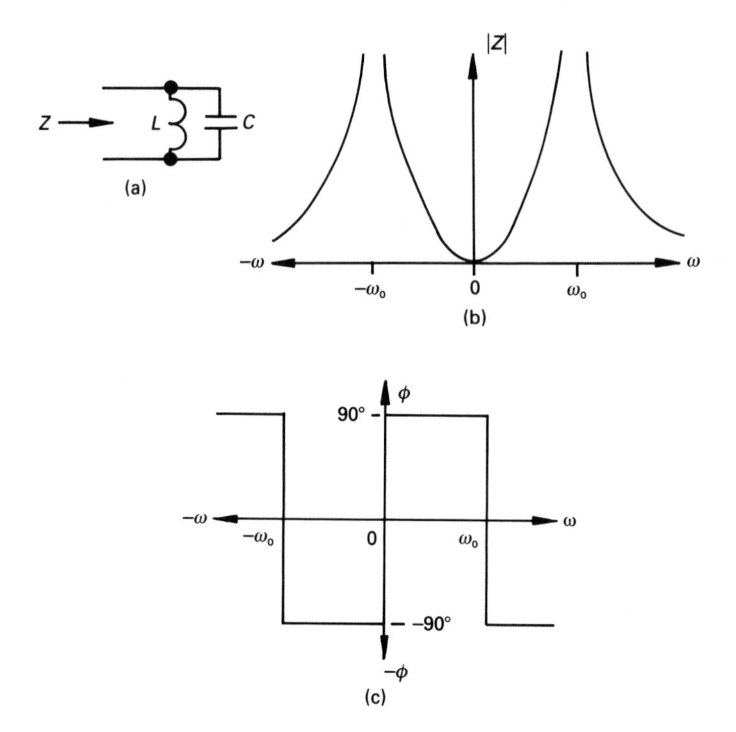

Figure 11.4 *(a) Parallel* LC *circuit, (b) the magnitude response and (c) the phase response.*

$$\omega_0 = 1/\sqrt{(LC)}$$

This is shown in figure 11.4.

At this frequency, which is known as the *resonant frequency*, ω_0, the denominator of the impedance equation is zero, and the magnitude of the impedance is infinity. When $\omega > \omega_0$, then $\omega^2LC > 1$ and the magnitude of the impedance begins to diminish until, at $\omega = \infty$, $|\mathbf{Z}(j\omega)| = 0$.

Also, when $\omega^2LC > 1$ the sign of the denominator of $\mathbf{Z}(j\omega)$ becomes negative, resulting in the phase shift becoming $-90°$.

The corresponding values for $-\omega$ are shown in figure 11.4.

Poles and zeros

At this point we will define two critical values of complex frequency *s*, one being known as a zero and the other as a pole, which determine both the transient and the steady-state behaviour of a circuit. A *zero* is a value of *s* which *makes the value of a complex function equal to zero*, and a *pole* is a value of *s* which *makes the value of a complex function equal to infinity*.

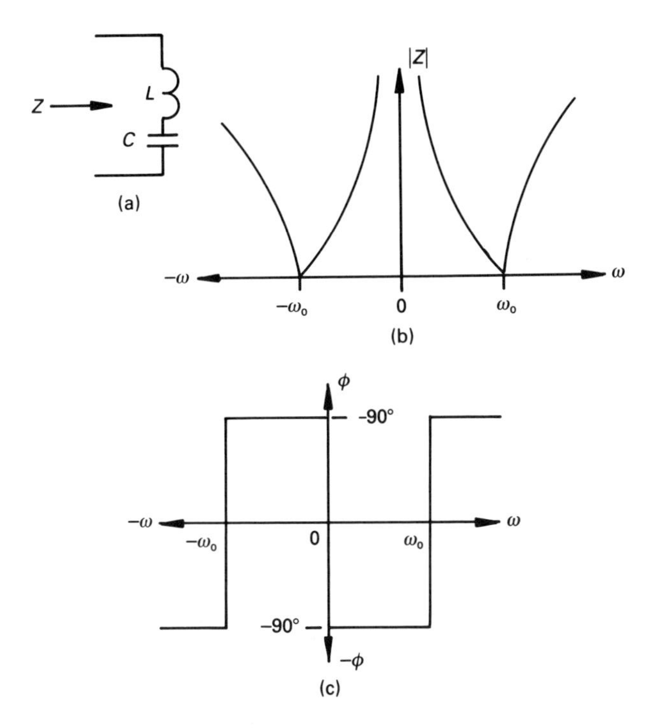

Figure 11.5 *(a) Series resistanceless* LC *circuit, (b) the magnitude response of the impedance of the circuit and (c) the phase response.*

Zeros occur in the numerator of the expression, and poles occur in the denominator. Since the complex impedance in this problem is

$$\mathbf{Z}(s) = \frac{Ls}{1 + LCs^2}$$

there is a zero at $s = 0$, and a pair of poles at $s = \omega_0 = 1/\sqrt{(LC)}$ and $s = -\omega_0 = -1/\sqrt{(LC)}$. The two poles form conjugate points on the s-plane (see section 11.4).

Worked example 11.6.2

Repeat worked example 11.6.1 but for a series circuit containing a pure inductor and a pure capacitor.

Solution

The circuit is shown in figure 11.5(a), and the circuit impedance is

$$Z(s) = Ls + \frac{1}{Cs}$$

or, replacing s by jω

$$Z(j\omega) = j\omega L - \frac{j}{\omega C} = j\left(\omega L - \frac{1}{\omega C}\right) = -j\left(\frac{1}{\omega C} - \omega L\right)$$

$$= \left(\frac{1}{\omega C} - \omega L\right) \angle -90° \ \Omega$$

Clearly, when $\omega = 0$, $|Z(j\omega)| = \infty$ (see diagram (b)), and the phase shift is $-90°$, as shown in diagram (c) of figure 11.5. As the frequency $(+\omega)$ increases, $|Z(j\omega)|$ decreases in value and the phase angle remains constant at $-90°$ until some frequency ω_0, when $\omega_0 L = 1/\omega_0 C$. At this frequency ($\omega_0 = 1/\sqrt{(LC)}$), which is the *resonant frequency* of the series circuit, the modulus of the circuit impedance is zero.

When $\omega > \omega_0$, $Z(j\omega)|$ increases with ω, finally approaching infinity as ω approaches infinity. At the same time, the phase shift becomes $+90°$, as shown in figure 10.5.

Corresponding values for $-\omega$ are also shown in the figure.

11.7 Transfer functions

Often we need to look at the frequency response of the ratio of the forced function, for example, the output voltage or current, to the forcing function which produces it, which may also be a voltage or current. The ratio of the forced function to the forcing function is known as the *transfer function* $H(s)$ or $H(j\omega)$. That is

$$H(s) = \frac{\text{response } (s)}{\text{forcing function } (s)} = \frac{\text{output } (s)}{\text{input } (s)}$$

or

$$H(j\omega) = \frac{\text{response } (j\omega)}{\text{forcing function } (j\omega)} = \frac{\text{output } (j\omega)}{\text{input } (j\omega)}$$

We can, for example, look on impedance as a transfer function in which

$$H(s) = Z(s) = \frac{V}{I}(s)$$

or

$$H(j\omega) = Z(j\omega) = \frac{V}{I}(j\omega)$$

where $\boldsymbol{V}$ is the response and $\boldsymbol{I}$ the forcing function. In general, a transfer function may take on one of many dimensions. For example, it may be dimensionless, as is the case when we are considering the voltage gain of an operational amplifier; it may have the dimensions of impedance, as in the case considered above; it may have the dimensions of, say, radians per volt where we are considering the operation of a voltage-driven position control system, etc.

Worked example 11.7.1

Evaluate the transfer function $\frac{\boldsymbol{V}_2}{\boldsymbol{V}_1}(s)$ for the circuit in figure 11.6.

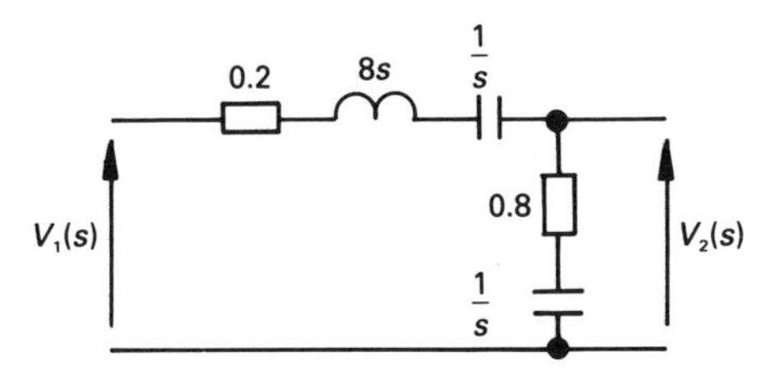

Figure 11.6 *Figure for worked example 11.7.1.*

Solution

Bearing in mind that the two 1 F capacitors are connected in series (giving an effective capacitance of 0.5 F), the transfer function is

$$\boldsymbol{H}(s) = \frac{\boldsymbol{V}_2(s)}{\boldsymbol{V}_1(s)} = \frac{0.8 + 1/s}{1 + 8s + 1/0.5s}$$

$$= \frac{0.125(1 + 0.8s)}{s^2 + 0.125s + 0.25}$$

$$= \frac{0.125(1 + 0.8s)}{(s + 0.0625 + \mathrm{j}0.5)(s + 0.0625 - \mathrm{j}0.5)}$$

Worked example 11.7.2

Derive an expression for $\boldsymbol{H}(s) = \boldsymbol{V}_0(s)/\boldsymbol{V}_1(s)$ for the operational amplifier circuit in figure 11.7. The operational amplifier is ideal. This circuit is the basis of many *active filter* networks.

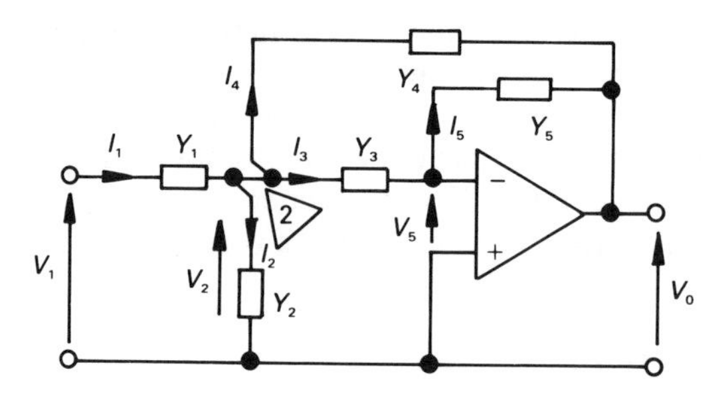

Figure 11.7 *Figure for worked example 11.7.2.*

Solution

The circuit is that of an active filter, and is applicable to a low-, a high- or a band-pass filter, according to the position of resistors and capacitors in the circuit (see also problem 11.11).

At node 2

$$I_1 = I_2 + I_3 + I_4 = (V_1 - V_2)Y_1$$

The individual currents are

$$I_2 = V_2Y_2 \qquad I_3 = (V_2 - V_5)Y_3$$

$$I_4 = (V_2 - V_0)Y_4 \qquad I_5 = (V_5 - V_0)Y_5$$

Since the op-amp is ideal, V_5 is practically zero, so that no current flows into the amplifier, and

$$I_3 = V_2Y_3$$

$$I_5 = -V_0Y_5$$

Inserting the above expressions for current into the first equation gives

$$V_2Y_2 + V_2Y_3 + V_2Y_4 - V_0Y_4 = (V_1 - V_2)Y_1$$

and, after a little more algebraic manipulation we get

$$H(s) = \frac{V_0}{V_1} = \frac{-Y_1Y_3}{Y_5(Y_1 + Y_2 + Y_3 + Y_4) + Y_3Y_4}$$

At this stage we are dealing with an academic exercise, and the reader should note that the admittance values should be of an electrically acceptable form. The reader should refer to problem 11.11 for typical values for a practical circuit.

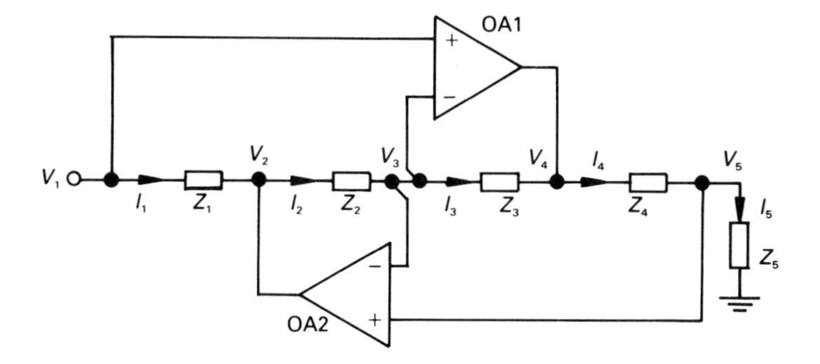

Figure 11.8 *Figure for worked example 11.7.3.*

Worked example 11.7.3

Derive an expression for the input impedance, $\boldsymbol{Z}_{in}(s)$, for the *generalised impedance convertor* (GIC) in figure 11.8. The op-amps are ideal.

Solution

The purpose of a GIC is to convert an impedance of one kind into an impedance of another kind, that is, it makes a capacitor 'look' to the external circuit as though it were an inductor. This is particularly useful in integrated circuit applications where inductors are difficult to manufacture. The circuit layout in figure 11.8 looks fairly involved, but it is drawn in this way to make the analysis as simple as possible.

Since the op-amps are ideal, the potential difference between the amplifier input terminals is practically zero. We may therefore say that $\boldsymbol{V}_1 \approx \boldsymbol{V}_3$ and $\boldsymbol{V}_3 \approx \boldsymbol{V}_5$; hence, we can use $\boldsymbol{V}_1$ to represent both $\boldsymbol{V}_3$ and $\boldsymbol{V}_5$. Also, since the op-amp input current is practically zero, we may say that $\boldsymbol{I}_4 = \boldsymbol{I}_5$, therefore

$$\frac{\boldsymbol{V}_4 - \boldsymbol{V}_5}{\boldsymbol{Z}_4} = \frac{\boldsymbol{V}_5}{\boldsymbol{Z}_5}$$

or, using the above simplification

$$\frac{\boldsymbol{V}_4 - \boldsymbol{V}_1}{\boldsymbol{Z}_4} = \frac{\boldsymbol{V}_1}{\boldsymbol{Z}_5} \quad \text{or} \quad \boldsymbol{V}_4 = \boldsymbol{V}_1\left(1 + \frac{\boldsymbol{Z}_4}{\boldsymbol{Z}_5}\right)$$

Also $\boldsymbol{I}_2 = \boldsymbol{I}_3$, or

$$\frac{\boldsymbol{V}_2 - \boldsymbol{V}_1}{\boldsymbol{Z}_2} = \frac{\boldsymbol{V}_1 - \boldsymbol{V}_4}{\boldsymbol{Z}_3} = \frac{\boldsymbol{V}_1(1 - [1 + \boldsymbol{Z}_4/\boldsymbol{Z}_5])}{\boldsymbol{Z}_3}$$

that is

$$V_2 = V_1\left(1 - \frac{Z_2Z_4}{Z_3Z_5}\right)$$

The input impedance of the circuit is

$$Z_{in}(s) = \frac{V_1}{I_1} = \frac{V_1}{(V_1 - V_2)/Z_1}$$

$$= \frac{V_1Z_1}{V_1(1 - [1 - Z_2Z_4/Z_3Z_5])} = \frac{Z_1}{Z_2Z_4/Z_3Z_5}$$

$$= \frac{Z_1Z_3Z_5}{Z_2Z_4}$$

If $Z_1 = Z_2 = Z_3 = Z_5 = R$ and $Z_4 = 1/sC$, then

$$Z_{in}(s) = \frac{R \times R \times R}{R/sC} = (R^2C)s = Ls$$

That is, the capacitor C in the Z_4 position appears as though it were an inductor L at the input of the circuit, where $L = R^2C$ H. If $R = 1$ kΩ and $C = 1$ μF, then $L = 1000^2 \times 10^{-6} = 1$ H.

11.8 Bode diagrams

In order to understand more fully the way in which the magnitude and phase shift of a transfer function of a given circuit changes with frequency, we need a simple method of estimating the shape of the curves. Accurate curves can be plotted after some more-or-less frantic manipulations with a calculator (or a programmable calculator), or they can be obtained directly by an all-out assault using a computer (see also chapter 14, where the SPICE package is described). However, in the short term, we need a fairly quick method of obtaining the response curves; one method has been provided by Hendrik Bode of the Bell Telephone Laboratories, and is described here.

We have already met with frequency response in section 11.6, where simple frequency response diagrams were described. Here we look in a little detail at the frequency response of more general forms of transfer function in electrical circuits.

Many transfer functions combine elements having an equation similar to the following

$$H(s) = K\frac{(s + z_1)(s + z_1)\ldots}{s^n(s + p_1)(s + p_1)\ldots}$$

where K is a constant gain (or attenuation) factor which is independent of frequency, z is a frequency at which a zero occurs in the complex frequency plane, p is a frequency at which a pole occurs in the complex frequency plane and n is an integer.

Finally, there may be quadratic terms in the denominator, similar to the one shown below, which can give rise to a resonant effect.

$$1/\left(1 + 2\zeta\left[\frac{s}{\omega_0}\right] + \left[\frac{s}{\omega_0}\right]^2\right)$$

First, we must look at the way in which we can estimate the overall magnitude and phase response from the many parts of the transfer function of the system.

A Bode diagram is an asymptotic plot of the magnitude and phase of the transfer function to a base of frequency, so that we can write the above equation as a function of ω in the following form

$$\boldsymbol{H}(\mathrm{j}\omega) = |\boldsymbol{H}(\mathrm{j}\omega)|\angle\boldsymbol{H}(\mathrm{j}\omega)$$

$$= \frac{|\boldsymbol{H}_1(\mathrm{j}\omega)|\angle\boldsymbol{H}_1(\mathrm{j}\omega) \times |\boldsymbol{H}_2(\mathrm{j}\omega)|\angle\boldsymbol{H}_2(\mathrm{j}\omega)\ldots}{|\boldsymbol{H}_3(\mathrm{j}\omega)|\angle\boldsymbol{H}_3(\mathrm{j}\omega) \times |\boldsymbol{H}_4(\mathrm{j}\omega)|\angle\boldsymbol{H}_4(\mathrm{j}\omega)\ldots}$$

$$= \frac{|\boldsymbol{H}_1| \times |\boldsymbol{H}_2| \times \ldots}{|\boldsymbol{H}_3| \times |\boldsymbol{H}_4| \times \ldots}\arg\left(\angle\boldsymbol{H}_1 + \angle\boldsymbol{H}_2 + \ldots - \{\angle\boldsymbol{H}_3 + \angle\boldsymbol{H}_4 + \ldots\}\right)$$

The final expression tells us that the overall magnitude is obtained by multiplying (or dividing) individual magnitudes of each part of the transfer function; the overall phase shift is obtained by adding (or subtracting) the individual phase angles of the different parts.

This may seem a relatively complex procedure, but it can be reduced to a series of simple steps when we look at the basic techniques for expressions commonly found in electrical and electronic circuits.

To simplify the process of drawing the diagrams, both the magnitude and phase curves are drawn to a logarithmic scale for the abscissa. The phase scale is drawn linearly; to simplify the multiplication and division of the magnitude values, the magnitude scale is in logarithmic units called *decibels* (dB), where the decibel is defined as follows

$$H_{\mathrm{dB}} = 20 \log |\boldsymbol{H}(\mathrm{j}\omega)|$$

and logarithms to base 10 are used.

The Bode diagram yields two graphs. One is the magnitude or gain expressed in dB plotted to a base of frequency (or to a function of frequency, which may be dimensionless) on a logarithmic scale. The other is a graph of phase shift (usually in degrees) plotted to the same logarithmic frequency base.

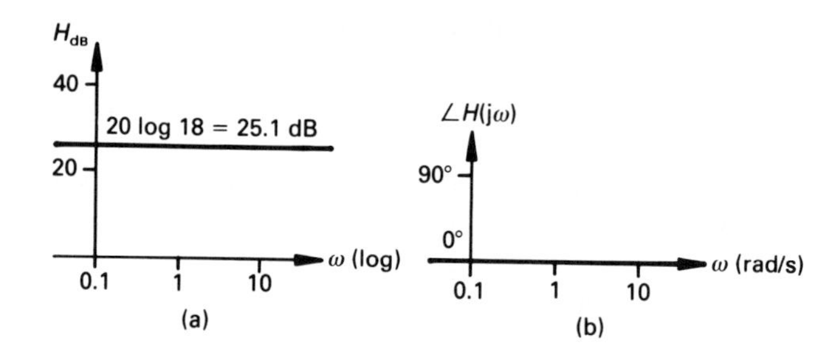

Figure 11.9 *Bode diagram (a) magnitude plot, (b) phase plot for* **H**($j\omega$) = K.

Normally both graphs are drawn on the same sheet of log–linear paper. In this way it is possible, at a glance, to review the frequency response of the system.

11.9 Bode diagram of $H(j\omega)$ = K

If $K = 18$, the decibel magnitude is

$$H_{dB} = 20 \log 18 = 25.1 \text{ dB}$$

for all ω, and results in the graph in figure 10.9(a).

The phase shift for a simple numeric gain factor is zero for all ω, as shown in figure 11.9(b).

11.10 Bode diagram of $H(j\omega) = (j\omega\tau)^n$

In this case the dB gain is

$$\begin{aligned} H_{dB} &= 20 \log |j\omega\tau|^n = 20n \log\sqrt{(\omega\tau)^2} \\ &= 20n \log \omega\tau \end{aligned}$$

If $n = 1$, the way in which the magnitude changes with $\omega\tau$ is shown in table 11.1. We see from the table that the graph is a straight line having a slope of 6 dB/octave (strictly speaking, it is 6.02 dB/octave), meaning that there is a change of 6 dB in the magnitude each time the frequency changes by a factor of 2. Alternatively, the slope is 20 dB/decade (that is, a change of 20 dB for each ten-fold change in frequency), as shown in figure 11.10(a). A look at the mathematics shows that the slope of the magnitude plot, for any integer n, is $6n$ dB/octave (approx.) or $20n$ dB/decade.

The phase angle associated with $(j\omega\tau)^n$ is

$$\angle(j\omega\tau)^n = \angle(j)^n = 90n°$$

Table 11.1

$\omega\tau$	H_{dB}	*Slope of graph*	
0.1	−20	6 dB/octave	20 dB/decade
0.2	−14 (approx.)		
1.0	0	6 dB/octave	20 dB/decade
2.0	6 (approx.)		
10.0	20		20 dB/decade
100.0	40		

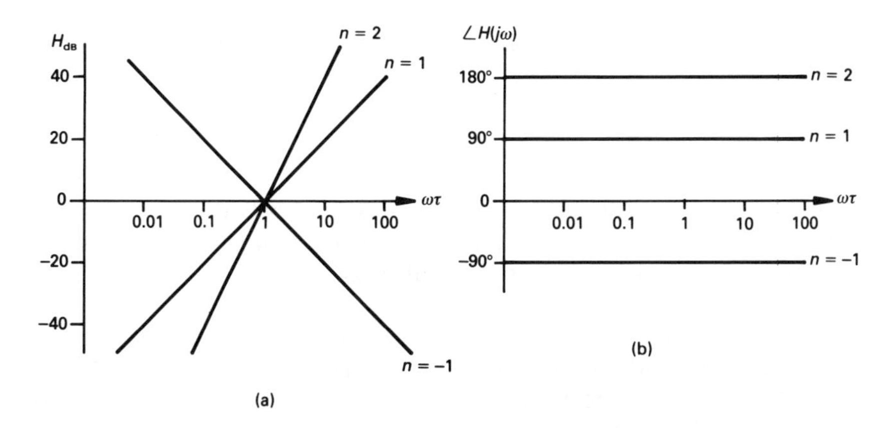

Figure 11.10 *(a) Magnitude plot of* $H(j\omega) = (j\omega\tau)^n$ *for* $n = -1$, *1 and 2, (b) corresponding phase plot.*

This is illustrated in figure 11.10(b) for values of n of -1, 1 and 2.

11.11 Bode diagram of $H(j\omega) = (1 + j\omega\tau)$

Gain–frequency plot

The expression for the gain in dB of this expression is

$$H_{dB} = 20 \log |1 + j\omega\tau| = 20 \log \surd\,(1 + \omega^2\tau^2)$$

We will look at the value of this expression over two ranges of frequency, namely (1) when $(\omega\tau)^2 << 1$ and (2) when $(\omega\tau)^2 >> 1$.

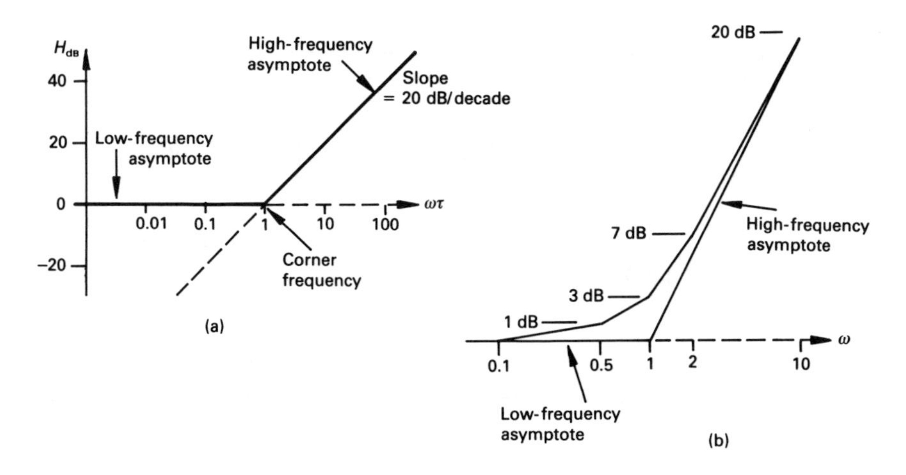

Figure 11.11 *(a) Asymptotic magnitude–frequency plot for* H($j\omega$) = *(1 +* $j\omega\tau$*), (b) a means of sketching the curve in the region of the corner frequency.*

1. $(\omega\tau)^2 << 1$
   In this case $H_{dB} = 20 \log \sqrt{(1 + \omega^2\tau^2)} \approx 20 \log 1 = 0$ dB for all ω within this frequency range. This results in the *low-frequency asymptote* in figure 11.11(a), which lies on the 0 dB axis.
2. $(\omega\tau)^2 >> 1$
   Here $H_{dB} = 20 \log \sqrt{(1 + \omega^2\tau^2)} \approx 20 \log \omega\tau$, which is a straight line of slope 20 dB/decade cutting the 0 dB axis when $\omega\tau = 1$ (see also section 10.11). This gives the *high-frequency asymptote* (see figure 11.11(a)) which cuts the $\omega\tau$ axis at $\omega\tau = 1$ or where $\omega = 1/\tau$.

Clearly, the two asymptotes intersect when the high-frequency asymptote has the value $H_{dB} = 0$, that is, when $\omega\tau = 1$. This frequency is known as a *corner frequency*, ω_C.

Occasionally the asymptotes do not give sufficiently accurate information to sketch the curve in the region of the corner frequency, and it is necessary to plot a few more points. In particular, values at the corner frequency together with frequencies which are one-half and double the corner frequency are usually sufficient to allow a fairly accurate curve to be sketched, as follows.

Magnitude at $\omega = \omega_C$

The actual gain at this frequency is

$$H_{dB} = 20 \log \sqrt{(1 + 1)} = 3 \text{ dB (strictly 3.01 dB)}$$

Magnitude at $\omega = \omega_C/2$ *and* $\omega = 2\omega_C$

The dB gain at $\omega_C/2$ is 1 dB (which is 1 dB away from the low-frequency asymptote), and at $2\omega_C$ is 7 dB (which is 1 dB away from the high-frequency asymptote).

Magnitude at $\omega \leqslant (\omega_C/10)$ *and* $\omega \geqslant 10\omega_C$

The dB gain at these frequencies almost lie on the low- and high-frequency asymptotes, respectively.

Taking these values into account, a plot of the gain curve in the region of the corner frequency is shown in figure 11.11(b).

Phase–frequency plot

The phase–frequency response is calculated from the

$$\angle \boldsymbol{H}\,(j\omega) = \angle(1 + j\omega\tau) = \tan^{-1}\omega\tau$$

This is a smoother curve than the magnitude–frequency asymptotic curve but, nevertheless, we can draw a simple asymptotic response as follows

1. *When* $\omega\tau$ *is small*

Here, very approximately, we take this to mean $0 \leqslant \omega\tau < 0.1$.

$$\angle \boldsymbol{H}\,(j\omega) \approx \tan^{-1} 0 = 0°$$

2. *When* $\omega\tau$ *is large*

Here we can take this to mean $10 \leqslant \omega\tau < \infty$.

$$\angle \boldsymbol{H}\,(j\omega) \approx \tan^{-1} \infty = 90°$$

3. *When* $0.1 < \omega\tau < 10$

The plot approximates in this region to a straight line which joins the low- and high-frequency phase asymptotes. This is a line of slope 45°/decade.

The resulting asymptotic phase–frequency curve is shown in full line figure 11.12. The actual phase curve is approximately 5° above the junction of the low- and mid-frequency asymptotes, and is 5° below the junction of the mid- and high-frequency asymptotes. The actual curve and the mid-frequency asymptotic line intersect when $\omega\tau = 1$, where $\angle \boldsymbol{H}(j\omega) = 45°$.

Worked example 11.11.1

A series circuit contains a resistor of 25 Ω resistance and an inductor of 0.25 H inductance. Write down an expression for the *s*-domain impedance of the circuit, and sketch the frequency response curve showing how the impedance varies with frequency.

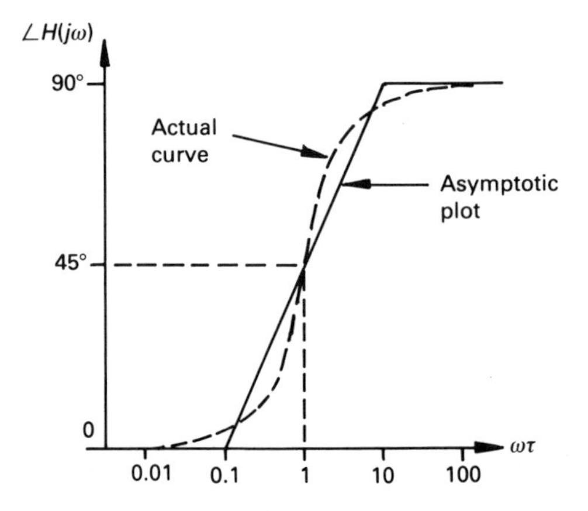

Figure 11.12 *The simplified asymptotic phase angle response for* $\mathbf{H}(j\omega) = (1 + j\omega\tau)$. *The actual curve is shown dotted.*

Solution

The expression for the *s*-domain impedance is

$$\mathbf{Z}(s) = \boldsymbol{H}(s) = R + Ls = 25 + 0.25s = 25(1 + 0.01s)$$

that is

$$\begin{aligned}\mathbf{Z}(\mathrm{j}\omega) &= \boldsymbol{H}(\mathrm{j}\omega) = R + \mathrm{j}\omega L = 25 + \mathrm{j}\omega 0.25 \\ &= 25(1 + \mathrm{j}\omega 0.01)\end{aligned}$$

The final step in the impedance equations has been taken in order to convert the expression into the standard form developed in section 11.11.

Initially, the equation is reduced to one containing the factors described earlier, namely a constant (= 25) and a term of the form $(1 + \mathrm{j}\omega\tau)$, that is, $(1 + \mathrm{j}\omega 0.01)$. The magnitude in dB of each is drawn separately in figure 11.13(a) and, finally, the two are added to give the complete asymptotic magnitude plot, which is drawn in full line in figure 11.13(b). The $K = 25$ factor gives a constant gain of 28 dB, and the $(1 + \mathrm{j}\omega 0.01)$ term gives a break point at $0.01\omega_C = 1$ or $\omega_C = 100$ rad/s. In this case, the Bode diagram has been plotted to a logarithmic base of frequency. The factor $(1 + \mathrm{j}\omega 0.01)$ indicates that there is a simple zero at $s = -100$ on the *s*-plane.

There is a correction of +3 dB to be made at the corner frequency on the magnitude plot, and the corrected curve is shown in dotted line in figure 11.13(b).

An inspection of the magnitude plot indicates that the impedance of the circuit changes from about 25 Ω (corresponding to 28 dB) at 10 rad/s (1.592 Hz) to about 250 Ω (or 48 dB) at 1000 rad/s (159.2 Hz).

Next we look at the phase plot. The constant term of 25 in the $\boldsymbol{H}(\mathrm{j}\omega)$

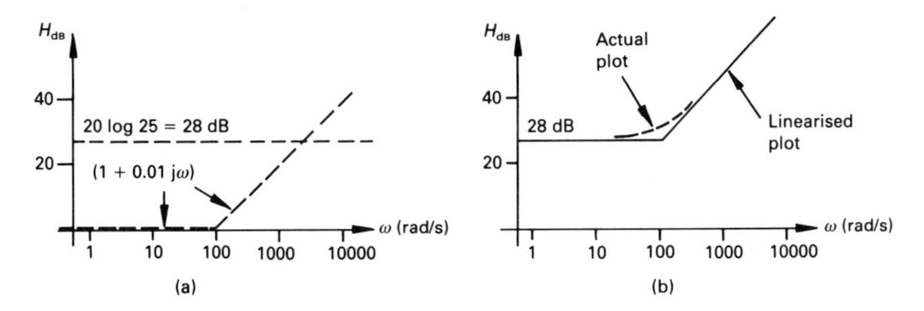

Figure 11.13 *(a) Plot of the magnitude factors of* **H**$(j\omega) = 25(1 + j\omega 0.01)$, *(b) the complete Bode magnitude plot.*

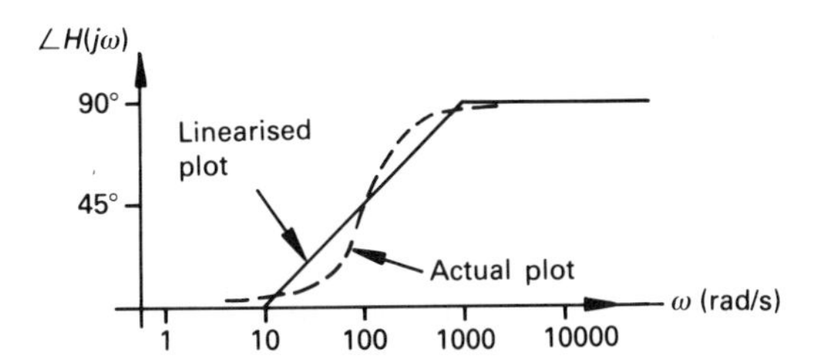

Figure 11.14 *The Bode phase plot for* $25(1 + j\omega 0.01)$.

expression imparts zero phase shift, so that the overall phase shift is equal to that of the $(1 + j\omega 0.01)$ term. This produces corner frequency on the asymptotic phase plot (see figure 11.14) at 100 rad/s. If necessary, the asymptotic phase plot can be corrected to give an accurate curve, as shown in broken line in the figure.

Normally, both the magnitude and phase plots are shown on one diagram.

11.12 Bode diagram of $H(j\omega) = (1 + j\omega\tau)^n$

As with the case of the $\boldsymbol{H}(j\omega) = (1 + j\omega\tau)$ response, the low-frequency gain is approximately zero dB up to a frequency of $\omega\tau = 1$, that is, $\omega = 1/\tau$. At this point the high-frequency asymptote assumes a slope of $20n$ dB/decade (or approximately $6n$ dB/octave), as shown in figure 11.15(a); the reader will find it an interesting exercise to verify the value of the slopes.

Once again, the actual magnitude curve will deviate from the associated asymptotes, but in this case the deviation will be n times greater than in the case for the $\boldsymbol{H}(j\omega) = (1 + j\omega\tau)$ curve. That is, at the corner frequency, the deviation is $3n$, dB, and at twice and one-half of the corner frequency the deviation is n dB from the associated asymptote.

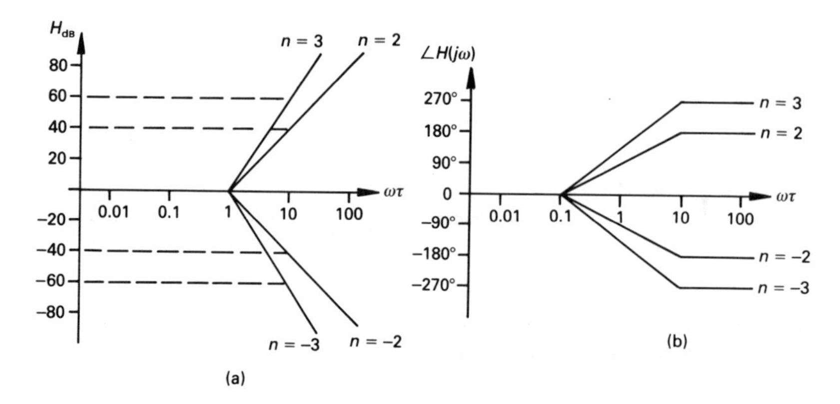

Figure 11.15 *(a) The magnitude–frequency and (b) the phase-frequency asymptotic response for* $\mathbf{H}(j\omega) = (1 + j\omega\tau)^n$.

The phase shift (see figure 11.15(b)) is, as might be expected, n times greater at each frequency than it was for the $\boldsymbol{H}(j\omega) = (1 + j\omega\tau)$ curve. Once again, the angular deviation of the actual curve from the start of the low-frequency phase asymptote, and the end of the high-frequency asymptote in figure 11.15 is $5n°$.

A reasonably accurate phase curve (to within about 1° accuracy) can be obtained with a little more trouble, by drawing straight-line segments between the points listed in table 11.2.

Table 11.2

$\omega\tau$	*Phase angle*
0.01	0°
0.1	$5n°$
0.5	$25n°$
2.0	$65n°$
10.0	$85n°$
100.0	$90n°$

Worked example 11.12.1

The voltage gain, V_{out}/V_{in}, of a transistor amplifier is

$$\boldsymbol{H}(s) = \frac{\boldsymbol{V}_{out}}{\boldsymbol{V}_{in}}(s) = \frac{-2.5s}{(1 + s/100)(1 + s/10\,000)}$$

Draw the Bode diagram for the amplifier.

Solution

The expression for the voltage gain can be written in the form

$$\boldsymbol{H}(\mathrm{j}\omega) = \frac{\boldsymbol{V}_{\mathbf{out}}}{\boldsymbol{V}_{\mathbf{in}}}(\mathrm{j}\omega) = \frac{-0.5\mathrm{j}\omega}{(1 + 0.01\mathrm{j}\omega)(1 + 10^{-5}\mathrm{j}\omega)}$$

We can look at the solution in terms of the factors -0.5, $\mathrm{j}\omega$, $(1 + 0.01\mathrm{j}\omega)^{-1}$, and $(1 + 10^{-5}\mathrm{j}\omega)^{-1}$, as follows:

1. The gain associated with the constant factor is

   $$H_{\mathrm{dB}} = 20 \log |-0.5| = -6 \text{ dB}$$

   and is shown as the broken line (1) in figure 11.16(a). The reader should note that the negative sign associated with the numerator is a 'phase' factor of 180° rather than a 'gain' factor.
2. The factor $\mathrm{j}\omega$ results in a gain which increases at a constant rate of 20 dB per decade, and passes through 0 dB at $\omega = 1$ rad/s, shown by broken line (2) in figure 11.16(a).
3. The factor $(1 + 0.01\mathrm{j}\omega)^{-1}$ gives a corner frequency at $\omega = 1/0.01 = 100$ rad/s, shown in broken line (3) in figure 11.16(a).
4. The broken line (4) in figure 11.16(a) corresponds to the factor $(1 + 10^{-5}\mathrm{j}\omega)^{-1}$, and has a corner frequency of $1/10^{-5} = 10^5$ rad/s.

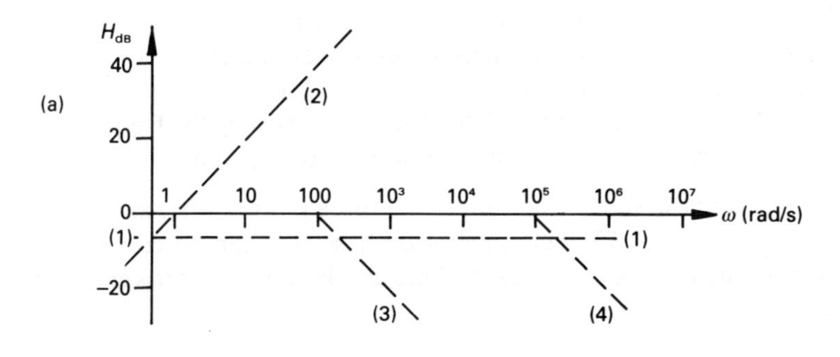

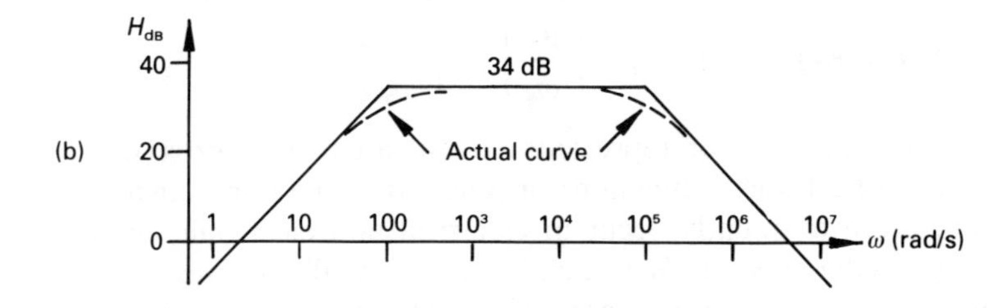

Figure 11.16 *The gain curve for worked example 11.12.1.*

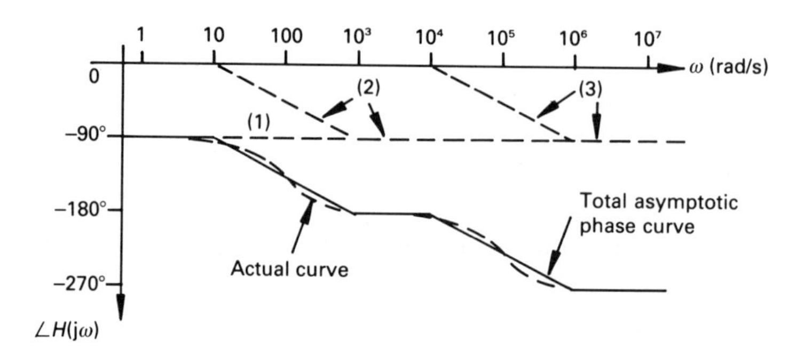

Figure 11.17 *The phase curve for worked example 11.12.1.*

The overall magnitude response is the sum of the four separate elements, and is shown in figure 11.16(b). The corrections to the corner frequencies are shown in broken lines in figure 11.16(b). The mid-band gain is seen to be 34 dB, corresponding to a numerical gain of 50.

When dealing with the phase shift, we can divide the transfer function into three factors, namely $(-0.5j\omega)$, $(1 + 0.01j\omega)^{-1}$ and $(1 + 10^{-5}j\omega)^{-1}$, as follows:

a. The phase shift associated with the factor $(-0.5j\omega)$ is $\angle -j = -90°$, shown as broken line (1) in figure 11.17.
b. The term $(1 + 0.01j\omega)^{-1}$ gives a phase lag which increases from zero (or thereabouts) to $-90°$ (or thereabouts) over the range 10 rad/s to 10^3 rad/s, shown as broken line (2) in figure 11.17.
c. This is similar to (2), but the phase shift occurs over the frequency range 10^4 to 10^6 rad/s, shown as broken line (3) in the figure.

The total phase shift is the sum of the three phase shifts, and is drawn in full line in figure 11.17. The linearised phase plot can be corrected at the corner frequencies, as outlined earlier, and is shown in broken line in the figure.

11.13 Plot of $H(j\omega) = 1 + 2\zeta \left[\frac{j\omega}{\omega_0}\right] + \left[\frac{j\omega}{\omega_0}\right]^2$

This quadratic expression represents a conjugate pair of zeros (if it is in the numerator of a transfer function) or poles (if in the denominator) on the *s*-plane; these frequently occur in electrical and electronic circuits. The quantity ζ is the *damping factor*, and ω_0 (as we shall see later) is the corner frequency of the asymptotic response for this function; when $\zeta = 0$, ω_0 is the *resonant frequency* of the system.

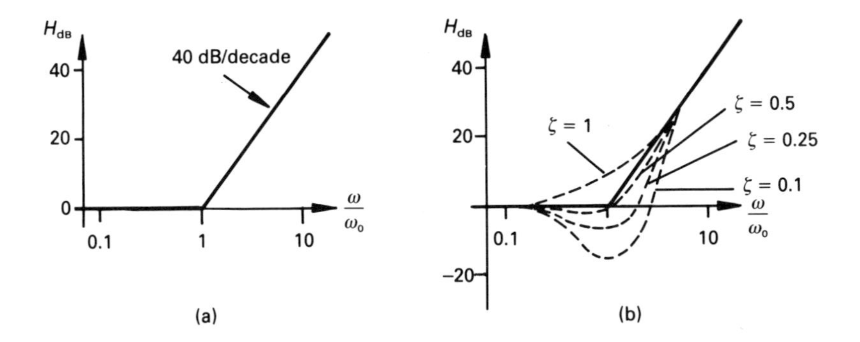

Figure 11.18 *(a) Bode asymptotic magnitude plot for* $H_{dB} = 20 \log \left| 1 + j2\zeta \left[\frac{\omega}{\omega_0}\right] - \left[\frac{\omega}{\omega_0}\right]^2 \right|$
(b) Details at the corner frequency for different values of damping factor.

Magnitude Response

The expression for the magnitude is

$$H_{dB} = 20 \log \left| 1 + j2\zeta \left[\frac{\omega}{\omega_0}\right] - \left[\frac{\omega}{\omega_0}\right] \right|$$

1. *When* $\omega/\omega_0 << 1$

$$H_{dB} \approx 20 \log 1 = 0 \text{ dB}$$

   This gives the *low-frequency asymptote*, which is a straight line on the 0 dB axis.

2. *When* $\omega/\omega_0 >> 1$

$$H_{dB} \approx 20 \log \left| -\left[\frac{\omega}{\omega_0}\right]^2 \right| = 40 \log \frac{\omega}{\omega_0}$$

The resulting *high-frequency asymptote* is a line of slope 40 dB/decade. The two asymptotes intersect (which is the *corner frequency* of the plot) when $(40 \log (\omega/\omega_0)) = 0$, that is when $\omega/\omega_0 = 1$ or $\omega = \omega_0$, as shown in figure 11.18(a).

In order to get an accurate impression of the magnitude curve, we must make corrections at the corner frequency. To a first approximation we may say that, whatever the value of ζ in the range 0.1 to 1 (which is the range in which most engineers are interested), the actual magnitude curve begins to leave the low-frequency asymptote at about $\omega/\omega_0 = 0.2$, and rejoins the high-frequency asymptote at about $\omega/\omega_0 = 2$ (see figure 11.18(b)). We need one other cardinal point at a given frequency in order to sketch the

Table 11.3

ζ	H_{dB}
0.1	−14
0.25	−6
0.5	0
1	6

magnitude curve for a given value of ζ; an appropriate frequency is $\omega/\omega_0 = 1$, or $\omega = \omega_0$. The value of H_{dB} at this frequency is calculated from

$$H_{dB} = 20 \log |1 + j2\zeta - 1| = 20 \log |2\zeta|$$

A list of values for various damping factors is given in table 11.3, and the associated magnitude curves are shown in figure 11.18(b).

The reader should note that, when there is negative peak in the magnitude curve, it occurs at a frequency which is a little less than $\omega/\omega_0 = 1$.

Phase response

The phase shift associated with $\boldsymbol{H}(\mathbf{j}\omega) = 1 + 2\zeta\left[\frac{j\omega}{\omega_0}\right] + \left[\frac{j\omega}{\omega_0}\right]^2$ is

$$\angle \boldsymbol{H}(j\omega) = \tan^{-1}\frac{2\zeta\ \omega/\omega_0}{1 - (\omega/\omega_0)^2}$$

1. *When* $\omega/\omega_0 << 1$

$$\angle \boldsymbol{H}(j\omega) \approx \tan^{-1}(2\zeta(\omega/\omega_0)) = 0°$$

This gives the *low-frequency asymptote*, which lies on the 0° axis. Within reasonable accuracy, this can be through to span from a very low frequency up to about $\omega/\omega_0 = 0.1$.

2. *When* $\omega/\omega_0 >> 1$

$$\angle \boldsymbol{H}(j\omega) \approx \tan^{-1}(-2\zeta/(\omega/\omega_0)) = 180°$$

That is, the *high-frequency asymptote* is a straight line at a phase shift of 180°. This spans from about $\omega/\omega_0 = 10$ up to infinite frequency.

3. *When* $\omega/\omega_0 = 1$

$$\angle \boldsymbol{H}(j\omega) = \angle j2\zeta = 90°$$

The asymptotic phase curve therefore consists of three straight lines, as shown in figure 11.19(a); the mid-frequency line which joins the low- and high-frequency asymptotes has a slope of 90°/decade, and has a phase shift of 90° when $\omega/\omega_0 = 1$.

Finally, we need to calculate a few values of ω/ω_0 which will allow us,

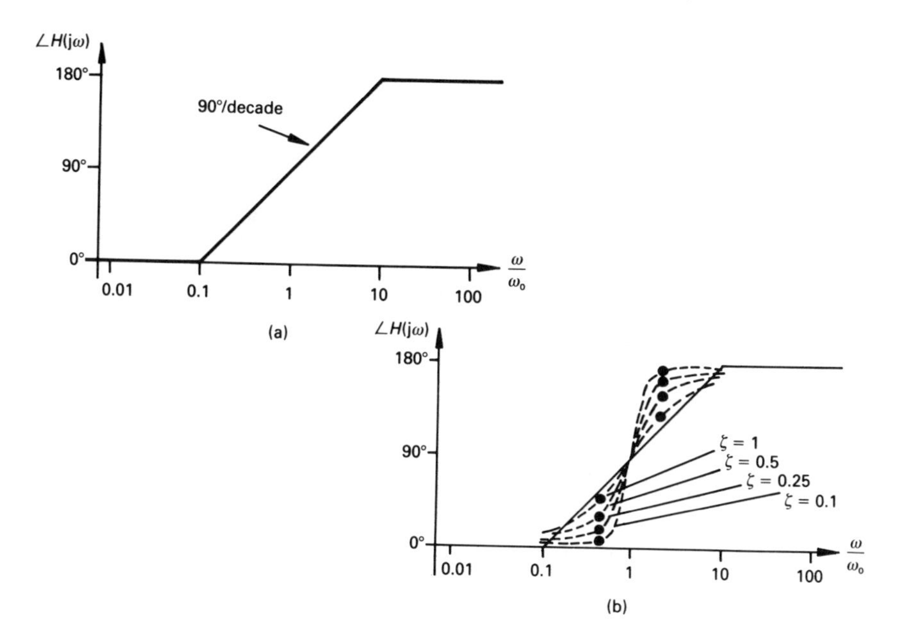

Figure 11.19 *(a) Linearised approximation to the phase response of* $\mathbf{H}(j\omega) = 1 + 2\zeta\left[\frac{j\omega}{\omega_0}\right] + \left[\frac{j\omega}{\omega_0}\right]^2$ *and (b) the estimated curves for a range of selected values of damping factor.*

with a reasonable degree of accuracy, to sketch the phase curve for any selected value of ζ. We will select the values $\omega/\omega_0 = 0.5$, 1 and 2.

1. When $\omega/\omega_0 = 0.5$, the overall phase shift is

$$\angle \boldsymbol{H}(\mathrm{j}\omega) = \tan^{-1}(1.3333\zeta)$$

2. When $\omega/\omega_0 = 1$, the overall phase shift is 90°.
3. When $\omega/\omega_0 = 2$, the overall phase shift is

$$\angle \boldsymbol{H}(\mathrm{j}\omega) = \tan^{-1}(-1.3333\zeta)$$

The corresponding values for various values of ζ are listed in table 11.4. The corresponding phase curves are drawn in broken line in figure 11.19(b); the points in table 11.4 are shown on the curves as large dots.

If the quadratic expression is in the denominator of the transfer function (see, for example, worked example 11.13.1), both the gain and phase curves are inverted or reversed. That is, the high-frequency asymptotic gain curve has a slope of −40 dB/decade, commencing at $\omega/\omega_0 = 1$; the asymptotic phase plot has a slope of −90°/decade, commencing at $\omega/\omega_0 = 0.1$, causing the phase to change from 0° to −180° by $\omega/\omega_0 = 10$.

Table 11.4

Damping factor ζ	*Phase shift (deg.) at* $\omega/\omega_0 = 0.5$	$\omega/\omega_0 = 2$
0.1	7.6	180 − 7.6 = 172.4
0.25	18.4	180 − 18.4 = 161.6
0.5	33.7	180 − 33.7 = 146.3
1	53.1	180 − 53.1 = 126.9

Worked example 11.13.1

The transfer function of an operational amplifier circuit is

$$\boldsymbol{H}(s) = \frac{\boldsymbol{V}_{\mathbf{out}}}{\boldsymbol{V}_{\mathbf{in}}}(s) = \frac{-2000s}{10\ 000 + 20s + s^2}$$

Draw the Bode diagram for the amplifier.

Solution

This expression can be re-written in the form

$$\boldsymbol{H}(s) = \frac{-0.2s}{1 + 2 \times 10^{-3}s + 10^{-4}s^2}$$

hence

$$\boldsymbol{H}(\mathrm{j}\omega) = \frac{-0.2\mathrm{j}\omega}{1 + 2 \times 10^{-3}\mathrm{j}\omega + 10^{-4}(\mathrm{j}\omega^2)}$$

The numerator contains a constant together with a simple linear function of ω, and the denominator is a quadratic expression. The linearised gain curve is plotted as follows.

1. The decibel gain of the numerator constant is

$$H_{\mathrm{dB}} = 20 \log 0.2 = -14 \text{ dB}$$

which is shown in figure 11.20(a) as the broken line (1). The reader should note that the negative sign associated with the numerator term is a 'phase' factor of 180°, rather than a 'gain' factor.

2. The gain associated with the numerator $\mathrm{j}\omega$ term is a line of slope 20 dB/decade, crossing the 0 dB axis at $\omega = 1$ rad/s. It is shown as broken line (2) in figure 11.20(a).

3. The quadratic term in the denominator has a corner frequency at $\omega = \surd(1/10^{-4}) = 100$ rad/s. The damping factor is determined from the fact that $2\zeta/\omega_0 = 2 \times 10^{-3}$, or $\zeta = 0.1$. The resulting curve is shown as broken curve

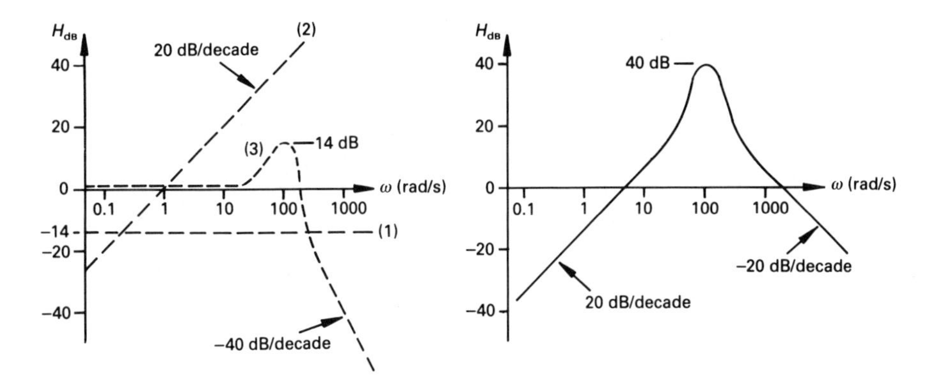

Figure 11.20 *(a) Gain plots for the individual factors of the transfer function in worked example 11.13.1 and (b) the overall gain plot.*

(3) in figure 11.20(a). The maximum gain for a damping factor of 0.1 is +14 dB in the region of the corner frequency of 100 rad/s.

The three asymptotic plots in figure 11.20(a) are combined to give the complete magnitude plot in figure 11.20(b). The maximum is 40 dB at the corner frequency of 100 rad/s.

The phase curve is plotted in two parts. The −j (= 1/j) term in the numerator contributes a constant phase shift of −90° (see broken line (1) in figure 11.21), and the quadratic term in the denominator gives a phase shift which increases from a low value (about −5°) to about −180° between the frequencies of 100 rad/s and 10^4 rad/s (see broken line (2)). The overall phase shift is the sum of the two; the asymptotic phase curve is shown in full line in figure 11.21, and the actual phase curve is in broken line.

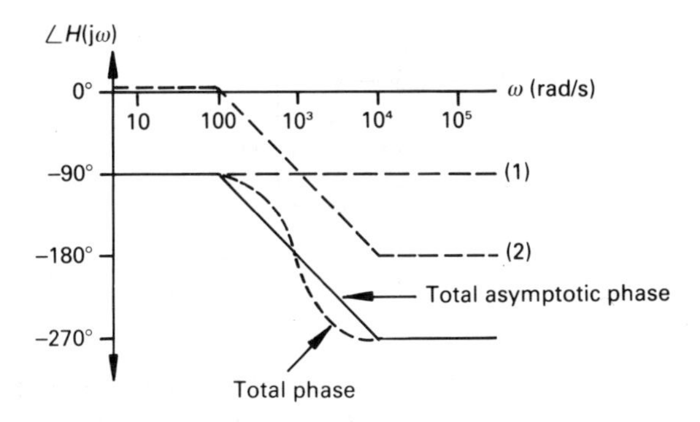

Figure 11.21 *Phase plot for worked example 11.13.1.*

Unworked problems

11.1. State the complex frequencies in the following time functions: (a) $5 \cos 8t$, (b) $(5 + 4e^{-6t} \cos 8t$, (c) $5 + 4e^{-6t} \cos 8t$, (d) $4e^{-6t} \cos (8t - 20°)$, (e) $4 \cos 3t - 5 \sin 2t$, (f) $8 \cos 2t.\ \sin 3t$
[(a) j8, −j8; (b) j8, −j8, −6 + j8, −6 − j8; (c) −6 + j8, −6 − j8; (d) −6 + j8, −6 − j8; (e) j3, −j3, j2, −j2; (f) j1, −j1, j5, −j5]

11.2. Write down the function of time having the following complex frequencies: (a) −6, (b) 7, (c) 2, −3, (d) 0, −6 + j4, −6 − j4; (e) −5, 6, 7 + j2, 7 − j2.
[(a) Ae^{-6t}; (b) Ae^{7t}; (c) $Ae^{2t} + Be^{-3t}$; (d) $A + Be^{-6t} \cos (4t + \phi)$; (e) $Ae^{-5t} + Be^{6t} + Ce^{7t} \cos (2t + \phi)$]

11.3. A voltage V_1 is defined by

$$200 = (2 + 3s)V_1 + \frac{(3 + 4s)}{s} V_1$$

Determine an expression for V_1, and calculate its value if $s = -3 + j4$. What is its value at $t = 0.2$ s?
[$200s/(3s^2 + 6s + 3)$; 16.67 V; 6.37 V]

11.4. At what value of s is $Z(s)$ zero in figure 11.6. At what value of s is $Z(s) = 7.9 = j0.06\ \Omega$?
[−0.0625 ± j0.348; −5 − j5]

11.5. Determine an expression for Z_{in} of figure 11.22 as a function of σ? Hence determine the frequencies of the poles and zeros. What is the circuit impedance at (a) $\sigma = 0$, (b) $\sigma = \infty$?
[$2(\sigma + 3.732)(\sigma + 0.268)/(1 + \sigma)$; zeros at −3.732 and −0.268, and a pole at −1; $Z(0) = 2\ \Omega$; $Z(\infty) = \infty$]

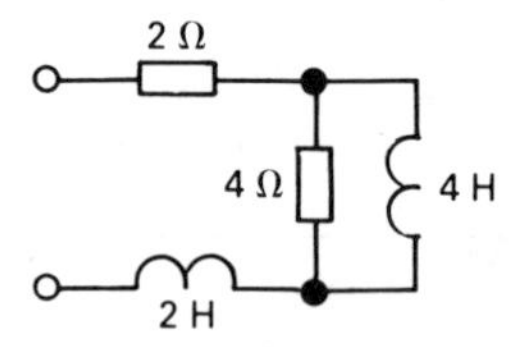

Figure 11.22

11.6. The complex-frequency impedance of a circuit has simple zeros at $s = -6$ and $s = -10$, and a pair of poles at $s = -4 \pm j8$. If its impedance at infinite frequency is 5 Ω, deduce an expression for

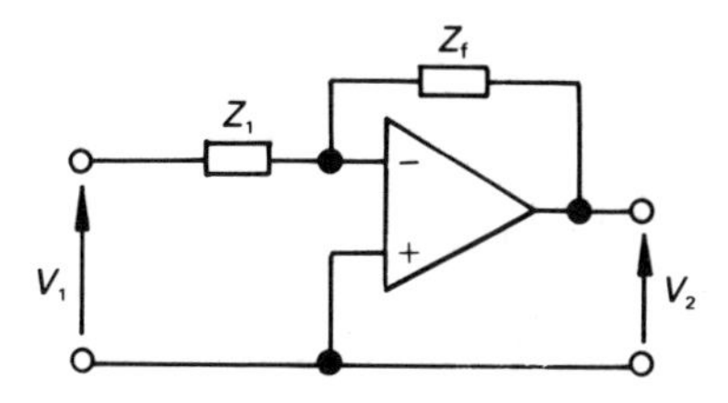

Figure 11.23

$\boldsymbol{Z}(s)$. If a forcing voltage $v(t) = 10e^{-2t} \cos(5t + 60°)$ V is applied to the circuit, determine the current $i(t)$ flowing into the network.
[$5(s + 6)(s + 10)/(s^2 + 8s + 80)$; $1.57e^{-2t} \cos(5t + 1.6°)$ A]

11.7. If the op-amp in figure 11.23 is ideal, deduce the transfer function $\boldsymbol{H}(s) = \boldsymbol{V}_2(s)/\boldsymbol{V}_1(s)$ as a ratio of two polynomials in s. Determine $\boldsymbol{H}(s)$ if (a) $\boldsymbol{Z}_1 = R_1$, $\boldsymbol{Z}_f = \boldsymbol{R}_f$ in parallel with C_f, (b) $\boldsymbol{Z}_1 = R_1$ in series with C_1, $\boldsymbol{Z}_f = R_f$.
[$-\boldsymbol{Z}_f(s)/\boldsymbol{Z}_1(s)$; (a) $-R_f/(R_1(sR_fC_f + 1))$; (b) $-sC_1R_f/(1 + sR_1C_1)$]

11.8. In problem 11.7, given that $R_f = 10$ kΩ, calculate values of R_1 and C_f so that $\boldsymbol{H}(s)$ is (a) -10, (b) $-250/(s + 50)$.
[$C_f = 0$ (open-circuit), $R_1 = 1$ kΩ; (b) $C_f = 2$ μF, $R_1 = 2$ kΩ]

11.9. The asymptotic magnitude–frequency Bode diagrams of a number of systems are described below. Draw the magnitude–frequency and phase–frequency diagrams, and deduce the transfer function for each.
(a) Constant gain of 40 dB up to 2 rad/s, after which the gain falls at the rate of 20 dB/decade.
(b) Constant gain of 6 dB up to 20 rad/s, after which it increases at the rate of 6 dB/octave.
(c) Constant gain of 10 dB up to 5 rad/s, after which it falls at the rate of 6 dB/octave up to 10 rad/s, after which the gain remains constant.
(d) The gain falls at 6 dB/octave up to 4 rad/s, after which it falls at 40 dB/decade. The gain at 1 rad/s is 60 dB.
(e) The gain falls at 40 dB/decade up to 20 rad/s, when it falls at 6 dB/octave up to 2000 rad/s, when it falls again at 12 dB/octave. At 2 rad/s the gain is 60 dB.
(f) The gain falls at 6 dB/octave up to 5 rad/s, after which the reduction in gain becomes 40 dB/decade. At 10 rad/s the fall in gain reduces to 6 dB/octave again until, at 100 rad/s, the gain reduction becomes 40 dB/decade again. The gain at 1 rad/s is 60 dB.
[(a) $10/(1 + 0.5j\omega)$; (b) $2(1 + 0.05j\omega)$;

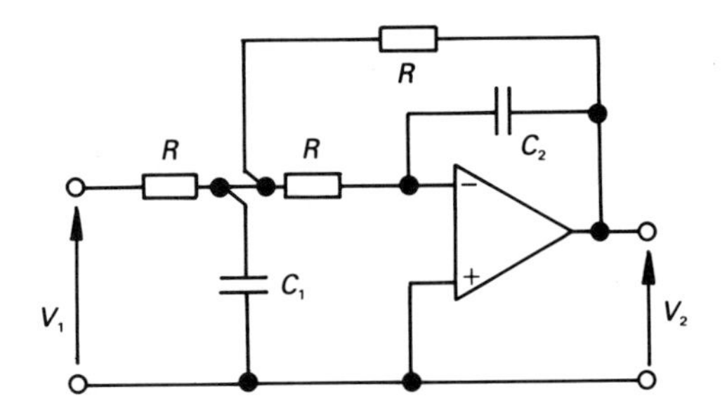

Figure 11.24

(c) $3.162(1 + 0.1j\omega)/(1 + 0.2j\omega)$; (d) $1000/(j\omega(1 + 0.25j\omega))$; (e) $4000(1 + 0.05j\omega)/((j\omega)^2(1 + 0.0005j\omega))$; (f) $1000/(1 + 0.1j\omega)/(j\omega(1 + 0.2j\omega)(1 + 0.01j\omega))$]

11.10. Draw the Bode diagram for each of the following transfer functions: (a) $10(1 + 0.2j\omega)$, (b) $20(1 + 0.02j\omega)^2$, (c) $1/(5\,(1 + 0.2j\omega)^2)$, (d) $10/(j\omega(1 + 0.05j\omega))$, (e) $100(1 + 0.2j\omega)/(j\omega(1 + 0.05j\omega))$, (f) $20(1 + 0.1j\omega)/((j\omega)^2(1 + 0.05j\omega))$]

11.11. Deduce the transfer function $\boldsymbol{H}(s) = \boldsymbol{V}_2(s)/\boldsymbol{V}_1(s)$ for the amplifier in figure 11.24. Plot the Bode diagram for the circuit if $R = 1$ kΩ, $C_1 = 2.388 \times 10^{-3}$ F, $C^2 = 10.613$ μF.
[$1/((R^2C_1C_2)s^2 + 3RC_2s + 1)$; with the values given, the peak response on the gain curve occurs at a frequency of 1 Hz, when the phase shift is 90° and the gain is 14 dB. Note: *the reader may find it advantageous to refer to worked example 11.7.2*

12

Resonance

12.1 Introduction

It was shown earlier that as the frequency increases, inductive reactance increases and capacitive reactance reduces. Because of this change in the magnitude of the reactances, and of the opposing phase angle of the two forms of reactance, there may be a frequency at which X_L and X_C in a given circuit are equal in magnitude and opposite in phase. When this occurs, we reach a condition of *electrical resonance* in the circuit.

Depending on the circuit, this can result in a large current in the circuit and a large voltage across part of the circuit. This is of particular interest to electrical engineers.

The resonance phenomenon is not limited to electrical engineering, and can exist in mechanical systems, as witness high amplitude vibrations at the *natural frequency* of mechanical systems. Perhaps the most famous of these was the destruction of the Tacoma Narrows Bridge in Washington State, USA, when the bridge was forced into a state of resonance by a pulsating gale.

12.2 The resonant condition

Resonance in a two-terminal network containing at least one inductor and one capacitor is defined as the condition which causes the *input impedance at one frequency to be purely resistive*. At this frequency, the voltage across the circuit is in phase with the current through it. In some circuits there may be more than one resonant frequency.

A wide variety of resonant conditions occurs in electrical circuits including series resonance, parallel resonance, resonance with selected frequencies, resonance between magnetically coupled circuits, etc,.

Resonance may or may not be desirable, depending on the circuit in which it occurs.

We have briefly looked at resonance in chapter 11 (see worked examples 11.6.1 and 11.6.2) when we considered the frequency response of idealised parallel and series circuits. In this chapter we will analyse practical circuits which include resistance.

12.3 Series resonance

We will look at a number of points of significant interest to electrical and electronic engineers.

12.3.1 Resonance frequency and frequency response

The impedance of the series *RLC* circuit in figure 12.1 is

$$\mathbf{Z} = R + \mathrm{j}(\omega L - 1/\omega C)$$

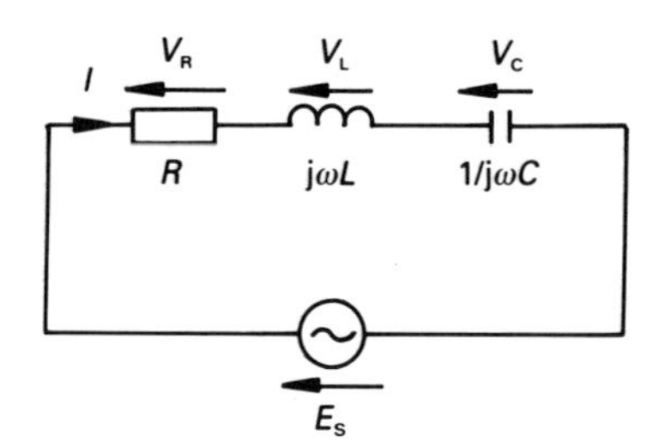

Figure 12.1 *A series* RLC *circuit.*

When $\omega L = 1/\omega C$, that is, when $X_L = X_C$, the impedance of the circuit is purely resistive. This occurs at the *resonant frequency*, ω_0, when $\omega_0 L = 1/\omega_0 C$ or

$$\omega_0 = 1/\sqrt{(LC)} \text{ rad/s}$$

and

$$f_0 = \omega_0/2\pi = 1/(2\pi\sqrt{(LC)}) \text{ Hz}$$

At resonance, the circuit impedance is

$$\mathbf{Z}_0 = R + \mathrm{j}0 = R\angle 0° \ \Omega$$

and is the *minimum impedance* of the circuit. The current at resonance is

$$\mathbf{I}_0 = \mathbf{E}_s/R \text{ A}$$

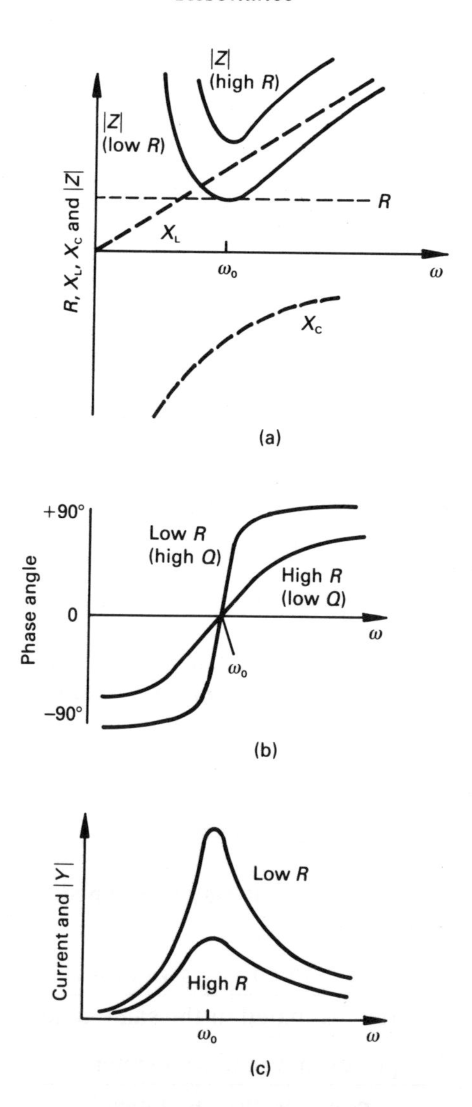

Figure 12.2 *The effect of frequency change in a series circuit on (a) reactance and impedance, (b) phase angle and (c) current and modulus of admittance.*

and is therefore the *maximum current* in the circuit. If R is reduced in value, the current at resonance is increased. At the resonant frequency, the supply voltage is

$$\boldsymbol{E}_s = \boldsymbol{I}_0 R$$

The frequency response of the series circuit is shown in figure 12.2.

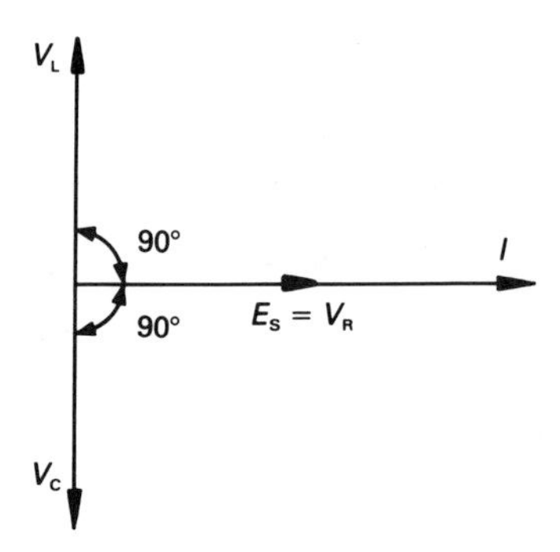

Figure 12.3 *Phasor diagram for a series circuit containing ideal elements at resonance.*

At resonance $X_L = X_C$, so that the magnitude of the voltage across each reactive element is the same at resonance. That is

$$I_0X_L = I_0X_C$$

A phasor diagram for a series circuit containing ideal elements at resonance is shown in figure 12.3.

However, if the coil in the circuit has some resistance (as it will in practice), then *the voltage across the coil at resonance* is not equal to the voltage across the capacitor (see worked example 12.3.2), and is not at 90° to the current.

12.3.2 The quality factor, Q_s, of the series circuit

The *quality factor* of a series circuit may be defined in any one of several ways, including the following

$$Q_S = \frac{\text{energy stored in a given time}}{\text{energy dissipated in the same time}}$$

$$= \frac{\text{power in a reactive element}}{\text{power in the resistance}}$$

A high value of Q_S indicates a low level of energy dissipated for a given energy stored in a reactive element. If we use the inductor as the reactive element, then

$$Q_S = \frac{I^2X_L}{I^2R} = \frac{X_L}{R} = \frac{\omega_0L}{R}$$

or, using the capacitive element

$$Q_S = \frac{I^2X_C}{I^2R} = \frac{X_C}{R} = \frac{1}{\omega_0CR}$$

Also, we may say that

$$Q_s = \frac{X_L}{R} = \frac{|I|X_L}{|I|R} = \frac{\text{modulus of the voltage across } L \text{ at resonance}}{\text{modulus of the voltage across } R \text{ at resonance}}$$

and

$$Q_s = \frac{X_C}{R} = \frac{|I|X_C}{|I|R} = \frac{\text{modulus of the voltage across } C \text{ at resonance}}{\text{modulus of the voltage across } R \text{ at resonance}}$$

or

$$Q_s = \frac{|\text{voltage across either } L \text{ or } C \text{ at resonance}|}{|E_s|}$$

Now, for the series circuit at resonance

$$Q_s = \frac{\omega_0 L}{R} = \frac{1}{\sqrt{(LC)}}\frac{L}{R} = \frac{1}{R}\sqrt{(L/C)}$$

Typically, in a well-designed high-quality (high-Q) communications circuit, the value of Q may be 50 or greater. It has, of course, a much lower value in a power–frequency circuit. In power circuits, where resistance values are much lower than they are in electronic circuits, series resonance can give rise to dangerously high values of current. Moreover, these currents can produce very high values of voltage across the reactive elements, possibly causing damage to the insulation of the elements, and excessive heating in all the elements.

Sometimes the only effective circuit resistance is in the winding of the coil, so that we can refer to the Q-factor of the coil, which is

$$Q_s = \frac{X_L}{R} = \frac{\omega_0 L}{R}$$

Worked example 12.3.1

A series circuit contains a resistor of 4 Ω resistance, a pure inductor of 0.1 mH inductance, and a 1 μF capacitor. Calculate (a) the resonant frequency of the circuit, (b) the net impedance of the circuit at resonance, (c) the current in the circuit at resonance if $\boldsymbol{E_s} = 10\angle 0°$ V, (d) the voltage across each element in the circuit at resonance and (e) the Q-factor of the circuit at resonance.

Solution

(a) The resonant frequency is

$$\begin{aligned} \omega_0 &= 1/\sqrt{(LC)} = 1/\sqrt{(0.1 \times 10^{-3} \times 1 \times 10^{-6})} \\ &= 100\,000 \text{ rad/s or } 15\,915.5 \text{ Hz} \end{aligned}$$

(b) At resonance

$$X_L = \omega L = 100\,000 \times 0.1 \times 10^{-3} = 10\ \Omega$$

and

$$X_C = 1/\omega C = 1/(100\,000 \times 1 \times 10^{-6}) = 10\ \Omega$$

hence the impedance of the circuit at resonance is

$$\boldsymbol{Z_0} = R + \mathrm{j}(X_L - X_C) = 4 + \mathrm{j}(10 - 10) = 4\ \Omega$$

(c) At the resonant frequency

$$\boldsymbol{I_0} = \boldsymbol{E_s}/\boldsymbol{Z_0} = 10\angle 0°/4 = 2.5\angle 0° \text{ A}$$

(d) voltage across each element in the circuit is

$$\begin{aligned} \boldsymbol{V_R} &= \boldsymbol{I_R} = 2.5 \times 4 = 10 \text{ V} = \boldsymbol{E_s} \\ \boldsymbol{V_L} &= \boldsymbol{IZ_L} = 2.5 \times 10\angle 90° = 25\angle 90° \text{ V} \\ \boldsymbol{V_C} &= \boldsymbol{IZ_C} = 2.5 \times 10\angle -90° = 25\angle -90° \text{ V} = -\boldsymbol{V_L} \end{aligned}$$

(e) We can calculate Q_S from any of the equations in section 12.3.2 as follows

$$Q_S = X_L/R = 10/4 = 2.5$$

Note: $|\boldsymbol{V_L}| = Q_S \times |\boldsymbol{E_s}| = 2.5 \times 10 = 25$ V

Worked example 12.3.2

If, in worked example 12.3.1, the whole resistance of the circuit is in the coil, calculate the voltage across the coil at resonance.

Solution

Since all the resistance is in the coil, the impedance of the coil is $(4 + \mathrm{j}10) = 10.77\angle 68.2°\ \Omega$. At the resonant frequency the current is 2.5 A, and the voltage across the coil is

$$\boldsymbol{I_0 Z_{coil}} = 2.5\angle 0° \times 10.77\angle 68.2° = 26.93\angle 68.2° \text{ V}$$

That is, the voltage across the coil at resonance does not lead the current by 90°; this is due to the effect of the resistance of the coil. However, the quadrature voltage across the coil is $25\angle 90°$ V, which is of opposite phase angle but equal magnitude to the voltage across the capacitor.

12.3.3 The peak voltage across **R, L** *and* **C**

One would imagine from the foregoing that, since the current rises to its peak value at ω_0, the p.d. across R, L, and C would peak at this frequency. However, it is shown in the following that this is not quite the case.

Since the voltage across R is proportional to the current then, in fact, $\boldsymbol{V}_\mathbf{R}$ does peak at ω_0. However, since the capacitive reactance commences at a high value and reduces with increasing frequency, I rises at a faster rate than X_C falls. It can be shown that $|\boldsymbol{V}_C|$ peaks just before resonance, and $|\boldsymbol{V}_L|$ peaks at a frequency just above resonance (the reader will find it an interesting exercise to verify this fact). The two voltages are equal in value at ω_0.

For values of Q_S greater than about 5, the frequency difference at which the peak voltage occurs across R, L and C is generally insignificant.

12.3.4 Bandwidth and selectivity of a series **RLC** *circuit*

If we plot the variation in power consumed in a series RLC circuit to a base of frequency, we get a graph of the type in figure 12.4.

At frequencies ω_1 and ω_2, the power consumed is one-half the maximum power consumed (which occurs at the resonant frequency), and these frequencies are known as the *half-power frequencies* or *half-power points*.

The sharpness of the peak in the resonance curve is described by the *bandwidth*, B, and is given by

$$B = \omega_2 - \omega_1$$

The smaller the bandwidth, the 'sharper' the amplitude response. The values of ω_1 and ω_2 can be calculated as follows. At a half-power frequency

$$\text{power consumed} = \frac{P_0}{2} = \frac{|I_0|^2 R}{2} = \left[\frac{I_0}{\sqrt{2}}\right]^2 R = \left[\frac{E_S}{R\sqrt{2}}\right]^2 R$$

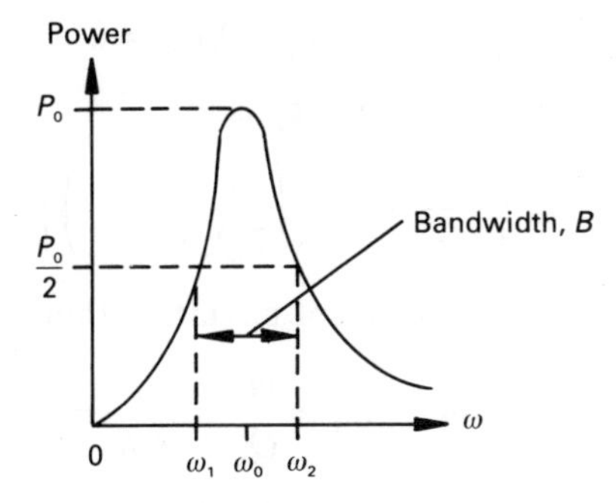

Figure 12.4 *Variation in power with frequency in a series* RLC *circuit.*

That is, the modulus of the impedance of the circuit at a half-power frequency is

$$|\mathbf{Z}| = R\sqrt{2}\ \Omega$$

The impedance of the circuit at any frequency is

$$\begin{aligned}\mathbf{Z} &= R + \mathrm{j}(\omega L - 1/\omega C)\\ &= R + \mathrm{j}\left[\omega L \times \frac{\omega_0 L}{\omega_0 L} - \frac{1}{\omega C} \times \frac{\omega_0 C}{\omega_0 C}\right]\\ &= R + \mathrm{j}\left[\frac{\omega}{\omega_Q}\omega_0 L - \frac{\omega_0}{\omega}\,\frac{1}{\omega_0 C}\right]\\ &= R + \mathrm{j}\left[\frac{\omega}{\omega_0}RQ_\mathrm{S} - \frac{\omega_0}{\omega}RQ_\mathrm{S}\right] = R\left[1 + \mathrm{j}Q_\mathrm{S}\left(\frac{\omega}{\omega_0} - \frac{\omega_0}{\omega}\right)\right]\end{aligned}$$

Since $|\mathbf{Z}| = R\sqrt{2}$ at a half-power frequency, it is clear that at a half-power frequency the quadrature component must have a value of unity, or

$$Q_\mathrm{S}\left(\frac{\omega}{\omega_0} - \frac{\omega_0}{\omega}\right) = Q_\mathrm{S}\left(\frac{\omega^2 - \omega_0^2}{\omega_0\omega}\right) = \pm 1$$

In the case where its value is −1 then

$$Q_\mathrm{S}\left(\frac{\omega^2 - \omega_0^2}{\omega_0\omega}\right) = -1 \quad \text{or} \quad Q_\mathrm{S}\omega^2 + \omega_0\omega - Q_\mathrm{S}\omega_0^2 = 0$$

Solving the quadratic equation for ω gives a positive value and a negative value. Selecting the positive value of ω gives the half-power frequency

$$\omega_1 = \omega_0\left\{-\frac{1}{2Q_\mathrm{S}} + \sqrt{\left(\left[\frac{1}{2Q_\mathrm{S}}\right]^2 + 1\right)}\right\}$$

In the case where its value is −1, we get

$$Q_\mathrm{S}\omega^2 - \omega_0\omega - Q_\mathrm{S}\omega_0^2 = 0$$

and the resulting half-power frequency is

$$\omega_2 = \omega_0\left\{\frac{1}{2Q_\mathrm{S}} + \sqrt{\left(\left[\frac{1}{2Q_\mathrm{S}}\right]^2 + 1\right)}\right\}$$

The bandwidth of the series resonant circuit is, therefore

$$B = \omega_2 - \omega_1 = \frac{\omega_0}{Q_\mathrm{S}} = \frac{1}{CR} = \frac{R}{L}$$

The *selectivity* of a circuit describes the sharpness of the response curve, and defines the ability of the circuit to discriminate in favour of frequencies near resonance. Its value is equal to the Q-factor of the circuit, and is

$$\text{selectivity} = Q = \omega_0/B$$

Worked example 12.3.3

Calculate the half-power frequencies and the bandwidth of the series circuit in worked example 12.3.1. Determine also complex impedance (a) 10 per cent below and (b) 20 per cent above the resonant frequency.

Solution

The reader is reminded, in worked example 12.3.1, that $\omega_0 = 100\ 000$ rad/s and $Q_S = 2.5$. From the foregoing

$$\omega_1 = \omega_1\left[-\frac{1}{2Q_S} + \sqrt{\left(\left[\frac{1}{2Q_S}\right]^2 + 1\right)}\right]$$

$$= 100\ 000\left[-\frac{1}{2 \times 2.5} + \sqrt{\left(\left[\frac{1}{2 \times 2.5}\right]^2 + 1\right)}\right]$$

$$= 100\ 000\ [-0.2 + \sqrt{1.04}] = 81\ 980 \text{ rad/s}$$

and

$$\omega_2 = \omega_0\left[\frac{1}{2Q_S} + \sqrt{\left(\left[\frac{1}{2Q_S}\right]^2 + 1\right)}\right]$$

$$= 100\ 000\ [0.2 + \sqrt{1.04}] = 121\ 980 \text{ rad/s}$$

hence

$$B = \omega_2 - \omega_1 = 40\ 000 \text{ rad/s}$$

or

$$B = \omega_0/Q = 100\ 000/2.5 = 40\ 000 \text{ rad/s}$$

The reader should note at this point that the resonant frequency is not midway between the two half-power frequencies (in fact, it is the geometric mean of the two frequencies, that is $\omega_0 = \sqrt{(\omega_1\omega_2)}$).

The complex impedance at any frequency can be calculated from the expression

$$\mathbf{Z} = R\left[1 + jQ_S\left(\frac{\omega}{\omega_0} - \frac{\omega_0}{\omega}\right)\right]$$

(a) *When* $\omega = 0.9\omega_0$

$$\mathbf{Z} = 4\left[1 + j2.5\left(\frac{0.9\omega_0}{\omega_0} - \frac{\omega_0}{0.9\omega_0}\right)\right]$$

$$= 4[1 + j2.5 \times (-0.2111)] = 4.52\angle{-27.82°}\ \Omega$$

(b) *When* $\omega = 1.2\omega_0$

$$\mathbf{Z} = 4\left[1 + j2.5\left(\frac{1.2\omega_0}{\omega_0} - \frac{\omega_0}{1.2\omega_0}\right)\right] = 5.44\angle 42.5°\ \Omega$$

12.4 Parallel resonance

In this case, the attention of the reader is directed towards circuits containing G, L and C in parallel.

12.4.1 *Resonant frequency and frequency response of an ideal parallel circuit*

An ideal parallel circuit contains a pure inductor, a pure capacitor and a pure conductance in parallel, as shown in figure 12.5. The admittance of the circuit is

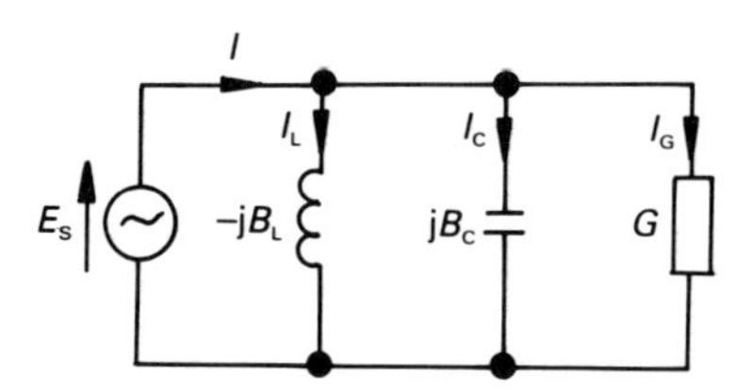

Figure 12.5 *Parallel circuit containing ideal elements.*

$$\mathbf{Y} = G + j(B_C - B_L)$$
$$= G + j\left(\omega C - \frac{1}{\omega L}\right)\ \text{S}$$

Once again, the *circuit is resonant when the reactive element of the admittance is zero, which occurs at the resonant frequency* ω_0, when $\omega_0 C = 1/\omega_0 L$, that is

$$\omega_0 = 1/\sqrt{(LC)}\ \text{rad/s} \quad \text{or} \quad f_0 = 1/(2\pi\sqrt{(LC)})\ \text{Hz}$$

Since, at resonance, the reactive element of the admittance is zero, the current at resonance is

$$\mathbf{I}_0 = \mathbf{I}_G = \mathbf{E}_S G$$

Clearly, G can have any value, including zero (corresponding to an infinite resistance shunting the circuit), so that *the current at resonance represents a minimum value of current.*

Since, at resonance, $Y_0 = G$, then the circuit impedance at resonance is

$$Z_0 = 1/G$$

The value of Z_0 is known as the *dynamic impedance* or *dynamic resistance* of the circuit.

At resonance, the parallel LC section of the ideal parallel circuit (known as a *tank circuit*), behaves as though it were an open-circuit, that is, no current is drawn from the supply. The reason for this is that the magnitude of the current in both the L and the C branches has the same value (this is because they both have the same value of reactance at resonance), but are 180° out of phase with one another.

The frequency response for the circuit is shown in figure 12.6, and the phasor diagram at resonance is in figure 12.7

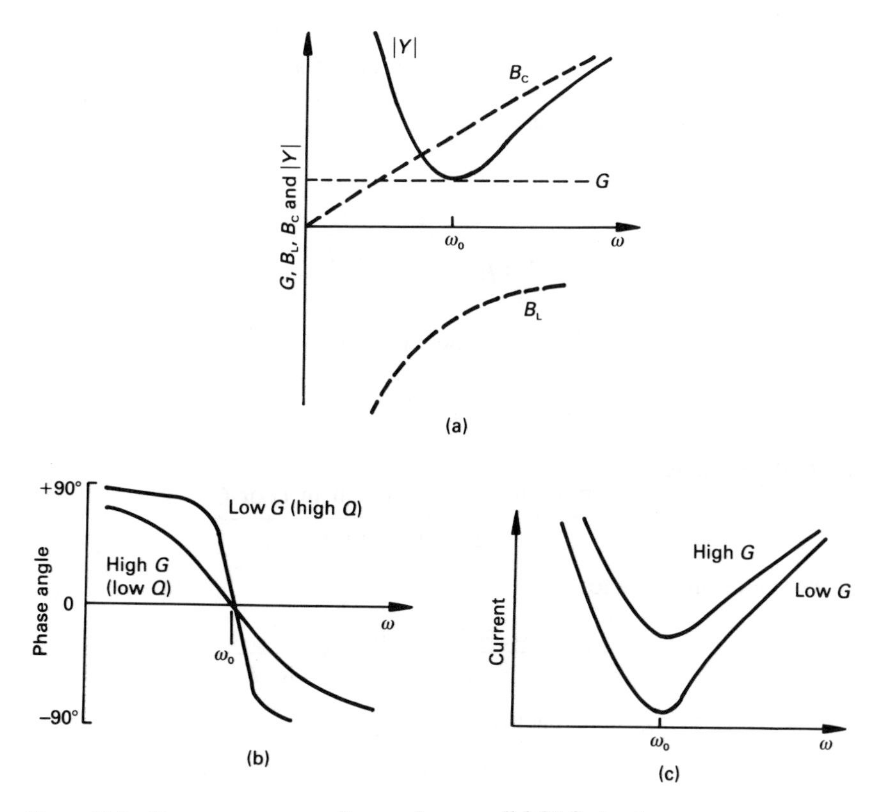

Figure 12.6 *Frequency response diagram for a parallel* GLC *circuit.*

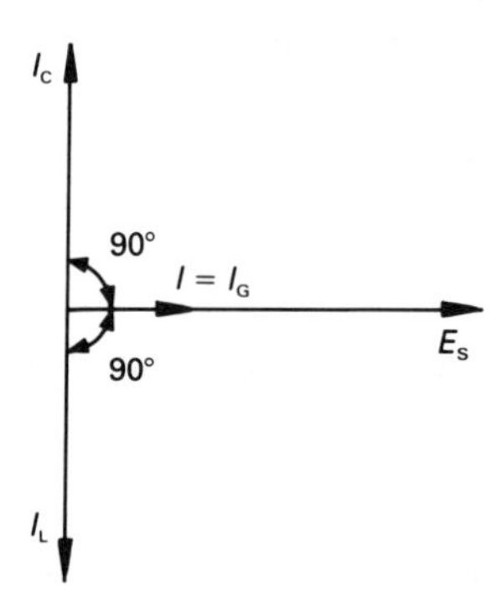

Figure 12.7 *Phasor diagram for a parallel* GLC *circuit at resonance.*

12.4.2 The quality factor, Q_P of an ideal parallel circuit

As with the series circuit, the Q-factor is defined as the ratio of the energy stored at resonance in one of the reactive elements to the energy dissipated in the circuit (which is in the shunting conductance). Alternatively, it is the ratio of the reactive power in a reactive element to the power dissipated in the shunting conductance. Selecting the inductive branch gives

$$Q_P = \frac{I_L^2 X_L}{I_R^2 R} = \frac{\dfrac{E_S^2}{X_L^2} X_L}{\dfrac{E_S^2}{R^2} R} = \frac{R}{X_L} = \frac{R}{\omega_0 L} = \frac{1}{G\omega_0 L}$$

where $R = 1/G$. Similarly, it may be shown that

$$Q_P = \omega_0 CR = R\sqrt{(C/L)} = \frac{\sqrt{(C/L)}}{G}$$

Also at resonance

$$Q_P = \frac{\text{modulus of the current in } L \text{ or } C}{\text{supply current}}$$

If we select the inductor

$$Q_P = \frac{E_S/\omega_0 L}{E_S/R} = \frac{R}{\omega_0 L} = \frac{1}{G\omega_0 L}$$

Similarly, selecting the capacitive branch gives

$$Q_P = \omega_0 CR$$

12.4.3 Bandwidth and selectivity of a parallel circuit

As with the series circuit, the ideal parallel circuit consumes maximum power at resonance (the reader will find it an interesting exercise to argue this point both quantitatively and qualitatively), and the bandwidth is equal to the band of frequencies between the two half-power frequencies.

The admittance of the parallel circuit at any frequency is

$$Y = \frac{1}{R} + \mathrm{j}\left(\omega C - \frac{1}{\omega L}\right) = \frac{1}{R}\left[1 + \mathrm{j}Q_P\left(\frac{\omega}{\omega_0} - \frac{\omega_0}{\omega}\right)\right]$$

where $G = 1/R$, and at each half-power frequency $|Y| = \sqrt{2}/R$, which occurs when the quadrature term has unity value, that is $Q_P\left(\frac{\omega}{\omega_0} - \frac{\omega_0}{\omega}\right) = \pm 1$. Considering the case for $+1$ and -1 separately, we obtain the two half-power frequencies

$$\omega_1 = \omega_0\left[-\frac{1}{2Q_P} + \sqrt{\left(\left[\frac{1}{2Q_P}\right]^2 + 1\right)}\right]$$

$$\omega_2 = \omega_0\left[\frac{1}{2Q_P} + \sqrt{\left(\left[\frac{1}{2Q_P}\right]^2 + 1\right)}\right]$$

The reader should note that ω_1 and ω_2 are given by the same formulas as for the series circuit (see section 12.3.4). The bandwidth of the circuit is

$$B = \omega_2 - \omega_1 = \omega_0/Q_P = \omega_0 X_L/R = \omega_0 X_C/R$$

also

$$\text{selectivity} = Q_P$$

12.4.4 Resonance in a practical parallel circuit

A simple practical parallel circuit has two branches (see figure 12.8(a)), each containing a reactive element and some resistance (the resistance in the capacitive branch may, alternatively, be shown as a leakage resistance in parallel with the capacitor). A typical phasor diagram at resonance is shown in figure 12.8(b).

If we convert the circuit in figure 12.8(a) into the equivalent ideal circuit in figure 12.8(c), we can use the equations developed earlier to determine various factors associated with the circuit.

Initially we will convert the capacitive branch of figure 12.8(a) into its equivalent parallel circuit. The admittance of this branch at frequency ω is

$$Y_1 = \frac{1}{R_C + 1/\mathrm{j}\omega C} = \frac{\mathrm{j}\omega C}{1 + \mathrm{j}\omega C R_C} = \frac{\mathrm{j}\omega C(1 - \mathrm{j}\omega C R_C}{1 + (\omega C R_C)^2}$$

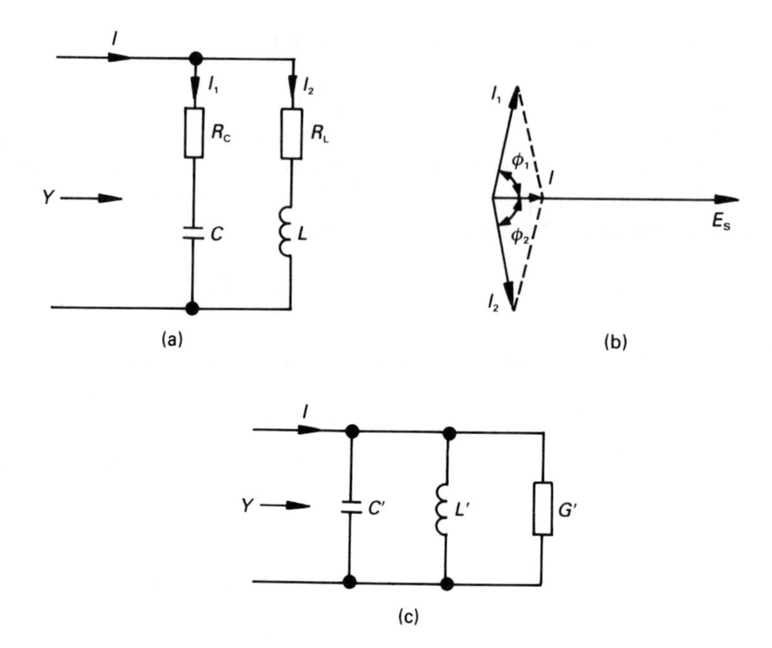

Figure 12.8 *(a) A practical parallel circuit, (b) a typical phasor diagram at resonance and (c) the equivalent ideal parallel circuit.*

$$= \frac{R_C(\omega C)^2}{1 + (\omega CR_C)^2} + \mathrm{j}\frac{\omega C}{1 + (\omega CR_C)^2} = G'_C + \mathrm{j}\omega C'$$

where $C' = C/[1 + (\omega CR_C)^2]$ and G'_C is the equivalent conductance in parallel with C'. In the case of the inductive branch of figure 12.8(a) we get

$$\boldsymbol{Y}_2 = \frac{1}{R_L\mathrm{j} + \mathrm{j}\omega L} = \frac{R_L - \mathrm{j}\omega L}{R_L{}^2 + \omega^2 L^2}$$

$$= \frac{R_L}{R_L^2 + \omega^2 L^2} - \mathrm{j}\frac{\omega L}{R_L^2 + \omega^2 L^2} = G'_L - \mathrm{j}\frac{1}{\omega L'}$$

where $L' = (R_L^2 + \omega^2 L^2)/\omega^2 L$, and G'_L is the equivalent conductance in parallel with L'.

The equivalent ideal parallel circuit in figure 12.8(c) comprises the two above branches in parallel, having an effective conductance of

$$G' = G'_C + G'_L = \frac{R_C(\omega C)^2}{1 + (\omega CR_C)^2} + \frac{R_L}{R_L^2 + \omega^2 L^2}$$

We can therefore say that, for the circuit in figure 12.8(a), the resonant frequency is

$$\omega_0 = 1/\surd(L'C') = \frac{1}{\surd(LC)}\sqrt{\left[\frac{L/C - R_L^2}{L/C - R_C^2}\right]} \text{ rad/s}$$

In the more usual case where $R_C = 0$, the resonant frequency is

$$\omega_0 = \frac{1}{\surd(LC)}\sqrt{\left[\frac{L - CR_L^2}{L}\right]} \text{ rad/s}$$

and if, at the same time $R_L^2 << (\omega_0 L)^2$ (or $Q_L >> 1$), the dynamic impedance at resonance is

$$Z_0 = L/CR_L$$

Worked example 12.4.1

A transistor amplifier with an output resistance of 50 kΩ is connected to a load comprising a two-branch parallel circuit which is resonant at 60 kHz; the parallel circuit contains a coil of inductance 1mH and resistance 15 Ω. If the bandwidth of the complete circuit is to be 3 kHz, calculate the other components in the circuit (assume that the capacitor has no leakage resistance).

Solution

The Q-factor of the complete circuit is

$$Q = f_0/R = 60\ 000/3000 = 20$$

Since the output impedance of the amplifier is 50 kΩ, it is effectively connected across the resonant circuit as shown in figure 12.9(a). The idealised equivalent circuit is shown in figure 12.9(b), and we will calculate the value of L' and C' in the following. From the work in the section above

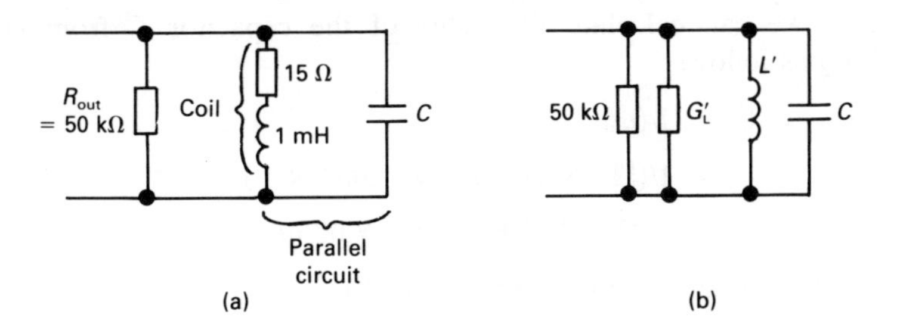

Figure 12.9 *(a) The equivalent output circuit for worked example 12.4.1 and (b) the idealised circuit.*

$$L' = (R_L^2 + \omega_0^2 L^2)/\omega_0^2 L)$$

$$= \frac{(15^2 + [2\pi \times 60\,000]^2 \times 1 \times 10^{-6})}{(2\pi \times 60\,000)^2 \times 1 \times 10^{-3}}$$

$$= 1.002 \times 10^{-3}\text{ H} \quad \text{or} \quad 1.002\text{ mH}$$

or, at the resonant frequency

$$X_L' = \omega_0 L' = 377.75\ \Omega$$

and

$$G_L' = R_L/(R_L^2 + \omega_0^2 L^2)$$

$$= 15/(15^2 + (2\pi \times 60\,000)^2 \times 1 \times 10^{-6})$$

$$= 1.0537 \times 10^{-4}\text{ S} \quad \text{or} \quad 0.10537\text{ mS}$$

The effective conductance shunting the $L'C$ section of the circuit is

$$G'' = G_L' + 1/50\,000 = 1.2537 \times 10^{-4}\text{ S}$$

hence the Q-factor for the curcuit in figure 12.9(b) is

$$Q = 1/(G''X_L') = 1/(1.2537 \times 10^{-4} \times 377.75)$$

$$= 21.12$$

This is larger than the required value of 20 (see the beginning of the solution), and we can reduce the overall Q-factor to the required value by shunting the circuit with another conductance. The total conductance needed to reduce the Q-factor to 20 is

$$G = 1/QX_L') = 1/(20 \times 377.75) = 1.324 \times 10^{-4}\text{ S}$$

hence the extra shunting conductance is

$$G - G'' = (1.324 - 1.2537) \times 10^{-4} = 0.0703 \times 10^{-4}\text{ S}$$

which corresponds to a shunting resistance of 142.25 kΩ.

Finally we can calculate the value of the capacitor C from $\omega_0 = 1/\sqrt{(L'C)}$ as follows

$$C = 1/(\omega_0^2 L')$$

$$= 1/([2\pi \times 60\,000]^2 \times 1.002 \times 10^{-3}]$$

$$= 7.02 \times 10^{-9}\text{ F} \quad \text{or} \quad 7.02\text{ nF}$$

The resulting circuit is shown in figure 12.10.

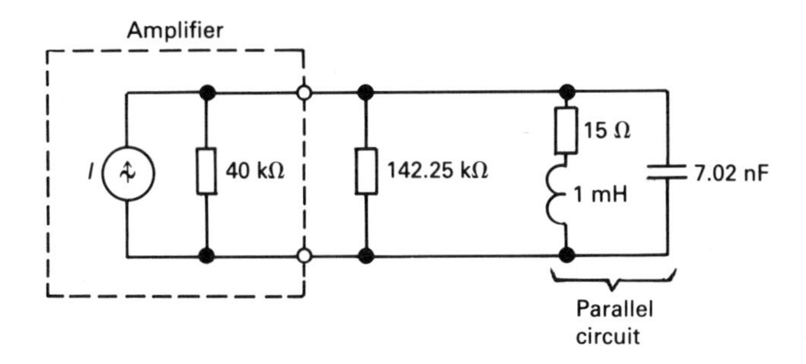

Figure 12.10 *The complete circuit for worked example 12.4.1.*

12.5 Scaling electrical circuits

When analysing a particular circuit, the values used may be difficult to handle or to visualise. It may, for example, be easier to think in terms of a resonant frequency of 1 rad/s rather than 10367.6 rad/s, or an impedance level of 1 Ω rather than 579.7 Ω. Using standard equations, we can 'scale' a circuit so that it appears, initially, to have relatively simple values; after initial calculations, we can return the circuit to its original form by the inverse of the scaling process.

Scaling is achieved in two stages, namely impedance scaling and frequency scaling as follows.

12.5.1 Impedance or magnitude scaling

To magnitude-scale an impedance by a factor K_m, we need to multiply the *impedance* of the element by a scaling factor K_m. Thus a resistor of R Ω is scaled by K_m to a value $K_m R$ Ω. An L H inductance of impedance $j\omega L$ is scaled to an impedance of $j\omega K_m L$ Ω; that is, the scaled value of inductance is $K_m L$ H. A capacitor of C F capacitance has an unscaled impedance of $1/j\omega C$ Ω, and this is scaled to $K_m/j\omega C$ Ω; that is, the scaled capacitance is C/K_m. Thus, the magnitude scaling changes are

$$R \text{ becomes } K_m R$$
$$L \text{ becomes } K_m L$$
$$C \text{ becomes } C/K_m$$

Similarly, it can be shown that voltage values are multiplied by K_m, and current values are divided by K_m. Ratios of similar quantities, such as the Q-factor of a resonant circuit, are unchanged by magnitude scaling. Other non-impedance factors, such as resonant frequency, are unchanged by impedance or magnitude scaling.

Consider the following values from worked example 12.3.1: $R = 4\ \Omega$, $L = 0.1$ mH, $C = 1\mu$F. If we magnitude scale these by a factor of $K_m = 10^4$, then the scaled values are $R' = 4 \times 10^4\ \Omega$, $L' = 1$ H, $C' = 1 \times 10^{-10}$ F. The original values refer to a series circuit having a resonant frequency of 100 000 rad/s, and a Q-factor of 2.5. The corresponding values for the scaled circuit are

$$\begin{aligned}\omega_0' &= 1/\surd(L',C') = 1/\surd(1 \times 1 \times 10^{-10}) \\ &= 100\ 000 \text{ rad/s}\end{aligned}$$

and

$$Q_S' = \omega_0' L'/R' = 100\ 000 \times 1/(4 \times 10^4) = 2.5$$

both of which are unchanged by magnitude scaling.

12.5.2 Frequency scaling

In this case, each frequency-dependent function is multiplied by a scaling factor K_f; for example, the scaled frequency ω becomes $K_f\omega$. Consequently, the resonant frequency and bandwidth of a tuned circuit are increased by K_f.

Clearly, the resistance of a resistor is independent of frequency, and its value is unchanged by frequency scaling. In the case of an inductor, its reactance is proportional to frequency, so that the frequency-scaled inductive impedance is related to the original value by

$$j\omega L = jK_f\omega L'$$

where L' is the frequency-scaled inductance; that is, the effective frequency-scaled inductance is $L' = L/K_f$. Similarly, a frequency-scaled capacitance C' is C/K_f. That is, the frequency scaling changes are

$$\begin{aligned}&R \text{ becomes } R \\ &L \text{ becomes } L/K_f \\ &C \text{ becomes } C/K_f\end{aligned}$$

Since $X_L = \omega L = K_f\omega \times L/K_f = \omega$L (and similarly for a capacitive reactance), impedance levels are unchanged by frequency scaling. Following this line of discussion, it will be seen that the Q-factor of a resonant circuit is unchanged by frequency scaling.

Suppose we were to frequency scale the values in worked example 12.3.1 by $K_f = 10^{-5}$, then

$$L' = L/K_f = 0.1 \times 10^{-3}/10^{-5} = 10 \text{ H}$$

$$C' = C/K_f = 10^{-6}/10^{-5} = 0.1 \text{ F}$$

and

$$\omega_0' = 1/\sqrt{(L'C')} = 1/\sqrt{(10 \times 0.1)} = 1 \text{ rad/s}$$

That is, the resonant frequency has been multiplied by $K_f = 10^{-5}$. The Q-factor of the circuit is

$$Q' = X_L'/R' = \omega_0' L'/R' = 1 \times 10/4 = 2.5$$

which is unchanged.

12.5.3 Combined scaling

Many problems involve both magnitude and frequency scaling, and the result involved both scaling factors. For R, L and C the effect is as follows

R becomes $K_m R$
L becomes $K_m L/K_f$
C becomes $C/K_m K_f$

12.5.4 Universal resonant circuits

Universal resonant circuits are scaled circuits with $R = 1\ \Omega$ and $\omega_0 = 1$ rad/s, that is, $LC = 1$. All conventional passive resonant circuits, that is, series RLC and parallel GLC circuits, can be designed from this type of circuit.

Worked example 12.5.1

Calculate the impedance and frequency scaling factors for a parallel resonant circuit, in which the resonant frequency is 10 000 rad/s, and the LC section is shunted by a 10 kΩ resistor.

Solution

The resistance R′ which shunts the resonant section is

$$R' = K_m R = K_m \times 1 = 10\,000$$

hence $K_m = 10\,000$

Also $\omega_0' = K_f \omega_0 = 10\,000 \times 1 = 10\,000$

that is $K_f = 10\,000$

Worked example 12.5.2

A series *RLC* circuit containes the following elements: $R = 100\ \Omega$, $L = 1.6$ mH and $C = 0.01\ \mu$F. Scale the circuit so that it contains a 2 H inductor and a 2 F capacitor.

Solution

From the above

$$L' = 2 = 1.6 \times 10^{-3}\ K_m/K_f$$

$$C' = 2 = 0.01 \times 10^{-6}/K_m K_f$$

Multiplying the two together gives

$$L'C' = 4 = 1.6 \times 10^{-11}/K_f^2$$

or

$$K_f = 2 \times 10^{-6}$$

From the expression for L' we get

$$K_m = 2K_f/1.6 \times 10^{-3} = 2.5 \times 10^{-3}$$

hence

$$R' = K_m R = 100K_m = 0.25\ \Omega$$

$$L' = K_m L/K_f = 1.6 \times 10^{-3} K_m/K_f = 2\ \text{H}$$

$$C' = C/K_m K_f = 0.01 \times 10^{-6}/K_m K_f = 2\ \text{F}$$

12.6 Passive and active filters

Passive series- and parallel-resonant circuits can be used in both band-pass and band-stop filters. A *band-pass filter* is a circuit which will transmit frequencies within a designated range known as the *pass-band*, and attenuates other frequencies. A *band-stop filter* attenuates frequencies within the *stop-band*, and allows other frequencies to be transmitted.

Both *passive* and *active* filters are widely used. A passive filter is one using only *R* (or *G*), *L* and *C* elements.

12.6.1 Passive band-pass filter

The basic characteristic of a band-pass filter is shown in figure 12.11(a). This characteristic can be achieved by the series circuit in figure 12.11(b) which, at resonance gives $V_0 = V_1$ (that is if the resistance of the coil is small compared with the value of R).

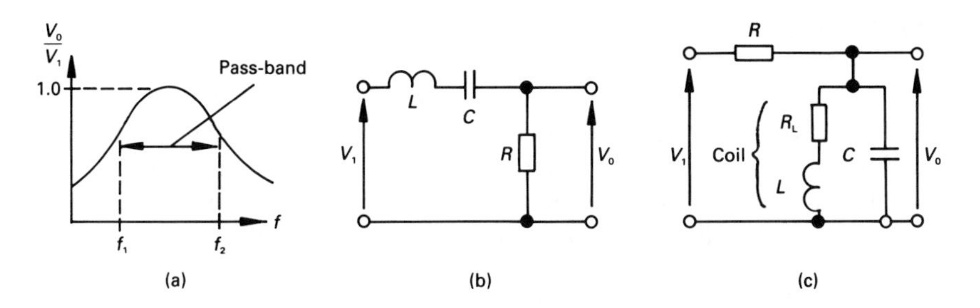

Figure 12.11 *(a) A band-pass filter characteristic, (b) and (c) simple passive circuits.*

The parallel circuit in figure 12.11(c) gives a similar frequency response since, at resonance, the dynamic resistance of the parallel circuit is much higher than the resistance of R.

In fact, at resonance, both circuits have an output resistance of R Ω (approximately). The frequencies f_1 and f_2 are the *cut-off frequencies* of the filter.

12.6.2 Passive band-stop filter

The general characteristic of this type of filter is shown in figure 12.12(a). The series circuit in figure 12.12(b) can achieve this characteristic since, at resonance, the net voltage, V_0, across L and C is zero (or nearly so); this, of course, assumes that the resistance of the coil is small compared with that of R.

Similarly, the parallel circuit in figure 12.12(c) gives a similar characteristic because, at resonance, the impedance of the parallel circuit is very high (ideally infinite) when compared with R (which is low).

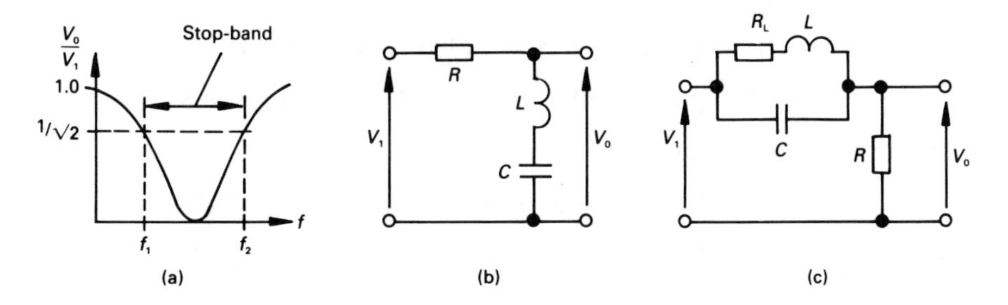

Figure 12.12 *(a) Band-stop filter characteristic, (b) and (c) simple passive circuits.*

12.6.3 Active RC *filters*

These are a class of filters utilising operational amplifiers as active elements in conjunction with passive resistors and capacitors. They can be used to provide almost any filter characteristic, that is, low-pass, high-pass, band-pass, band-stop, etc. and are the subject of many specialised texts.

We will not dwell on their operation here, and it is sufficient to say that we have already analysed a circuit which is the basis of many forms of active filter. The reader should refer to worked example 11.7.2 and problem 11.11 for this type of circuit.

12.7 Selective resonance

A periodic voltage which is non-sinusoidal can be shown mathematically to contain sinusoids of many frequencies (see chapter 13 for details). It may happen that a circuit produces *selective resonance* with one of the frequencies, which may result in a large value of voltage or current at the selected frequency in the circuit.

This is one reason why supply authorities discourage consumers from connecting apparatus which generate high values of harmonic voltage or current as, for example, may occur when high-power machines are controlled by certain types of thyristor-controlled equipment.

Worked example 12.7.1

A voltage of $100(\sin \omega t + \sin 2\omega t + \sin 4\omega t)$ is applied to a series circuit comprising a resistance of 10 Ω, a capacitor of 10 μF, and a 0.625 H inductance. If $\omega = 200$ rad/s, calculate the current in the circuit due to each harmonic frequency.

Solution

For the 200 rad/s frequency

$$\begin{aligned} \mathbf{Z}_{200} &= 10 + \mathrm{j}(200 \times 0.625 - 1/(200 \times 10 \times 10^{-6})) \\ &= 10 + \mathrm{j}(125 - 500) = 10 - \mathrm{j}375 = 375.1\angle{-88.5^\circ}\ \Omega \end{aligned}$$

For the 400 rad/s frequency

$$\begin{aligned} \mathbf{Z}_{400} &= 10 + \mathrm{j}(400 \times 0.625 - 1/(400 \times 10 \times 10^{-6})) \\ &= 10 + \mathrm{j}(250 - 250) = 10 - \mathrm{j}0 = 10\angle 0^\circ\ \Omega \end{aligned}$$

For the 800 rad/s frequency

$$Z_{800} = 10 + j(800 \times 0.625 - 1/(800 \times 10 \times 10^{-6}))$$
$$= 10 + j(500 - 125) = 10 + j375 = 375.1\angle 88.5° \ \Omega$$

Hence the current for the 200 rad/s frequency component is

$$I_{200} = 100/375.1\angle -88.5° = 0.267\angle 88.5° \ \text{A}$$

The current for the 400 rad/s frequency component is

$$I_{400} = 100/10\angle 0° = 10\angle 0° \ \text{A}$$

and for the 800 rad/s frequency component is

$$I_{800} = 100/375.1\angle 88.5° = 0.267\angle -88.5° \ \text{A}$$

In this case, the circuit selectively resonates with the second harmonic frequency component, and passes a much larger current at that frequency.

12.8 Tuned coupled circuits

Many communications circuits contain a tuned band-pass filter comprising two parallel circuits, both tuned to the same frequency, in which the inductive elements in the two circuits are magnetically coupled.

The general arrangement of the coupled circuit is as shown in figure 12.13(a), and a corresponding equivalent circuit is in diagram (b). The individual resistance of each circuit is shown but, in practice, the value of R_1 and R_2 are very much lower than the circuit reactance at the operating frequency. We can, to simplify calculations and without significant loss of accuracy, ignore the two resistance values at this stage. In this case, the mesh currents are

$$0 = j(\omega L_1 - 1/\omega C_1)I_1 - j\omega M I_2$$
$$0 = -j\omega M I_1 + j(\omega L_2 - 1/\omega C_2)I_2$$

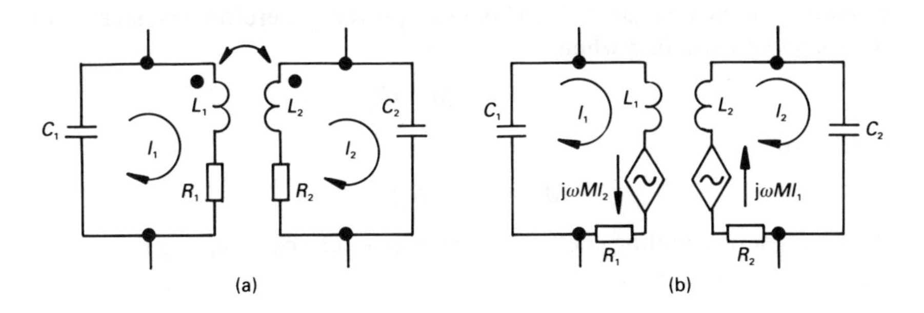

Figure 12.13 *(a) Inductively coupled tuned circuits, (b) an equivalent circuit.*

Substituting the value of I_2 from the second equation into the first equation gives

$$\left((\omega L_1 - 1/\omega C_1) - \frac{\omega^2 M^2}{\omega L_2 - 1/\omega C_2}\right) I_1 = 0$$

and since I_1 is not zero, then

$$(\omega L_1 - 1/\omega C_1)(\omega L_2 - 1/\omega C_2) - \omega^2 M^2 = 0$$

Since each circuit has the same resonant frequency, it follows that

$$\frac{1}{\omega_0^2} = L_1 C_1 = L_2 C_2$$

giving, after some manipulation

$$\frac{1}{\omega^4} - \frac{2}{\omega^2 \omega_0^2} + \frac{1}{\omega_0^4}(1 - k^2) = 0$$

where $k = M/\surd(L_1 L_2)$, and is the magnetic coupling coefficient between the two inductors. The above equation is a quadratic in $(1/\omega^2)$, and solving gives

$$\omega = \omega_0^2/\surd(1 \pm k)$$

That is to say, there are two resonant peaks in the response, one being at frequency $\omega_0/\surd(1 + k)$ and the other at $\omega_0/\surd(1 - k)$.

Critical coupling

When maximum power is transferred in figure 12.13(a) from the primary winding to the secondary winding, we say that *critical coupling* has been achieved between the two circuits. If the mutual inductance needed to give this coupling is M_C then, at this coupling, the effective impedance of the secondary circuit at resonance is R_2, and this impedance referred to the primary winding is $(\omega_0 M_C)^2/R_2$. Maximum power is therefore transferred to the secondary winding when

$$R_1 = (\omega_0 M_C)^2/R_2$$

that is when

$$M_C = \surd(R_1 R_2)/\omega_0$$

If the magnetic coupling coefficient which gives critical coupling is k_C, then $M_C = k_C \surd(L_1 L_2)$, hence

$$k_C = M_C/\surd(L_1 L_2) = \sqrt{\left(\frac{R_1}{\omega_0 L_1} \frac{R_2}{\omega_0 L_2}\right)} = 1/\surd(Q_1 Q_2)$$

Unworked problems

12.1. A series *RLC* circuit contains R = 20 Ω, L = 0.1 H and C = 0.1 μF. Calculate the resonant frequency, the Q-factor, the bandwidth and the half-power frequencies.
[10 000 rad/s; 50; 200 rad/s; 9900.5 rads/s; 10 100.5 rad/s]

12.2. In problem 12.1, if the applied voltage is 100 mV, calculate the current in the circuit and the voltage across C at (a) $0.8\omega_0$, (b) $1.3\omega_0$.
[(a) $2.22 \times 10^{-4}\angle 87.5°$ A; $0.28\angle -2.5°$ V; (b) $1.88 \times 10^{-4}\angle -87.8°$ A; $0.145\angle -177.8°$ V]

12.3. If, in the circuit problem in 12.1, the resistor is shunted by a dependent current source of $0.02V_L$, where V_L is the voltage across the inductor, calculate the new value of resonant frequency. The current in the dependent source flows in the same direction as that in the original circuit.
[12 910 rad/s]

12.4. Calculate the impedance at resonance of the circuit in problem 12.3, and determine the current in the circuit at 1.3 times the resonant frequency if the r.m.s. supply voltage is 0.1 V.
[(20 + j0) Ω; $0.243\angle -87.2°$ mA]

12.5. A 5 V a.c. source with an output resistance of 2 Ω is connected to a series circuit which resonates at 1 MHz. What values of inductance and capacitance are connected in the circuit if the resistance of the inductor is 3 Ω, and voltage across the capacitor at resonance is 100 V?
[15.9 μH; 1.59 nF]

12.6. An ideal parallel circuit of the type in figure 12.8(c) contains a resistance of 10 kΩ in one branch, a 10 mH inductance in the second branch, and a 1 μF capacitance in the third branch. Calculate (a) the resonant frequency of the circuit, (b) its Q-factor at resonance, (c) the bandwidth of the circuit and (d) the lower and upper half-power frequencies.
[(a) 10 000 rad/s; (b) 100; (c) 100 rad/s; (d) 9950.1 rad/s, 10 050.1 rad/s]

12.7. Calculate the impedance of the circuit in problem 12.6 at (a) ω_0, (b) $0.8\omega_0$, (c) $1.2\omega_0$.
[(a) 10 kΩ; (b) $222.2\angle 88.7°$ Ω; (c) $272.6\angle -88.4°$ Ω]

12.8. A coil of 10 Ω resistance and 0.2 H inductance is connected in parallel with a capacitor of 100 μF capacitance. Calculate the

resonant frequency of the circuit, and evaluate the current in each branch at resonance if the supply voltage is 1 V. What current is drawn from the supply?
[34.7 Hz; $I_{coil} = 0.0224\angle -77.1°$ A; $I_C = 0.0218\angle 90°$ A; 5 mA]

12.9. Convert the circuit in problem 12.8 into a *GLC* parallel circuit containing the ideal elements of the kind in figure 12.5.
[5 mS; 0.211 H; 100 μF]

12.10. A communications receiver circuit contains a 3-branch parallel circuit similar to that in figure 12.5. The circuit is tuned by a variable-capacitance capacitor over a broadcast band from 0.5 MHz to 1.6 MHz. If the value of Q_P does not exceed 40 and R is 25 kΩ, calculate the value of L, together with the maximum and minimum value of C.
[198.9 μH; 49.74 pF; 509.3 pF]

12.11. If, in the circuit in problem 12.6, a dependent voltage source is inserted in series with the capacitor so that the voltage generated assists the original current through the capacitor, determine an expression for the input admittance of the circuit if the magnitude of the dependent source is $1000\boldsymbol{I}_R$, where $\boldsymbol{I}_R$ is the current in the 10 kΩ resistor. Calculate the resonant frequency of the circuit.
[$10^{-4} + 1.1 \times 10^{-6}\, j\omega + 100/j\omega$; 1517 Hz]

12.12. Scale the series *RLC* circuit in worked example 12.3.1 so that it contains (a) a 1 H inductance and a 1 F capacitance, (b) it contains a 1 nF capacitance and is resonant at 1 MHz.
[(a) 0.4 Ω, 1 H, 1 F; (b) 4 kΩ, 1 nF, 23.3 μH]

12.13. In a band-pass filter of the type in figure 12.11(b), at a frequency of 20 kHz $X_L = 2.513$ kΩ, $X_C = 397.9$ Ω and $R = 100$ Ω; calculate (a) the centre frequency (f_0), (b) the half-power frequencies, (c) the Q-factor at resonance and (d) the Q-factor of the circuit if a resistance of 250 Ω is connected between the output terminals.
[(a) 7958 Hz; (b) 8366 Hz, 7570 Hz; (c) 10; (d) 14]

12.14. In problem 12.13, calculate the two half-power frequencies and the pass-band if a load resistance of 100 Ω is connected between the output terminals.
[7761 Hz, 8159 Hz; 398 Hz]

12.15. Two circuits tuned to a frequency of 1 kHz are magnetically coupled as shown in figure 12.13(a). Each circuit has a resistance of 100 Ω, and the Q-factor of the primary and secondary circuits are 50 and 100, respectively. Calculate (a) the critical coupling coefficient and (b) the lower and upper resonant frequencies.
[(a) 0.01414; (b) 0.993 kHz, 1.007 kHz]

13

Harmonics and Fourier Analysis

13.1 Introduction

Thus far we have thought almost exclusively of alternating waveforms as being pure sinusoids. However, there is a range of practical devices including inductors, semiconductors, etc., which result either in the flow of non-sinusoidal currents or in the production of non-sinusoidal voltages. It is the analysis of these which interests us now.

The *Fourier series*, developed by Baron Jean Fourier, is a series of terms that represent a non-sinusoidal waveform. The simplest form is the trigonometric series, in which the waveform is represented as the sum of a d.c. term together with a large number (theoretically infinity) of pure sinusoids. As a consequence of the principle of superposition, the effect of the waveform on circuits can be analysed using standard techniques.

The exponential form of the Fourier series is more compact than the trigonometric form, but is less easy to understand!

13.2 Harmonics

Sinusoids are, by far, the most frequent form in which *periodic waveforms* are met, that is, waveforms in which $f(t) = f(T + t)$, where T is the *periodic time* of the wave.

A waveform which is both periodic and non-sinusoidal is said to be a *complex wave*. A complex wave may be shown to be built up from a zero-frequency term (or d.c. term) and a series of sinusoids or *harmonic waves* whose frequency is an integral multiple of the *fundamental frequency* (or first harmonic frequency). The fundamental frequency is the basic waveform which establishes the general periodic time of the complex wave.

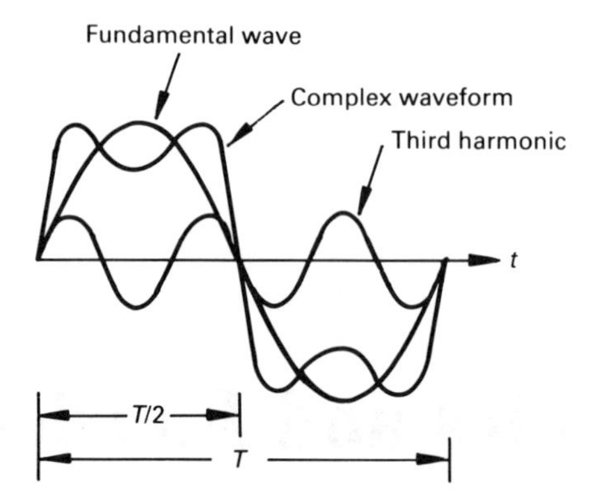

Figure 13.1 *A complex waveform which is the sum of the fundamental frequency and the third harmonic.*

Thus the second harmonic has a frequency which is twice the fundamental frequency, the third harmonic frequency is three times the fundamental, etc. A complex wave which is the sum of a fundamental frequency and a third harmonic frequency is shown in figure 13.1. In this case the third harmonic is in phase with the fundamental, and has a lower amplitude. In general, harmonics frequencies are not in phase with the fundamental, and do not have the same amplitude as the fundamental.

Harmonics are generally produced in electrical circuits by equipment having a non-linear characteristic. The simplest case is, for example, an iron-cored coil whose *B–H* curve is non-linear. A more complex case is a rectifier circuit, which allows flow of current in one-half of the supply voltage wave, but prevents it in the other half-cycle. Electronic devices such as thyristors and triacs give rise to even more complex waves.

The resulting harmonics produce problems in electrical circuits ranging from meter-reading errors to selective resonance.

13.3 Trigonometric Fourier series

A periodic waveform for which $f(t) = f(T + t)$, where T is the periodic time of the wave, can be represented by the *trigonometric Fourier series* of $f(t)$ in the form

$$f(t) = a_0 + a_1 \cos \omega t + a_2 \cos 2\omega t + \dots$$
$$+ b_1 \sin \omega t + b_2 \sin 2\omega t + \dots$$

or in the generalised form

$$f(t) = a_0 + \sum_{n=1}^{\infty} (a_n \cos n\omega t + b_n \sin n\omega t)$$

where $\omega = 2\pi/T$ rad/s, and $n\omega$ is the nth harmonic frequency. The coefficients a_0, a_n and b_n are the *trigonometric Fourier coefficients* of $f(t)$.

Alternatively, the series can be written in the form

$$f(t) = a_0 + \sum_{n=1}^{\infty} c_n \sin(n\omega t + \phi_n)$$

or

$$f(t) = a_0 + \sum_{n=1}^{\infty} c_n \cos(n\omega t + \theta_n)$$

where

$$\begin{aligned} c_n &= \sqrt{(a_n^2 + b_n^2)} \\ \phi_n &= \tan^{-1}(a_n/b_n) \\ \theta_n &= \tan^{-1}(-b_n/a_n) \end{aligned}$$

We are not concerned here with the proof of Fourier's theorem, but we have a deep interest in evaluating the Fourier coefficients. In order to do this we need to know the value of certain integrals, which are given below. Firstly, the average value of a sinusoid over a complete cycle is zero.

$$\int_0^T \sin \omega t \, dt = 0$$

$$\int_0^T \cos \omega t \, dt = 0$$

and the value of the following definite integrals is zero

$$\int_0^T \sin k\omega t \cos n\omega t \, dt = 0$$

$$\int_0^T \sin k\omega t \sin n\omega t \, dt = 0 \qquad k \neq n$$

$$\int_0^T \cos k\omega t \sin n\omega t \, dt = 0 \qquad k \neq n$$

also

$$\int_0^T \sin^2 n\omega t \, dt = \frac{T}{2}$$

$$\int_0^T \cos^2 n\omega t \, dt = \frac{T}{2}$$

The reader should note that the integration period can be any range of T, for example, the integration could be over the range 0 to T (as it is above), or from $-T/2$ to $T/2$, or from $-T/4$ to $3T/4$, etc.

We now turn our attention to a method of evaluating the Fourier coefficients.

The value of a_0

If we integrate the Fourier series over one complete cycle we get

$$\int_0^T f(t)\,dt = \int_0^T a_0\,dt + \int_0^T \sum_{n=1}^{\infty} (a_n \cos n\omega t + b_n \sin n\omega t)\,dt$$

and, since every term in the second expression is either a sine or a cosine, then

$$\int_0^T f(t)\,dt = a_0 T$$

or

$$a_0 = \frac{1}{T}\int_0^T f(t)\,dt$$

That is, the coefficient a_0 is the *average value of $f(t)$ over one complete cycle*. It is therefore the *d.c. component* of $f(t)$.

The value of a_n

To evaluate this term, we multiply both sides of the Fourier series by $\cos k\omega t$, and then integrate both sides over one cycle, as follows

$$\int_0^T f(t) \cos k\omega t\,dt = \int_0^T a_0 \cos k\omega t\,dt +$$

$$\int_0^T \sum_{n=1}^{\infty} (a_n \cos k\omega t \cos n\omega t\,dt + b_n \cos k\omega t \sin n\omega t\,dt)$$

Every term on the right-hand side of the equation is zero except for the case where $k = n$, when

$$\int_0^T f(t) \cos n\omega t\,dt = \frac{T}{2}\,a_n$$

or

$$a_n = \frac{2}{T}\int_0^T f(t) \cos n\omega t\,dt$$

That is

$$a_n = \text{twice the average value of } (f(t) \cos n\omega t) \text{ over one cycle}$$

The value of b_n

The procedure for evaluating b_n is generally similar to that for a_n, with the exception that we multiply both sides of the Fourier series by sin $k\omega t$ before integration. The final result is

$$b_n = \frac{2}{T}\int_0^T f(t) \sin n\omega t \, dt$$

or

$$b_n = \text{twice the average value of } (f(t) \sin n\omega t) \text{ over one cycle}$$

Using these relationships, we will take a look at the Fourier analysis of a rectangular wave and a rectified sinewave.

Worked example 13.3.1

Determine the Fourier series for the rectangular wave in figure 13.2.

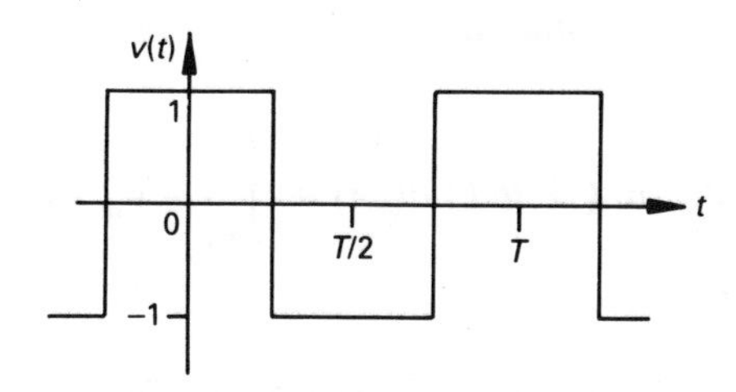

Figure 13.2 *Figure for worked example 13.3.1.*

Solution

In this case the graph is plotted to a base of time, and the integral will be evaluated over a range of T seconds. The equation of the wave is

$$v(t) = \begin{cases} V_m \text{ for } -T/4 < t < T/4 \\ -V_m \text{ for } T/4 < t < 3T/4 \end{cases}$$

By observation, the area above the axis in each cycle is equal to the area below it, hence

$$a_0 = 0$$

The value of a_n can be calculated as follows. Since $\omega T = 2\pi$, then $\omega = 2\pi/T$. Also, since the waveform repeats itself between $T/2$ and T, we shall analyse the wave from 0 to $T/2$ below.

$$a_n = \frac{2}{T/2}\int_0^{T/2} f(t) \cos n\omega t \, \mathrm{d}t$$

$$= \frac{4}{T}\left\{\int_0^{T/4} V_\mathrm{m} \cos 2n\pi t/T \, \mathrm{d}t + \int_{T/4}^{T/2} -V_\mathrm{m} \cos 2n\pi t/T \, \mathrm{d}t\right\}$$

$$= \frac{4V_\mathrm{m}}{T}\left\{\int_0^{T/4} \cos\frac{2n\pi t}{T}\,\mathrm{d}t - \int_{T/4}^{T/2} \cos\frac{2n\pi t}{T}\,\mathrm{d}t\right\}$$

$$= \frac{4V_\mathrm{m}}{T}\left\{\left[\frac{T}{2n\pi}\sin\frac{2n\pi t}{T}\right]_0^{T/4} - \left[\frac{T}{2n\pi}\sin\frac{2n\pi t}{T}\right]_{T/4}^{T/2}\right\}$$

$$= \frac{4V_\mathrm{m}}{n\pi}\sin n\pi/2$$

If n is even, $a_n = 0$. For odd values of n, a_n is finite. The value of b_n can be calculated as follows.

$$b_n = \frac{2}{T/2}\int_0^{T/2} f(t) \sin n\omega t \, \mathrm{d}t$$

$$= \frac{4}{T}\left\{\int_0^{T/4} V_\mathrm{m} \sin (2n\pi t/T) \, \mathrm{d}(\omega t) + \int_{T/4}^{T/2} -V_\mathrm{m} \sin (2n\pi t/T) \, \mathrm{d}t\right\}$$

$$= 0 \text{ for all } n$$

That is, there are no sine terms in the series. Hence, the Fourier series is of the form

$$v(t) = \frac{4V_\mathrm{m}}{\pi}\left(\cos \omega t - \frac{1}{3}\cos 3\omega t + \frac{1}{5}\cos 5\omega t - \ldots\right)$$

Worked example 13.3.2

Express the rectified current waveform in figure 13.3 as a Fourier series.

Solution

The waveform can be expressed over the range $0 \leqslant t \leqslant 0.1$ as

$$i(t) = \begin{cases} I_\mathrm{m} \sin \omega t & \text{for } 0 \leqslant t \leqslant 0.05 \\ 0 & \text{for } 0.05 \leqslant t \leqslant 0.1 \end{cases}$$

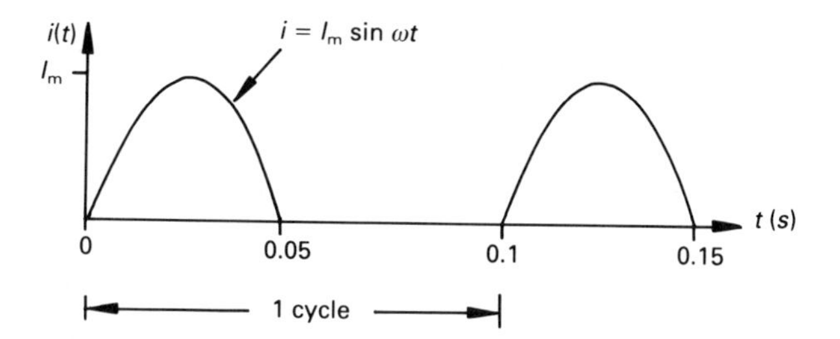

Figure 13.3 *Figure for worked example 13.3.2.*

Since $T = 0.1$ s, then $f = 1/T = 10$ Hz and $\omega = 20\pi$ rad/s.

The d.c. component or zero-frequency component is

$$a_0 = \frac{1}{T}\int_0^T f(t)\,dt$$

$$= \frac{1}{0.1}\left\{\int_0^{0.05} I_m \sin 20\pi t\,dt + \int_{0.05}^{0.1} 0\,dt\right\}$$

$$= I_m/\pi$$

The reader should note that the integration is broken down into intervals for which the functional form of $i(t)$ is known.

The value of a_n can be calculated as follows.

$$a_n = \frac{2}{T}\int_0^T f(t)\cos n\omega t\,dt$$

$$= \frac{2}{0.1}\left\{\int_0^{0.05} I_m \sin 20\pi t \cos 20\pi nt\,dt + \int_{0.05}^{0.1} 0\,dt\right\}$$

$$= 10I_m\int_0^{0.05} (\sin (n+1)20\pi t - \sin (n-1)20\pi t)\,dt)$$

We break the calculation at this point to note that

$$a_1 = 10I_m\int_0^{0.05} \sin (2 \times 20\pi t)\,dt) = 0$$

The calculation is continued below to include other values of n

$$a_n = \frac{10I_m}{20\pi}\left[-\frac{1}{n+1}\cos (n+1)20\pi t + \frac{1}{n-1}\cos (n-1)20\pi t\right]_0^{0.05}$$

$$= \frac{10I_m}{20\pi}\left[\frac{1}{n+1}(1 - \cos(n+1)\pi) - \frac{1}{n-1}(1 - \cos(n-1)\pi)\right]$$

Points to note here are that *if n is odd* (and $\neq 1$), then $(n \pm 1)$ is even, so that $a_n = 0$; that is $a_3 = a_5 = a_7 = \ldots = 0$. *If n is even* then $(n \pm 1)$ is odd and

$$a_n = \frac{10I_m}{20\pi}\left[\frac{2}{n+1} - \frac{2}{n-1}\right] = \frac{-2I_m}{(n+1)(n-1)\pi}$$

The coefficient b_n is calculated as follows

$$b_n = \frac{2}{T}\int_0^T f(t) \sin n\omega t \, \mathrm{d}t$$

$$= \frac{2}{0.1}\left\{\int_0^{0.05} I_m \sin 20\pi t \sin 20\pi n t \, \mathrm{d}t + \int_{0.05}^{0.1} 0 \, \mathrm{d}t\right\}$$

$$= 20I_m \int_0^{0.05} \sin 20\pi t \sin 20\pi n t \, \mathrm{d}t$$

If $n = 1$

$$b_1 = 20I_m \int_0^{0.05} \sin^2 20\pi t \, \mathrm{d}t = \frac{I_m}{2}$$

If $n \neq 1$, the calculation proceeds as follows

$$b_n = 20I_m \int_0^{0.05} \tfrac{1}{2}(\cos(n-1)20\pi t - \cos(n+1)20\pi t) \, \mathrm{d}t$$
$$= 0$$

Hence

$$i(t) = \frac{I_m}{\pi}\left(1 + \frac{\pi}{2}\sin 20\pi t - \frac{2}{3}\cos 40\pi t - \frac{2}{15}\cos 80\pi t - \frac{2}{35}\cos 120\pi t - \ldots\right)$$

13.4 Waveform symmetry

Many waveforms exhibit symmetry either about a particular point or a particular axis. A knowledge of the type of symmetry involved may tell us that certain coefficients in the Fourier series may be absent, allowing us (in certain cases) to simplify the calculations involved. If more than one form of symmetry exists, there will be more than one factor missing from the series.

When investigating the symmetry of a waveform, and if a d.c. component exists, the reader will find it useful to visualise the wave without its d.c. component.

The two most readily recognisable forms of symmetry are *even-function symmetry* (or *even symmetry*) and *odd-function symmetry* (or *odd symmetry*).

Even symmetry

An *even function* is defined as one for which $f(t) = f(-t)$, that is, it is *symmetrical about the y-axis*, as shown in figure 13.4(a). The Fourier series for the function contains only cosine terms (no sine terms exist): a_0 may exist, a_n exists and $b_n = 0$. The series is of the form

$$f(t) = a_0 + a_1 \cos \omega t + a_2 \cos 2\omega t + \ldots$$

Odd symmetry

An *odd function* is one in which $f(t) = -f(-t)$, that is it is *symmetrical about the origin* (see figure 13.4(b)). The Fourier series for this function contains only sine terms; to summarise, $a_0 = 0$, $a_n = 0$, b_n exists, and the Fourier series is of the form

$$f(t) = b_1 \sin \omega t + b_2 \sin 2\omega t + \ldots$$

Half-wave repetition

This is a wave in which $f(t) = f(t + T/2)$, as shown in figure 13.4(c). The Fourier series for this type of repetition contains only even terms; to summarise, a_0 may exist, and only even terms exist in a_n and b_n as follows. If the series has the general form

$$f(t) = c_0 + c_1 \sin(\omega t + \phi_1) + c_2 \sin(2\omega t + \phi_2) + c_3 \sin(3\omega t + \phi_3) + \ldots$$

then

$$f(t + T/2) = c_0 - c_1 \sin(\omega t + \phi_1) + c_2 \sin(2\omega t + \phi_2) - c_3 \sin(3\omega t + \phi_3) + \ldots$$

Hence if $f(t) = f(t + T/2)$, then the Fourier series for $f(t)$ contains only even harmonics.

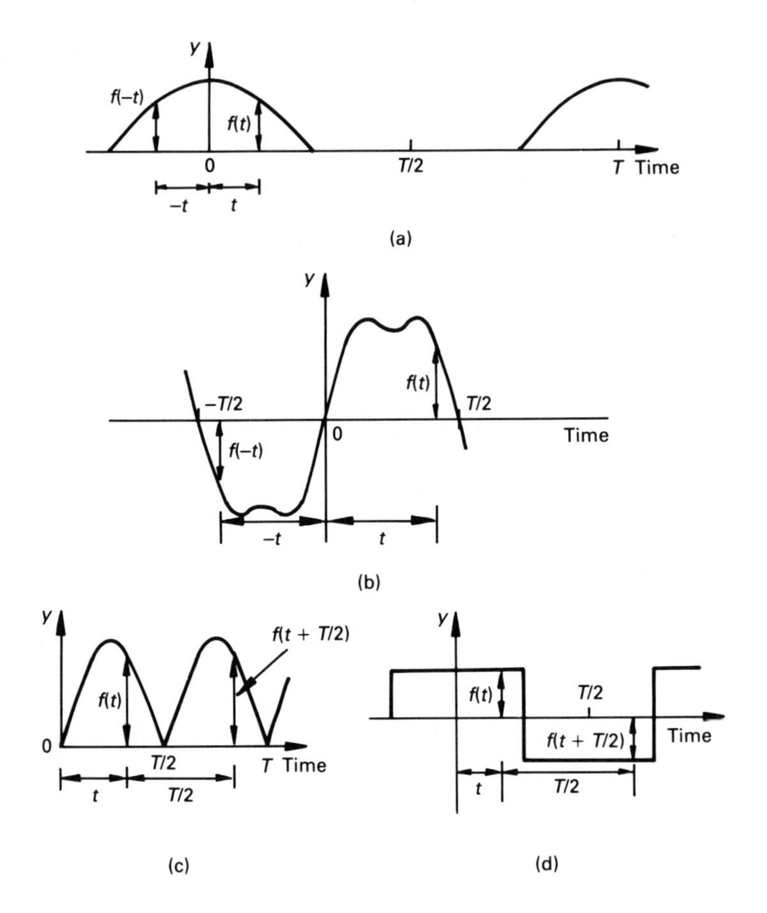

Figure 13.4 *Waveform symmetry.*

Half-wave inversion

In this type of wave, $f(t) = -f(t + T/2)$, as shown in figure 13.4(d). The Fourier series contains only odd terms; to summarise, $a_0 = 0$ and only odd terms exist in a_n and b_n.

If the series has the general form described in the Half-wave repetition section, it is clear that if $f(t) = -f(t + T/2)$, neither the 'd.c.' component nor even terms exist (see also worked example 13.4.1).

Worked example 13.4.1

Identify the symmetry of the waveforms in figure 13.5.

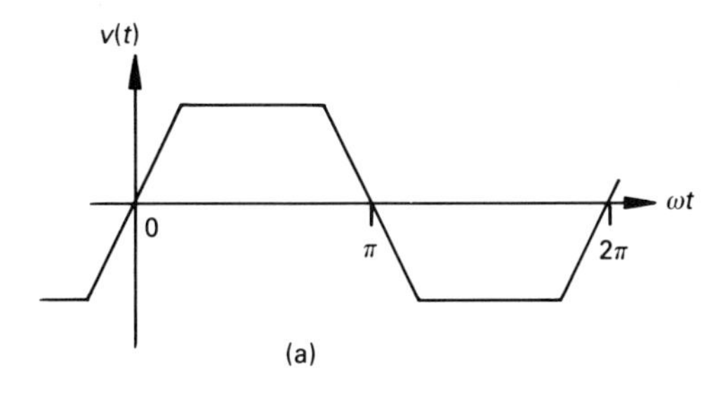

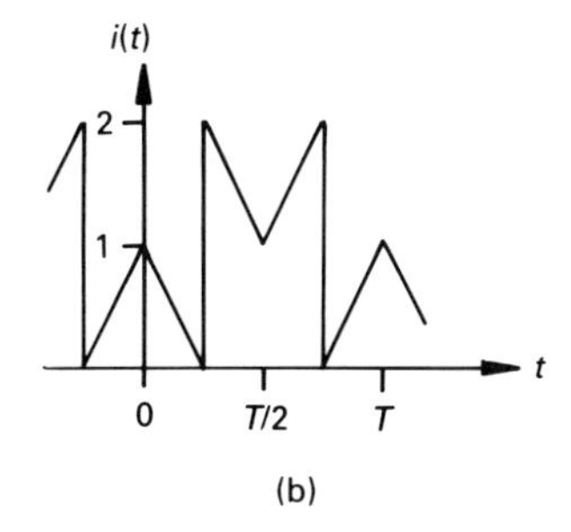

Figure 13.5 *Figure for worked example 13.4.1.*

Solution

The waveform in figure 13.5(a) corresponds to the idealised magnetic flux distribution in the air gap of an alternator. The waveform has odd symmetry and half-wave inversion. Consequently $a_0 = 0$, $a_n = 0$ and $b_{\text{even}} = 0$.

The wave in figure 13.5(b) has a 'd.c.' term which, by observation, has unity value, that is, $a_0 = 1$. If we visualise the 'a.c.' component without the 'd.c.' term, we see that the wave is an even function and has half-wave inversion. That is $b_n = 0$ and $a_{\text{even}} = 0$.

13.5 Line spectrum

A popular method of presenting the results of Fourier analysis is in the form of a *line spectrum* of the wave. In any given problem, there may be either one or two line spectrum diagrams. One diagram shows the amplitude of each frequency component in the form of a vertical line drawn at the corresponding frequency, the length of the line indicating the magnitude of the harmonic (this is the most usual form of line spectrum).

In addition, there may be a phase spectrum, the length of each vertical line representing the phase angle of the particular harmonic component (see worked example 13.5.1).

For a particular harmonic frequency $n\omega t$, we can combine the sine and cosine terms in the Fourier series as follows

$$a_n \cos n\omega t + b_n \sin n\omega t = \sqrt{(a_n^2 + b_n^2)} \sin (n\omega t + \tan^{-1}(a_n/b_n))$$

or

$$a_n \cos n\omega t + b_n \sin n\omega t = \sqrt{(a_n^2 + b_n^2)} \cos (n\omega t + \tan^{-1}(-b_n/a_n))$$

Using the above relationships, the magnitude and phase spectra can be drawn. Depending on whether the sine or cosine version is chosen, the phase shift between the two will differ by 90°.

If we replace the cosine and sine terms in the Fourier series by their exponential equivalent, we get the following simplified series (showing only a_0 and the nth harmonic).

$$\begin{aligned} f(t) &= a_0 + \ldots + a_n \cos n\omega t + \ldots + b_n \sin n\omega t + \ldots \\ &= a_0 + \ldots + \frac{a_n}{2}(\mathrm{e}^{\mathrm{j}n\omega t} + \mathrm{e}^{-\mathrm{j}n\omega t}) \\ &\quad + \ldots + \frac{b_n}{2\mathrm{j}}(\mathrm{e}^{\mathrm{j}n\omega t} - \mathrm{e}^{-\mathrm{j}n\omega t}) + \ldots \\ &= a_0 + \ldots + \frac{1}{2}(a_n - \mathrm{j}b_n)\mathrm{e}^{\mathrm{j}n\omega t} + \ldots \\ &\quad + \frac{1}{2}(a_n - \mathrm{j}b_n)\mathrm{e}^{\mathrm{j}n\omega t} + \ldots \end{aligned}$$

This series indicates that there is yet another version of the line spectrum, with spectral lines at frequencies $+n\omega$ and $-n\omega$, each of these lines being one-half the length of the original amplitude spectrum. The length of the d.c. term (a_0) is unchanged. If, for example, the trigonometric Fourier series of a wave is as shown in figure 13.6(a), the exponential series indicates that the alternative spectrum in figure 13.6(b) is also true for the wave.

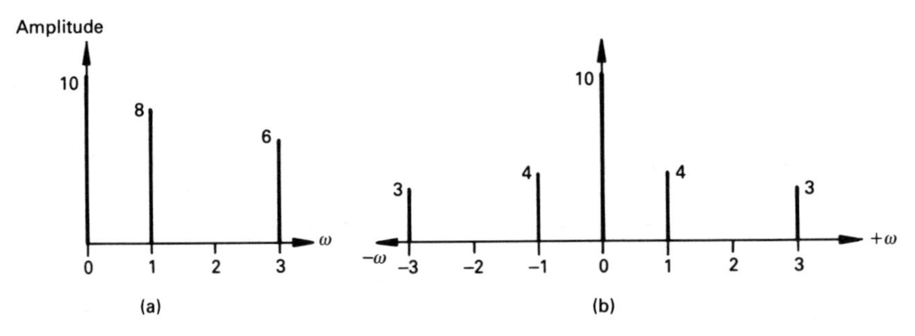

Figure 13.6 *Amplitude spectrum of a complex wave using (a) the trigonometric Fourier series, (b) the exponential Fourier series.*

Worked example 13.5.1

Draw the phase and magnitude spectra for a waveform with the following Fourier analysis

$$\begin{aligned} v(t) = 10 &- 3 \cos t + 2 \cos 2t - 5 \cos 3t \\ &+ 4 \sin t - 6 \sin 2t + 2 \sin 3t - 12 \sin 4t \end{aligned}$$

Solution

The line spectrum values can be calculated as follows; it is assumed that each term in the series can be expressed in the form $c_n \cos(n\omega t + \phi_n)$.

d.c. component

Amplitude = 10

ω = 1 rad/s component

Amplitude = $c_1 = \sqrt{(a_1^2 + b_1^2)} = \sqrt{(3^2 + 4^2)} = 5$

Phase shift = $\phi_1 = \tan^{-1}(-b_1/a_1) = \tan^{-1}(-4/-3) = -127°$

ω = 2 rad/s component

Amplitude = $c_2 = \sqrt{(a_2^2 + b_2^2)} = \sqrt{(2^2 + 6^2)} = 6.3$

Phase shift = $\phi_2 = \tan^{-1}(-b_2/a_2) = \tan^{-1}(6/2) = 71.5°$

ω = 3 rad/s component

Amplitude = $c_3 = \sqrt{(a_3^2 + b_3^2)} = \sqrt{(5^2 + 2^2)} = 5.4$

Phase shift = $\phi_3 = \tan^{-1}(-b_3/a_3) = \tan^{-1}(-2/-5) = -158.2°$

ω = 4 rad/s component

Amplitude = $c_4 = \sqrt{(a_4^2 + b_4^2)} = \sqrt{(0^2 + 12^2)} = 12$

Phase shift = $\phi_4 = \tan^{-1}(-b_4/a_4) = \tan^{-1}(12/0) = 90°$

The Fourier series can therefore be expressed in the form

$$v(t) = 10 + 5\cos(\omega t - 127°) + 6.3\cos(2\omega t + 71.5°)$$
$$+ 5.4\cos(3\omega t - 159.2°) + 12\cos(4\omega t + 90°)$$

and the corresponding line spectra are shown in figure 13.7.

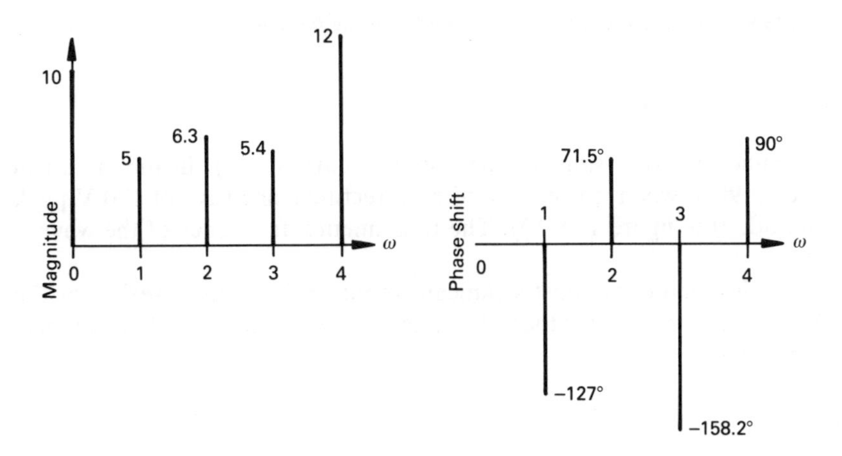

Figure 13.7 *Figure for worked example 13.5.1.*

13.6 Circuit response to a non-sinusoidal forcing function

A complex forcing voltage can be thought of as a direct voltage in series with an infinite number of cosine and sine voltages, as shown in figure 13.8. We can consider the response of a linear network to this voltage by calculating the current due to each independent harmonic of the complex voltage. By superposition, the resulting current is the sum of the individual terms, as shown in worked example 13.6.1.

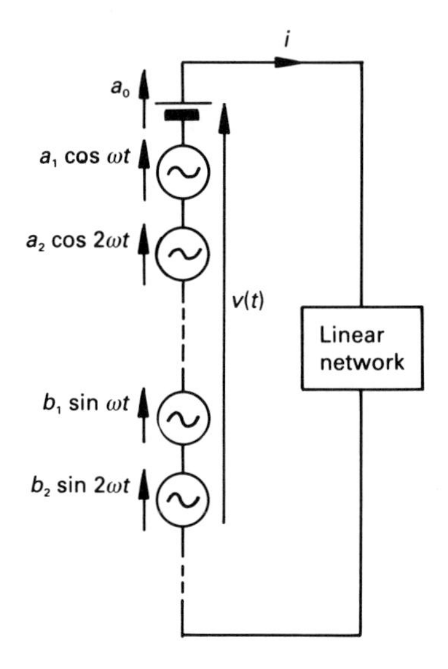

Figure 13.8 *Circuit response to a non-sinusoidal forcing function.*

Worked example 13.6.1

Determine an expression for the complex current, $i(t)$, in the circuit in figure 13.9(a), when it is energised by a rectified sinewave of 100 V peak amplitude (see figure 13.9(b)). The fundamental frequency of the wave is 1000 rad/s.

Also determine the most significant terms in the Fourier series for the voltage $v_R(t)$ and $v_L(t)$ for the voltage across the resistance and inductance, respectively.

Solution

Firstly we need the trigonometric Fourier expression for the complex wave; fortunately we have already deduced the first few terms in the series for half-wave rectified current in worked example 13.3.2, and the corresponding series for a voltage wave is

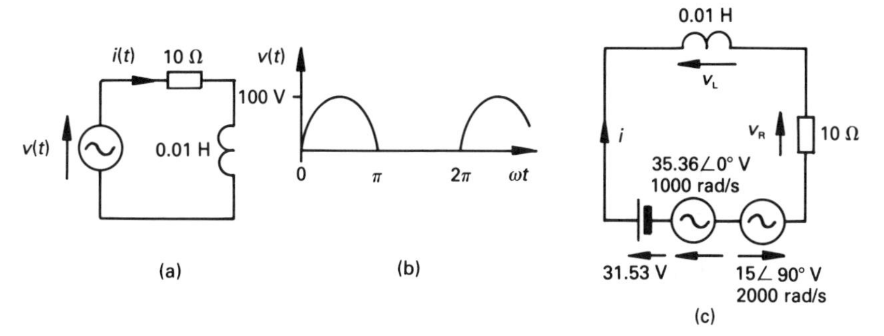

Figure 13.9 *Figure for worked example 13.6.1.*

$$v(t) = \frac{V_m}{\pi}\left(1 + \frac{\pi}{2}\sin\omega t - \frac{2}{3}\cos 2\omega t - \frac{2}{15}\cos 4\omega t - \frac{2}{35}\cos 6\omega t - \ldots\right)$$

Since the terms in the series diminish rapidly as the order of the harmonic increases, we shall consider only terms up to and including the second harmonic. Converting the second harmonic cosine term into a sine term, the expression for $v(t)$ is

$$v(t) = 31.83 + 50 \sin \omega t - 21.22 \sin (2\omega t + 90°) \text{ V}$$

When the maximum a.c. values are converted into r.m.s. values, the circuit diagram for the first three terms in the series is as shown in figure 13.9(c). We now consider each component in turn.

d.c. term

The current, I_0, produced by this term is

$$I_0 = a_0/R = 31.83/10 = 3.183 \text{ A}$$

The p.d. across the resistor produced by this voltage is

$$V_{R0} = RI_0 = 10I_0 = 31.83 \text{ V}.$$

and across the inductor (which has no resistance) is zero.

Fundamental frequency term

The r.m.s. voltage is $50/\sqrt{2} = 35.36$ V, hence $\mathbf{V}_1 = 35.36\angle 0°$ V. The circuit impedance at the fundamental frequency of 1000 rad/s is

$$\mathbf{Z}_1 = R + j\omega_1 L = 10 + j(1000 \times 0.01) = 10 + j10$$
$$= 14.14\angle 45° \ \Omega$$

hence

$$I_1 = V_1/Z_1 = 35.36\angle 0°/14.14\angle 45° = 2.5\angle -45° \text{ A}$$

and $I_{1max} = 3.54$ A. Hence

$$V_{R1} = RI_1 = 10 \times 2.5\angle -45° = 25\angle -45° \text{ V}$$

and

$$V_{L1} = jX_1I_1 = j10 \times 2.5\angle -45° = 25\angle 45° \text{ V}$$

Second harmonic term

The r.m.s value of this component is $21.22/\sqrt{2} = 15$ V and, since $\phi = 90°$, $V_2 = -15\angle 90° = 15\angle -90°$ V. The circuit impedance at a frequency of 2000 rad/s is

$$Z_2 = R + j\omega_2L = 10 + j(2000 \times 0.01) = 10 + j20$$
$$= 22.36\angle 63.4° \ \Omega$$

that is

$$I_2 = V_2/Z_2 = 15\angle -90°/22.36\angle 63.4° = 0.67\angle -153.4° \text{ A}$$

or $I_{2max} = 0.95$ A. The voltage across the circuit elements is

$$V_{R2} = RI_2 = 10 \times 0.67\angle -153.4° = 6.7\angle -153.4° \text{ V}$$

and

$$V_{L2} = jX_2I_2 = j20 \times 0.67\angle -153.4° = 13.4\angle -63.4° \text{ V}$$

Hence, the first three terms in the Fourier series for $i(t)$ are

$$i(t) = 3.183 + 3.54 \sin(1000t - 45°) + 0.95 \sin(2000t - 153.4°) \text{ A}$$

The first three terms in the Fourier series for $v_R(t)$ are

$$v_R(t) = 31.83 + 35.36 \sin(1000t - 45°) + 9.48 \sin(2000t - 153.4°) \text{ V}$$

and the terms in the Fourier series for $v_L(t)$ are

$$v_L(t) = 35.36 \sin(1000t + 45°) + 18.95 \sin(2000t - 63.4°) \text{ V}$$

At this point the reader will find it interesting to verify that $v(t) = v_R(t) + v_L(t)$.

13.7 T.M.S. value of a complex wave and the power supplied

The r.m.s. value of the complex wave

$$y(t) = a_0 + (a_1 \cos \omega t + \ldots + a_n \cos n\omega t + \ldots) \\ + (b_1 \sin \omega t + \ldots + b_n \sin n\omega t + \ldots)$$

is given by

$$Y_{\text{rms}} = \sqrt{(a_0^2 + \tfrac{1}{2}(a_1^2 + \ldots + a_n^2 + \ldots) + \tfrac{1}{2}(b_1^2 + \ldots + b_n^2 + \ldots))}$$

or, if A_n is the r.m.s. value of a_n, and B_n is the r.m.s. value of b_n, then

$$Y_{\text{rms}} = \sqrt{(A_0^2 + (A_1^2 + \ldots + A_n^2 + \ldots) + (B_1^2 + \ldots + B_n^2 + \ldots))}$$

The reader may like to verify that the r.m.s. voltage across the resistor in worked example 13.6.1 is 41 V, and the r.m.s. voltage across the inductor is 28.37 V. Accordingly, the r.m.s. supply voltage is

$$Vs = \sqrt{(41^2 + 28.37^2)} = 49.86 \text{ V}$$

which is in close agreement with the r.m.s. voltage of 49.88 V calculated in worked example 13.7.1.

If $v(t)$ and $i(t)$ are the respective voltage across and the current in the circuit, then the average power supplied by the nth harmonic is

$$\begin{aligned} P_n &= \frac{1}{2\pi} \int_0^{2\pi} v_n i_n \, \mathrm{d}(\omega t) \\ &= \frac{1}{2\pi} \int_0^{2\pi} V_n I_n \sin n\omega t \sin(n\omega t - \phi_n) \, \mathrm{d}(\omega t) \\ &= V_n I_n \cos \phi_n \end{aligned}$$

where ϕ_n is the phase angle between V_n and I_n.

The average power supplied by a voltage of one frequency (say $n\omega t$) and a current of another frequency (say $m\omega t$) is zero. Hence the total power supplied by a complex voltage (or current) is

$$\begin{aligned} P_{\text{T}} &= P_0 + P_1 + \ldots + P_n + \ldots \\ &= V_0 I_0 + V_1 I_1 \cos \phi_1 + \ldots + V_n I_n \cos \phi_n + \ldots \\ &= V_0 I_0 + \sum_{n=1}^{\infty} V_n I_n \cos \phi_n \end{aligned}$$

where V_n and I_n are r.m.s. values.

The *power factor* of a circuit to which a complex wave is applied is defined as

$$\text{power factor} = \frac{\text{power consumed}}{\text{r.m.s. voltage} \times \text{r.m.s. current}}$$

Worked example 13.7.1

Calculate, for worked example 13.6.1, (a) the r.m.s. current in the load, (b) the r.m.s. voltage across the load, (c) the power supplied by each component of the complex wave, (d) the total power supplied and (e) the power factor.

Solution

(a) The r.m.s. value of the current is

$$I = \sqrt{\left(I_0^2 + \frac{I_{1m}^2}{2} + \frac{I_{2m}^2}{2}\right)}$$

$$= \sqrt{\left(3.183^2 + \frac{3.54^2}{2} + \frac{0.95^2}{2}\right)} = 4.1 \text{ A}$$

(b) The r.m.s. value of the voltage across the load is

$$V = \sqrt{\left(V_0^2 + \frac{V_{1m}^2}{2} + \frac{V_{2m}^2}{2}\right)}$$

$$= \sqrt{\left(31.83^2 + \frac{50^2}{2} + \frac{21.22^2}{2}\right)} = 49.88 \text{ V}$$

(c) We can calculate the power consumed by each component of the complex wave as follows.
d.c. component

$$P_0 = V_0 I_0 = 31.83 \times 3.183 = 101.3 \text{ W}$$

or, alternatively $\quad P_0 = I_0^2 R = 3.183^2 \times 10 = 101.3 \text{ W}$
or $\quad P_0 = V_{0(R)}^2/R = 31.83^2/10 = 101.3 \text{ W}$

Fundamental frequency

$$P_1 = \text{Re}(\boldsymbol{V}_1\boldsymbol{I}_1^*) = \text{Re}(35.36\angle 0° \times 2.5\angle -45°)$$
$$= \text{Re}(62.5 - \text{j}62.5) = 62.5 \text{ W}$$

or $\quad P_1 = I_1^2 R = 2.5^2 \times 10 = 62.5 \text{ W}$
or $\quad P_1 = V_{1(R)}^2/R = 25^2/10 = 62.5 \text{ W}$

Second harmonic

$$P_2 = \text{Re}(\boldsymbol{V}_2\boldsymbol{I}_2^*) = \text{Re}(15\angle -90° \times 0.67\angle 153.5°) = 4.5 \text{ W}$$

or $\quad P_2 = I_2^2 R = 0.67^2 \times 10 = 4.49 \text{ W}$
or $\quad P_2 = V_{2(R)}^2/R = 6.7^2/10 = 4.49 \text{ W}$

(d) Total Power

$$P_T = P_0 + P_1 + P_2 = 168.3 \text{ W}$$

or

$$P_T = I^2R = 4.1^2 \times 10 = 168.1 \text{ W}$$

(e) The power factor of the circuit can be calculated from

$$\begin{aligned}\text{power factor} &= \text{power/(r.m.s. volt-amperes)}\\ &= 168.3/(49.88 \times 4.1) = 0.823\end{aligned}$$

13.8 Effect of harmonics in a.c. systems

Since resistance is independent of frequency, the voltage and current will, for each individual harmonic, be in phase with one another in a resistor.

In the case of an inductor, the reactance to the nth harmonic is n times greater than it is to the fundamental; consequently, the current produced (for each volt) by that harmonic is n times smaller than for the fundamental.

For a capacitor, the reactance to the nth harmonic is n times smaller than it is to the fundamental; consequently, the current produced (for each volt) by that harmonic is n times greater than for the fundamental.

Consequently, if measurements of impedance are made in a circuit containing reactive elements using a complex supply frequency, large errors can arise unless allowance is made for the individual harmonics.

Harmonics can also be produced in 3-phase systems. It can be shown that it is unusual for even harmonics to be present in these systems, but multiples of the third harmonic (called *triple-n harmonics*) can produce serious problems. In some cases, odd harmonics have a positive phase sequence (see also chapter 7), and others have a negative phase sequence.

13.9 Harmonic analysis

If we are provided with numerical data for a waveform (as we may from an oscillogram) then, using the theory in section 13.3, we can calculate the Fourier coefficients of the wave. The coefficients are calculated from ordinates taken at fixed intervals along the wave.

However, the reader should note that this method can, on occasions, give misleading results; for example, if the spacing of the ordinates is such that they are placed at the nodes of the nth harmonic, then the nth harmonic will apparently have zero value! Moreover, errors will arise if one tries to extend the theory to predicting high-order harmonics.

Worked example 13.9.1

A current waveform has the values in table 13.1 in its first half-cycle; they are repeated with a negative sign in the second half-cycle.

Calculate the maximum value of the fundamental frequency, together with the second and the third harmonics, and determine the phase angle between the fundamental and the third harmonic.

Solution

In this type of problem, the solution can be obtained in tabular form (see table 13.2) and, in so doing, the risk or error is reduced.

Table 13.1

$\theta°$	0	20	40	60	80	100	120	140	160
i (A)	−17.3	16.9	64.3	103.9	115.8	94.5	69.3	47	30.2

Table 13.2

$\theta°$	$f(t)$	$f(t)\cos\theta$	$f(t)\sin\theta$	$f(t)\cos 2\theta$	$f(t)\sin 2\theta$	$f(t)\cos 3\theta$	$f(t)\sin 3\theta$
0	−17.3	−17.3	0	−17.3	0	−17.3	0
20	16.9	15.9	5.8	12.9	10.9	8.5	14.6
40	64.3	49.3	41.3	11.2	63.3	−32.15	55.7
60	103.9	52.0	90.0	−52.0	90.0	−103.9	0
80	115.8	20.1	114.0	−108.8	39.6	−57.9	−100.3
100	94.5	−16.4	93.1	−88.8	−32.3	47.25	−81.8
120	69.3	−34.7	60.0	−34.7	−60.0	69.3	0
140	47.0	−36.0	30.2	8.2	−46.3	23.5	40.7
160	30.2	−28.4	10.3	23.1	−19.4	−15.1	26.2
Sum	524.6	4.5	444.7	−246.2	45.8	−77.8	−44.9

The calculations for the first half-cycle are shown in table 13.2, and for the second half-cycle the corresponding total values are

$f(t)$	$f(t)\cos\theta$	$f(t)\sin\theta$	$f(t)\cos 2\theta$	$f(t)\sin 2\theta$	$f(t)\cos 3\theta$	$f(t)\sin 3\theta$
−524.6	4.5	444.7	246.2	−45.8	−77.8	−44.9

The totals for the complete cycle are

$f(t)$	$f(t)\cos\theta$	$f(t)\sin\theta$	$f(t)\cos 2\theta$	$f(t)\sin 2\theta$	$f(t)\cos 3\theta$	$f(t)\sin 3\theta$
0	9	889.4	0	0	−155.6	−89.9

Thus a_0 = mean value of $\Sigma f(t) = 0/18 = 0$

and $a_n = 2 \times$ mean value of $\Sigma f(t) \cos n\theta$
$b_n = 2 \times$ mean value of $\Sigma f(t) \sin n\theta$

hence

$$\begin{aligned} a_1 &= 2 \times 9/18 = 1 \\ a_2 &= 2 \times 0/18 = 0 \\ a_3 &= 2 \times (-155.6)/18 = -17.3 \\ b_1 &= 2 \times 889.4/18 = 98.8 \\ b_2 &= 2 \times 0/18 = 0 \\ b_3 &= 2 \times (-89.8)/18 = -9.98 \end{aligned}$$

that is, the Fourier series for the wave is

$$\begin{aligned} f(t) &= \cos\theta + 98.8 \sin\theta - 17.3 \cos 3\theta - 9.98 \sin 3\theta \\ &= 98.8 \sin(\theta + 0.6°) + 19.97 \sin(3\theta - 120°) \end{aligned}$$

The reader should note that the $-120°$ phase angle of the third harmonic corresponds to a phase angle of $-40°$ at the fundamental frequency, so that the phase angle between the two sinusoids is 40.6°.

Unworked problems

13.1. A waveform has the following Fourier analysis:

$$f(t) = 10 + 6 \sin(20\pi t + 40°) + 8 \sin(40\pi t + 70°) + 2 \sin(60\pi t + 90°)$$

Calculate (a) the mean value of $f(t)$, (b) the r.m.s. value of $f(t)$, (c) the periodic time of the second harmonic and (d) the value of $f(t)$ at $t = 0.1$ s.
[(a) 10; (b) 12.33; (c) 0.05 s; (d) 23.38]

13.2. For the waveform in figure 13.10, evaluate (a) a_0, (b) a_1, (c) b_2.
[(a) -1.6; (b) -0.75; (c) 0.88]

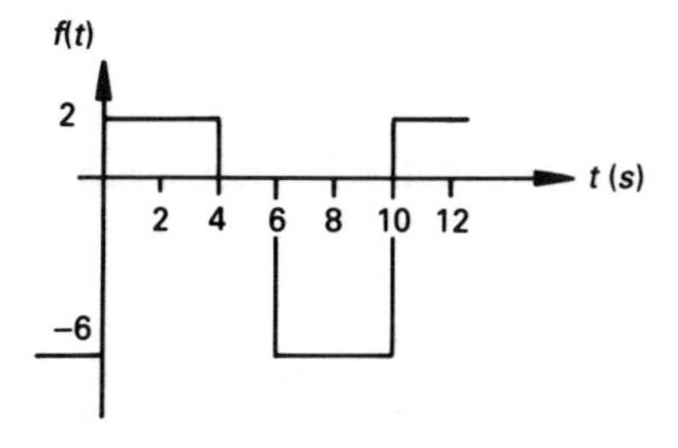

Figure 13.10

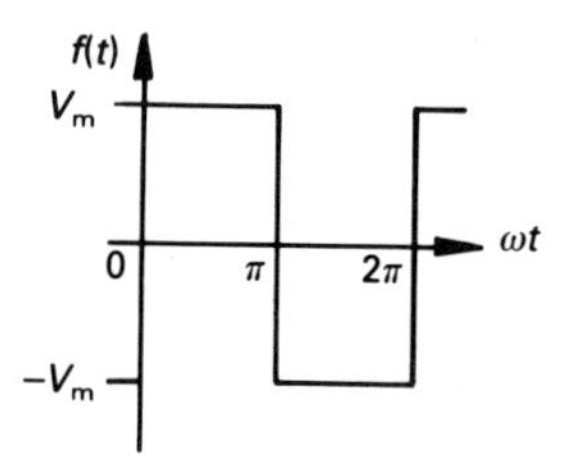

Figure 13.11

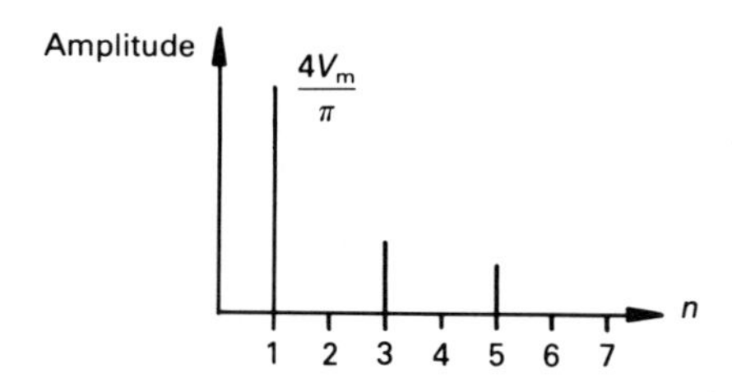

Figure 13.12

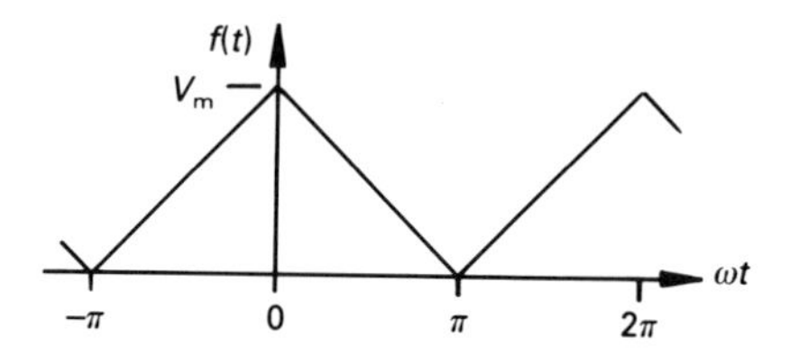

Figure 13.13

13.3. Evaluate the trigonometric Fourier series for the square wave in figure 13.11 and (b) draw the line spectrum for the wave.

$$\left[\text{(a) } f(t) = \frac{4V_m}{\pi}\left(\sin \omega t + \frac{1}{3}\sin 3\omega t + \frac{1}{5}\sin 5\omega t + \ldots\right);\right.$$

(b) see figure 13.12]

13.4. Determine the trigonometric Fourier series for the triangular waveform in figure 13.13.

$$\left[\frac{V_m}{2}\left(1 + \frac{8}{\pi^2}\cos \omega t + \frac{8}{(3\pi)^2}\cos 3\omega t + \frac{8}{(5\pi)^2}\cos 5\omega t + \ldots\right)\right]$$

13.5. A voltage waveform is defined by the following

$$v(t) = 200 \cos \omega t \text{ for } -\frac{\pi}{2} \leqslant \omega t \leqslant \frac{\pi}{2}$$

$$v(t) = 0 \text{ for } \frac{\pi}{2} \leqslant \omega t \leqslant \frac{3\pi}{2}$$

Deduce the Fourier series for the wave.

$$\left[\frac{200}{\pi}\left(1 + \frac{\pi}{2}\cos \omega t + \frac{2}{3}\cos 2\omega t - \frac{2}{15}\cos 4\omega t + \frac{2}{35}\cos 6\omega t - \ldots\right)\right]$$

13.6. If the periodic frequency of the wave in problem 13.5 is 300 rad/s, and is applied to a series circuit containing a 1 kΩ resistance and a 0.8 μF capacitance, determine the Fourier series up to the 6th harmonic for the current in the circuit.
[23.3 cos (ωt + 76.5°) + 18.4 cos ($2\omega t$ + 64.4°) + 5.9 cos ($4\omega t$ + 46.2°) + 3 cos ($6\omega t$ + 34.8°) mA]

13.7. For problem 13.6, calculate (a) the r.m.s. value of the voltage and the current, (b) the power supplied by each harmonic, (c) the total power supplied and (d) the power factor .
[(a) 99.84 V, 21.5 mA; (b) d.c. power = 0, fundamental power = 0.271 W, 2nd harmonic power = 0.169 W, 4th harmonic power = 17 mW, 6th harmonic power = 4.5 mW; (c) 0.462 W; (d) 0.215]

13.8. Write down the Fourier series for each of the following periodic functions, and determine the periodic time of the wave in each case: (a) $6 \sin^2 20t$, (b) $5 \cos^2 10t$,
(c) $5 + 4 \sin 100t + 8 \sin^2 100t + 10 \cos^2 100t$.
[(a) $3(1 - \cos 40t)$, 0.05π s; (b) $2.5(1 + \cos 20t)$, 0.1π s; (c) $14 + 4 \sin 100t + \cos 200t$, 0.02π s]

13.9. The trigonometric Fourier series for a periodic wave is

$$f(t) = 5 - 10 \cos 4t + 5 \cos 12t - 9 \cos 20t + 4 \sin 4t + 15 \sin 12t - 4.5 \sin 20t + 8 \sin 28t$$

Produce a table showing the magnitude and phase angle for each term in the series assuming that the expression for the nth harmonic is $c_n \sin (n\omega t + \phi_n)$. Calculate the r.m.s. value of the wave.

[The table is

Order	*Modulus*	ϕ
0	5	0°
1	10.77	−68.2°
3	15.81	18.4°
5	10.06	−116.6°
7	8	0°

; 17.05]

13.10. A Fourier series expressed in the form

$$f(t) = a_0 + \sum_{n=1}^{\infty} c_n \cos(n\omega t + \phi)$$

has magnitude and phase angle spectrum for the first five terms as follows.

Order of harmonic	0	1	3	5	7
Magnitude	10	8	10	4	3
ϕ°	0	0	−45	135	90

Deduce the Fourier series for the wave.
[$10 + 8 \cos \omega t + 7.07 \cos 3\omega t - 2.83 \cos 5\omega t + 7.07 \sin 3\omega t - 2.83 \sin 5\omega t - 3 \sin 7\omega t$]

13.11. The following values were obtained from measurements on a voltage wave.

ϕ°	0	30	60	90	120	150	180	210	240	270	300	330
voltage (V)	−134	−444	−326	−106	−94	−144	−66	144	126	306	494	344

Calculate the value of the parameters of the Fourier series up to and including the third harmonic.
[$-100 \cos 2\omega t - 34 \cos 3\omega t - 300 \sin \omega t - 173 \sin 2\omega t - 94 \sin 3\omega t$]

14

Computer Solution of Electric Circuits

14.1 Introduction

A range of software packages is available for the analysis of circuits using mainframe, mini and personal computers. The attention of the reader is directed here to one of the most popular of them, namely the *Simulation Program with Integrated Circuit Emphasis* or *SPICE*, developed at the University of California, Berkeley, which is widely available.

With the space available, only an introduction can be given here; for more details, the reader should study a copy of the *User Guide* for the version of SPICE employed in his computer system.

SPICE enables circuit analysis to be performed on d.c. and a.c. circuits (or a combination of the two) which include elements such as resistors, inductors, coupled circuits, capacitors, semiconductor devices, dependent and independent sources, transmission lines, etc. SPICE can also be used to perform frequency response analysis, Fourier analysis, transient analysis, and can handle problems involving electronic noise, temperature effects, etc.

The reader should note that some of these features may not be available in some versions of SPICE for personal computers.

The circuit to be analysed is described using a *text editor*, which allows the user to create and modify the *input file* (one program line per circuit component). The first line of the file normally contains a title statement, and the file is terminated by a '.END' statement. Such element as voltage sources, current sources, resistors, inductors, transformers, capacitors, diodes, transistors, etc., can be included in the input file.

Next, the input file is read by *analyser*, which checks the file for errors; if none are present, it performs the required analysis and supplies the results. On completion, control is returned to the editor.

Depending on the type of analysis required, the results may either be in numerical form, or graphical form, or both.

A brief introduction to SPICE circuit elements is given in section 14.3, and greater detail can be obtained from section 14.4, where programming examples are described.

SPICE provides a faster solution to circuit problems than can be obtained by hand or calculator, and is less error-prone than conventional methods. However, the reader is reminded that computer solution is not an aid to learning! Computer solution comes into its own when the solution of the problem is either tedious and/or complex, or when a range of solutions is needed for different circuit parameter values.

All the PLOTs or graphs in this chapter are typical 'text' type displays produced by a printer. Most versions of SPICE have a graphics post-processor option, which provides a very high quality output on a printer or plotter.

We will, at this stage, look at a very simple circuit comprising two series-connected resistors, R1 and R2, supplied by a battery V1. An input file describing the circuit may be as follows.

```
Simple Circuit
V1   1  0  DC   10; 10 V d.c. source
R1   1  2    5; R1 connected between nodes 1 and 2
R2   2  0   15; R2 connected between nodes 2 and 0
.END          ; this line MUST be included
```

Each input file MUST either commence with a 'title line' ('Simple Circuit' in this case), or a blank line; the purpose of this line is to enable the function of the file to be described in one line. This is followed by a series of 'element' lines, *which may be written in any order*. In this case we have chosen the order V1, R1, R2; we could have chosen the order R2, R1, V1, or R1, V1, R1, etc. The final line must be a '.END' line; the '.' at the beginning of the line tells SPICE that this line contains a control function. Other control lines are introduced in sections 14.3 and 14.4 and, apart from '.END', control lines may be included at any stage in the file (but before '.END').

In-line comments can be added following a ';' at the end of each line. A complete line of comments can be included in the file by commencing the 'comment' line with a '*' (see worked example 14.1).

SPICE symbolic values

Value	*SPICE symbolic form*	*Exponent form*
10^{-15}	F	1E–15
10^{-12}	P	1E–12
10^{-9}	N	1E–9
10^{-6}	U	1E–6
10^{-3}	M	1E–3
10^{3}	K	1E3
10^{6}	MEG	1E6
10^{9}	G	1E9
10^{12}	T	1E12

14.2 Circuit description

Each terminal of each element in the circuit being analysed is connected to a node, the nodes being numbered in the range 0, 1, 2, 3 . . . n; the value of n is only limited by the version of SPICE being used. Node 0 is the reference (or ground) node. A two-terminal element, such as a resistor, an inductor, a capacitor, etc., is connected between two nodes, and a three-terminal element such as a transistor is connected between three nodes.

The value of a component, such as the resistance of a resistor, can be specified either by entering its actual value (usually a maximum seven figures), or by expressing the value in exponent form, such as 1.5E–6, or by writing it in symbolic form as shown in the table above. Thus, a resistance of a 1200 ohm resistor may be specified either as 1200, or as 1.2E3, or as 1.2K.

14.3 Element description

In the following, the SPICE specifications of a number of the more important circuit elements are described.

Resistors

The general specification is of the form

RXXXXXXX N1 N2 VALUE

in which RXXXXXXX is the designation of the resistor in the circuit being analysed; the XXXXXXX part of the designation is an alphanumeric string from one to seven characters. For example, this could be R2, R23, or R3AC4XY, etc. N1 and N2 are the nodes between which the resistor is connected, and VALUE is the resistance of the resistor in ohms. A typical specification may be

R2 1 2 3.5

The reader should note that SPICE does not accept conductance values, and that a resistance of zero ohms is not allowed.

Capacitors

These are specified in the form

CXXXXXXX N+ N− VALUE ⟨IC = ICOND⟩

where N+ is the positive node, N− is the negative node, and VALUE is the capacitance in farads. The ⟨ ⟩ brackets indicate that the contents are optional and, where present, give the initial charge on the capacitor between N+ and N− in volts.

Examples include

C3 1 0 0.5E–6

C29 4 5 0.7U IC = − 2.3

In the case of C29, node 4 is charged to −2.3 V with respect to node 5.

Inductors

The general specification is

LXXXXXXX N+ N− VALUE ⟨IC = ICOND⟩

where N+ is the positive node, N− is the negative node, and VALUE is the inductance in henrys. Where present, the initial current (in A) flows *inside the inductor* from N+ and N−.

Examples include

L4 4 5 5.6M

L6 2 3 4 IC = 1.5

In the case of L6, an initial current of 1.5 A flows from node 2 to node 3 *through the inductor*.

Where an inductor is part of a coupled circuit (see the following section), node N+ is the end marked by a 'dot' using the dot notation (see chapter 9).

Coupled circuits

A coupled circuit is presented as separate inductors with a specified magnetic coupling coefficient between the windings in the form

KXXXXXXX LYYYYYYY LZZZZZZZ VALUE

where XXXXXXX, YYYYYYY and ZZZZZZZ are alphanumeric strings from one to seven characters in length. KXXXXXXX is the designation of the coupled circuit, such as K8, while LYYYYYYY and LZZZZZZZ are the designation of the windings of the coupled circuit; the designation of each of the windings is given separately. If the transformer has two windings of self-inductance L_1 and L_2, and a mutual inductance M, the coupling coefficient between the windings is $k = M/\sqrt{(L_1L_2)}$, where $0 < k \leqslant 1$; the value of the coupling coefficient is given in the VALUE section of the transformer specification. A complete specification of a two-winding coupled circuit may be

```
K3 L1      L2 0.65
L1  4   0   2
L2  5   0   4
```

This represents a transformer with a coupling coefficient of 0.65, having a coil L_1 (of inductance 2 H) connected between nodes 4 and 0, and coil L_2 (of inductance 4 H) connected between nodes 5 and 0. The 'dotted' end of coil L_1 is connected to node 4, and the 'dotted' end of L_2 to node 5.

Independent voltage source

The general specification of an independent constant (d.c.) voltage source has the form

VXXXXXXX N+ N− ⟨DC⟩ VALUE

for example V4 1 0 DC 7.5
and for a sinusoidal a.c. source it has the form

VXXXXXXX N+ N− ⟨AC ⟨ACMAG ⟨ACPHASE⟩⟩⟩

for example VS 3 4 AC 6.7
where VXXXXXXX designates the voltage source in the circuit, N+ is the positive node, and N− the negative node. In the case of the constant (d.c.) source, VALUE is the direct voltage produced by the source; in the a.c. case the voltage is given by ACMAG, and ACPHASE is the phase angle of

the source. If ACMAG is omitted, a value of one volt is assumed; if ACPHASE is omitted, a phase angle of zero is assumed.

SPICE also allows a selected range of other independent sources such as a pulse, an exponential wave, etc., to be used.

Linear dependent voltage sources

The general form of a *linear voltage-controlled* (or *voltage-dependent*) *voltage source* is

EXXXXXXX N+ N− NC+ NC− VALUE

where N+ and N− are the positive and negative nodes, respectively, of the controlled source, while NC+ and NC− are the respective positive and negative nodes of the dependent (or controlling) voltage source. VALUE is the voltage gain from the controlling to the controlled source.

The general form for a *linear current-controlled* (or *current-dependent*) *voltage source* is

HXXXXXXX N+ N− VNAM VALUE

where N+ and N− have the meaning given above, and VNAM is the name of the voltage source through which the controlling current flows. The direction of positive controlling current flow is from the positive node, through the source, to the negative node of VNAM. VALUE is the transresistance (ohms) between the controlling current and the controlled voltage.

Independent current source

For a constant (d.c.) source the general specification has the form

IYYYYYYY N+ N− ⟨DC⟩ VALUE

for example I5 1 0 DC 5.3
and for an a.c. source it has the form

IYYYYYYY N+ N− ⟨AC ⟨ACMAG ⟨ACPHASE⟩⟩⟩

for example I42 5 3 AC 5.6

IYYYYYYY designates the current source within the circuit, and the arrow designating the direction of current flow through the source is from node N+ to N−. VALUE is the value of the current. In the case of an a.c. source, if ACMAG is omitted a current of 1 A is assumed, and if ACPHASE is omitted a phase angle of zero is assumed.

Linear dependent current sources

The general form for a *linear voltage-dependent* (or *voltage-controlled*) *current source* is

GXXXXXXX N+ N− NC+ NC− VALUE

where N+ and N− are the positive and negative nodes, respectively, of the controlled source, NC+ and NC− are the respective positive and negative nodes of the controlling source, and VALUE is the transconductance (in S) between the two. Current flow is from N+, through the source, to N−.

The general form for a *linear current-dependent* (or *current-controlled*) *current source* is

FXXXXXXX N+ N− VNAM VALUE

where N+ and N− are as defined above, VNAM is the name of the voltage source through which the controlling current flows, and VALUE is the current gain between the two.

Print-out analysis

The general form of statement is

.PRINT PRTYPE VAR1 ⟨VAR2 . . . VAR8⟩

in which PRTYPE is the type of analysis to be performed, such as d.c., a.c., transient, noise, etc. For example

.PRINT DC V(1) V(3, 4)

In the above example, V(1) is the voltage at node 1 (with respect to node 0), and V(3, 4) is the voltage at node 3 with respect to node 4.

In the case of an a.c. analysis, five forms of output can be obtained as follows:

VR – real part
VI – imaginary part
VM – magnitude
VP – phase shift
VDB – $20 \times \log_{10}$ (magnitude)

As with the d.c. case, differential voltage can be printed, for example

.PRINT AC VP(1, 2) VR(4, 5)

The current flowing in an independent voltage source can be specified as follows

I(VXXXXXXX)

and the current in the voltage source can be printed, for example

.PRINT AC IM(VS) IP(VS)

A voltage source must be present in a branch before the current in the branch can be printed; it is sometimes useful to insert an independent voltage source of zero volts in a branch in order to determine the current in that branch.

Plotting

The general form of plot is produced by the .PLOT statement described below

.PLOT PLTYPE OV1⟨PLO1, PHI1⟩ ⟨OV2⟩⟨PLO2, PHI2⟩ . . . ⟨OV8⟩

Up to eight output variables (OV1 to OV8) may be plotted; the type of analysis is specified by PLTYPE (which may be DC, AC, or TRAN (SIENT)). The low and high boundaries for the independent variables may be specified by the user (PLO, PHI), but if the plot limits are not specified, the SPICE program will automatically determine suitable minimum and maximum scale values.

Table 14.22 shows a typical plot of two variables (gain in dB and phase) plotted against frequency. Both of the vertical scales were chosen automatically. Note also that the value of the first specified output variable (gain) has its values printed alongside the independent variable (frequency) values. There is no limit to the number of .PLOT lines specified for each type of analysis.

Transient analysis

The general form of statement is

.TRAN TSTEP TSTOP ⟨TSTART⟩ ⟨TMAX⟩ ⟨UIC⟩

where TSTEP is the plotting and/or printing increment of time for the output. TSTOP is the final value of time, and TSTART is the initial value of time used in computing the transient response; the default value of the latter is zero. Using TMAX enables the computing time step to be smaller than TSTEP. The optional parameter UIC means 'Use Initial Conditions', whose values are specified using IC = . . . with inductors or capacitors, or a .IC line to specify initial node voltages.

.PRINT TRAN and .PLOT TRAN are two lines which are respectively used to print out a table of node voltages, and to plot a graph of these as a function of time. For example, the lines

```
.TRAN 0.1M  5M
.PLOT  TRAN  V(4)  V(1)
```

produce a graph of the voltages at nodes 4 and 1 in a network, starting at zero time and incrementing in steps of 0.1 ms for a duration of 5 ms. The graph abscissa points will be those of time, with the corresponding values of node 4 voltages at these times. The scales of both voltages will be calculated and marked automatically.

14.4 Programming examples

In the following we will consider a series of worked examples associated with each chapter in the book as follows:

Worked example number	*Chapter number*
14.1	1
14.2	2
14.3	3
14.4	4
14.5	5
14.6	6
14.7	7
14.8	8
14.9	9
14.10	10
14.11 and 14.12	11
14.13	12
14.14	13

The worked examples in this chapter solve examples which have been worked out manually in other chapters, and illustrate typical techniques for the solution of circuits. The reader should make reference to the circuits mentioned in the text.

Worked example 14.1

Calculate the value of I_1, I_2, I_3 and V_p in figure 1.23 using the SPICE circuit analysis package (see also worked example 1.17.1, page 23).

Solution

Since this is the first example of the use of SPICE, we will describe the program in a little more detail than usual.

The *input file* in table 14.1 contains a description of the circuit; the first line in the input file is a *title line*, which does not contain program data. Following the title line are a series of comment lines which commence with a '*'; the purpose of these lines is to make the program readable. The reader should note that SPICE assumes that current flows *inside* any source from a '+' node to a '−' node.

The reader should refer to figure 1.23 when studying the input file.

SPICE does not include an 'ammeter' as a circuit element, but it does evaluate the current flowing through each voltage source. We can therefore insert a zero-value voltage source at any point in the circuit where an ammeter is needed; voltage sources V1, V2 and V3 are used as 'ammeters' in this example. The reader will recall that the positive direction of current through such a source is from the '+' node to the '−' node *inside the source*.

Each element line describes one element in the circuit; circuit elements can be given in any order. In this case, the circuit elements are taken from left to right as figure 1.23 is viewed. A current source is an 'I' element, a resistor is an 'R' element, and a voltage source a 'V' element.

At the end of the input file, it is usual to specify the type of analysis required. Each analysis type is preceded by a 'dot', for example '.DC' or '.AC', etc. In this case we have not specified an analysis type, since the information provided in the input file will be sufficient to allow SPICE to give the required output, namely the voltage at each node and the current in the voltage sources V1, V2 and V3.

The '.OPTIONS' control line enables the user to specify one or more options which will apply to his circuit and/or input file. The .OPTIONS control line is itself optional, and can be omitted. In this case (as in other cases in the book), we have included the NOPAGE option, which causes page ejects to be suppressed, concatinating the print-out, and saving considerable amount of waste paper being produced. Finally, the input file is terminated by a '.END' command line.

The reader will find it an interesting exercise to draw the circuit using the nodes listed in the input file because, in fact, we have added the three voltage sources V1, V2 and V3 for the reason given earlier.

Each independent current source is described as an 'I' element, and characters are added to the 'I' to describe the source, that is, I1 and I2. Each current source line is as follows

IXXXXXX N1 N2 DC VALUE

where IXXXXXX is the circuit definition of the source, such as I1, etc. The

Table 14.1

```
*** Worked Example 14.1 ***
*** 14 A d.c. current source ***
*source name
*|  (+) node
*|    |   (-) node
*|    |    |    type of source
*|    |    |      |    value
*|    |    |      |      |
I1    0    1     DC     14
*** 2 ohm and 4 ohm resistors ***
*element
*|    (+) node
*|    |   (-) node
*|    |    |    value
*|    |    |     |
R1    2    0     2
R2    3    0     4
*** 4 A d.c. current source ***
I2    1    0     DC     4
*** 5 ohm resistor ***
R3    4    0     5
* The following ZERO VALUE voltage sources act as ammeters *
*source name
*|  (+) node
*|    |    (-) node
*|    |     |  type of source
*|    |     |    |    value
*|    |     |    |      |
V1    1     2   DC      0
V2    1     3   DC      0
V3    1     4   DC      0
* The following line saves paper
.OPTIONS NOPAGE
.END
```

current flows *within the source* from node N1 to node N2. The 'DC' statement is optional, but is included in this case to remind us that it is a d.c. source. If it is an a.c. current source, it would be defined as an 'AC' source, the 'AC' being mandatory. Finally, we include the VALUE of the current.

Resistors are defined as 'R' elements, the two numbers following the element definition are the nodes between which the resistor is connected. For example, the line

R2 3 0 4

tells us that resistor R2 is connected between nodes 3 and 0, and its value is 4 Ω. *SPICE uses node 0 as the reference or zero-voltage node.*

Most versions of SPICE do not have 'ammeters', but they can give details of the current flowing in an independent voltage source. In this case, we insert zero-value voltage sources V1, V2 and V3 in series with resistors R1, R2 and R3, respectively, in order to determine the current in each of the resistors. The independent voltage sources are defined as 'V' sources as follows

VXXXXXX N+ N− DC VALUE

where VXXXXXX is the name given in the circuit to the source. Node N+ is the positive terminal of the source, and N− is the negative terminal. Once again, 'DC' is optional and, in each case here, the value of the source is zero. When SPICE was written, it was decided that *positive current enters the positive terminal of an independent voltage source*. If the reader draws the circuit using the node numbering in the input file, he will see that the current in each voltage source enters the positive node. Should the node numbering be reversed, the corresponding value of current printed in the output file has a negative sign!

The output file in table 14.2 shows that the voltage at each node is 10.526 V, which corresponds to the result in worked example 1.17.1 and, likewise, the current in resistors R1, R2 and R3, corresponding to the current in voltage sources V1, V2 and V3.

Table 14.2

```
NODE   VOLTAGE     NODE   VOLTAGE     NODE   VOLTAGE     NODE   VOLTAGE
(  1)   10.5260    (  2)   10.5260    (  3)   10.5260    ( 4)    10.5260

       VOLTAGE SOURCE CURRENTS
       NAME          CURRENT
       V1            5.263E+00
       V2            2.632E+00
       V3            2.105E+00
```

Table 14.3

```
.PARAM current = 10
I1    0    1    DC    {1.4*current}
R1    2    0    2
R2    3    0    4
I2    1    0    DC    {0.4*current}
R3    4    0    5
```

In many applications it is convenient to use a *parameter* instead of a numeric value, and the usefulness of parameters is greatly extended by allowing them to be combined into *arithmetic expressions*. The early part of the input file in table 14.1 is modified as shown in table 14.3 to include a parameter called 'current', the parameter first being defined in a '.PARAM' control line; the value of 'current' is arbitrarily set at 10 A. Next, the parameter is included in an arithmetic expression enclosed in curly brackets { }, in the lines for I1 and I2. The *value* of I1 is

$$\text{'current'} \times 1.4 = 14 \text{ A}$$

and in the case of I2 it is

$$\text{'current'} \times 0.4 = 4 \text{ A}.$$

A salutary note must be given here about the use of computer software for circuit solution. While the software may give the solution quickly, it does not necessarily help the reader to understand the engineering processes involved.

Worked example 14.2

Calculate the voltage at each node in figure 2.14 using the SPICE programming language (see also worked example 2.9.3 on page 46).

Solution

The input file describing the circuit is given in table 14.4 SPICE cannot accept conductance values, and each is converted into its equivalent resistance value. To specify each resistance in this case, each has been numbered by the nodes it is connected to. For example, the 4 S conductance in figure 2.14 is described as *resistance* R13 of value a 0.25 Ω (SPICE only accepts resistance values), connected between node 1 and 3. Similarly with the current sources; the 3 A source which drives current from node 2 to node 3 is described as I23, etc.

This problem introduces us to the SPICE version of the voltage-controlled current source, which is a 'G' source. Using the above convention, we call it G13, which drives current from node 1 to node 3. The voltage-controlled current source is defined as follows

GXXXXXX N+ N− NC+ NC− VALUE

where GXXXXXX is the circuit 'name' of the source, and current is driven *within the source* from N+ to N−, that is, current leaves terminal N−. Node NC+ is the positive controlling node, NC− is the negative controlling node and VALUE is the transconductance in Siemens of the source.

Table 14.4

```
*** Worked Example 14.2 ***
*** Resistance values ***
R01     0    1    0.1667
R12     1    2    0.5
R23     2    3    0.3333
R03     0    3    0.2
R13     1    3    0.25
** Independent current sources **
I02     0    2    DC    2
I23     2    3    DC    3
** Voltage-controlled current source **
*source name
*|     N(+) N(-) NC(+) NC(-) transconductance
*|      |    |    |     |     |
G13     1    3    3     2    1.5
.OPTIONS NOPAGE
.END
```

Table 14.5

```
NODE   VOLTAGE     NODE   VOLTAGE     NODE   VOLTAGE
(  1)    .0707     (   2)   .0174     (  3)    .3151
```

The relevant part of the output file is given in table 14.5, and the voltage at each node is seen to agree with the results of worked example 2.9.3

Worked example 14.3

Determine V_{AB} and the current in each generator in figure 3.11(a) (see also worked example 3.9.1 on page 71).

Solution

The circuit in figure 3.11 is described in the SPICE input file in table 14.6. The first three element lines specify the three independent direct voltage sources and the nodes to which they are connected. In this case we have chosen to enter the 'positive' direction of the voltage in the input file. That is, the positive terminal of V3 is listed in table 14.6 as being connected to node 0. The next three element lines describe the value and circuit connections of each resistor. The reader should draw the corresponding circuit diagram, and note that the common point to which the resistors are connected is node 4 (corresponding to node B in figure 3.11(a)).

Once again, it is not necessary to specify the type of analysis needed, because SPICE will perform an operating point analysis and will calculate the voltage at each node in the output file, which is given in table 14.7.

The reader will note that the voltage at nodes 1, 2 and 3 correspond to the voltage of the respective independent sources, and the voltage at node 4 corresponds to V_{BA} in worked example 3.9.1. SPICE also outputs the current in each voltage source but, as described in worked example 14.1, it is the current *flowing into the positive terminal of the corresponding voltage source*. That is, we need to multiply each current in table 14.7 by -1 to get the conventional current.

The current flowing in the voltage source V3 is also given a negative value but, since the positive node of V3 is connected to node 0, the current flows through this source from node 0 to node 1. Once again, we see that an engineering vision is necessary when interpreting the results of SPICE.

Table 14.6

```
*** Worked Example 14.3 ***
*** Independent Voltage Sources ***
V1    1    0    DC    10
V2    2    0    DC    20
V3    0    3    DC    25
*** Resistors ***
R1    1    4    20
R2    2    4    15
R3    3    4    10
.OPTIONS NOPAGE
.END
```

Table 14.7

```
****     SMALL SIGNAL BIAS SOLUTION          TEMPERATURE =   27.000 DEG C
(   1)  10.0000  (    2)   20.0000  (    3)  -25.0000  (    4)   -3.0769

     VOLTAGE SOURCE CURRENTS
       NAME            CURRENT
       V1             -6.538E-01
       V2             -1.538E+00
       V3             -2.192E+00

       TOTAL POWER DISSIPATION   9.21E+01   WATTS
```

Worked example 14.4

Using SPICE software, plot the waveforms of voltage across and current through the inductor in figure 4.11(a) (see also worked example 4.7.1 on page 86).

Solution

In this example we will introduce some additional features of the SPICE language, the input file being shown in table 14.8. In order to plot a graph of a variable, we must first perform a transient analysis on the circuit and, in order to do this, we must excite the system with a time-dependent source. That is, we must describe a source to SPICE in terms of the value of the signal and the time at which it occurs.

In this case the current source is described by means of a *Piece-Wise Linear* (PWL) function as follows

IXXXXXX N+ N− PWL(T1, V1 T2, V2 . . . Tn, Vn)

where IXXXXXX is the name used to describe the independent current source, and the current flows inside the source from node N+ to N−, that is, the current leaves node N−. Pairs of values inside the PWL brackets describe pairs of coordinates of the current waveform, Tn, Vn corresponding to the *n*th value of time (seconds) and the *n*th value of current (amperes). Pairs of points are corrected linearly (the comma between the time and value for each point is optional). Thus the three points in the PWL wave describe a triangular current wave.

Table 14.8

```
*** Worked Example 14.4 ***
*** The following current source is described by a ***
*** Piece-Wise Linear function (PWL) ***
*source name
*|  N(+)  N(-)       T1,V1  T2,V2  T3,V3
*|    |     |          |      |      |
I1    0     1     PWL(0,0    3,3    6,0)
* VZERO is used as an ammeter
VZERO 1    2    0
* 6 H Inductor
*inductor name
*|   N(+) N(-) value
*|    |    |    |
L1    2    0    6
.OPTIONS NOPAGE
*transient analysis
*|       Tstep  Tstop
*|         |      |
.TRAN    0.3    6.6
* PLOT the TRANsient analysis
.PLOT    TRAN    V(2)    I(VZERO)
.END
```

As stated earlier, current flows into the 'positive' pole of an independent voltage source, and SPICE can evaluate the current flowing in an independent voltage source. Therefore, in order to measure the current in the circuit, we introduce the concept of a zero-value voltage source, VZERO, in this circuit. The purpose is to use the source as an 'ammeter'. Finally, a 6 H inductor is connected to the PWL current source and the 'ammeter'.

The '.TRAN' command line causes SPICE to compute the transient response of the circuit every 0.3 s for a period of 6.6 s. Finally, the '.PLOT TRAN' line causes SPICE to plot the TRANsient response showing not only the voltage across the inductor (V(2)), but also the current in the circuit (I(VZERO)). The reason that we ask for I(VZERO) to be plotted is that, in order either to print or plot a current, *it must flow through a voltage source*.

Table 14.9

```
LEGEND:
*: V(2)
+: I(VZERO)
   TIME        V(2)
(*)------         -1.0000E+01  -5.0000E+00   0.0000E+00   5.0000E+00   1.0000E+01
(+)------         -4.1211E-30   1.0000E+00   2.0000E+00   3.0000E+00   4.0000E+00

   0.000E+00  0.000E+00 +            .            *            .           .
   3.000E-01  6.000E+00 .   +        .            .            .  *        .
   6.000E-01  6.000E+00 .       +    .            .            .  *        .
   9.000E-01  6.000E+00 .           +.            .            .  *        .
   1.200E+00  6.000E+00 .            .  +         .            .  *        .
   1.500E+00  6.000E+00 .            .      +     .            .  *        .
   1.800E+00  6.000E+00 .            .         +  .            .  *        .
   2.100E+00  6.000E+00 .            .            .+           .  *        .
   2.400E+00  6.000E+00 .            .            .    +       .  *        .
   2.700E+00  6.000E+00 .            .            .        +   .  *        .
   3.000E+00  6.000E+00 .            .            .            +  *        .
   3.300E+00 -6.000E+00 .         *  .            .        +   .           .
   3.600E+00 -6.000E+00 .         *  .            .    +       .           .
   3.900E+00 -6.000E+00 .         *  .            .+           .           .
   4.200E+00 -6.000E+00 .         *  .         +  .            .           .
   4.500E+00 -6.000E+00 .         *  .      +     .            .           .
   4.800E+00 -6.000E+00 .         *  .  +         .            .           .
   5.100E+00 -6.000E+00 .         * +.            .            .           .
   5.400E+00 -6.000E+00 .       + *  .            .            .           .
   5.700E+00 -6.000E+00 .   +     *  .            .            .           .
   6.000E+00 -6.000E+00 +         *  .            .            .           .
   6.300E+00  5.673E-20 +            .            *            .           .
   6.600E+00  0.000E+00 +            .            *            .           .
```

The corresponding section of the output file is shown in table 14.9. The reader will note that we have not specified the range over which the results are to be plotted, and we have left it to SPICE to decide. With the form of print-out used here, it is not possible to show the sudden transition in voltage across the inductor when the current changes in value, and the reader should not think that the current changes linearly between the points on the graph. Several suppliers of versions of SPICE offer a graphics post-processor to provide an accurate, high-quality graphics output.

Worked example 14.5

Solve the a.c. series circuit in worked example 5.10.1 (see page 112) using SPICE.

Solution

The input file is given in table 14.10, and it is left as an exercise for the reader to draw the corresponding circuit diagram. SPICE can accept voltages and currents in polar complex form, and can output data in either polar or rectangular form. The first element line in table 14.10 describes the voltage source as an a.c. source connected between node 1 and node 0, and having a magnitude of 10 V and a phase angle of 20°.

Unfortunately, SPICE can only accept information about inductance and capacitance, and cannot accept reactance data. However, there are several ways round this problem, and one is as follows. If we use a frequency of 1 rad/s (or 0.1592 Hz), then we get the correct result if we let $L = X_L$, and $C = 1/X_C$. This is done here.

Consider impedance Z1, which comprises a resistance of 7.071 Ω in series with an inductor $L_1 \equiv X_{L1} = 7.071$ H. We therefore show a resistance of 7.071 Ω connected between nodes 1 and 2, and an inductor of 7.071 H connected between nodes 2 and 3. The latter has a reactance of 7.071 Ω at the excitation frequency of 1 rad/s or 0.1592 Hz. Similarly for impedance Z3, where we have a resistance of 2.605 Ω in series with a capacitive reactance of 14.77 Ω. The corresponding 'capacitance' is $C_3 = 1/X_{C3} = 1/14.77 = 0.0677$ F = 67.7 mF. Finally we use an independent zero-value voltage source, Vam, as an ammeter.

The method of dealing with the frequency of the source is shown in the '.AC' control line, and is specified in the following line.

```
.AC DIST NOPOINTS STARTFREQ ENDFREQ
```

This control line is, strictly speaking, intended to deal with a range of frequencies for frequency response calculations; in this case we only need one frequency. Where the word DIST appears we enter the way in which the frequencies are distributed between the STARTFREQ frequency and the ENDFREQ frequency (both in Hz). There are three options, namely LIN (LINear), OCT (OCTave) and DEC (DECade); we choose LIN. Where NOPOINTS appears we insert the number of points to be calculated. Thus the line

.AC LIN 1 0.1592 0.1592

implies that we need one frequency between 1 rad/s and 1 rad/s!

There are two '.PRINT' command lines, each asking for the current through the 'ammeter' and the voltage of node 3 with respect to node 4 to be printed. The two lines do, however, ask for the data in a different form.

Table 14.10

```
*** Worked Example 14.5 ***
*** Voltage Source ***
*source name
*|    N(+) N(-) type of source
*|     |    |    | magnitude  phase (deg)
*|     |    |    |     |         |
VS     1    0    AC    10        20
*** Impedance Z1 ***
R1     1    2    7.071
* Let L1 = XL1
L1     2    3    7.071
*** Impedance Z2 ***
R2     3    4    10
*** Impedance Z3 ***
R3     4    5    2.605
* Let C3 = 1/XC3
C3     5    6    67.7M
*** Ammeter ***
Vam     6    0    AC    0
*analysis type
*|    LINear  number of frequencies
*|       |      | start freq  end freq
*|       |      |     |          |
.AC     LIN     1    0.1592    0.1592
.OPTIONS NOPAGE
* magnitude of I(Vam)          phase of I(Vam)
*                  |            |
.PRINT     AC    IM(Vam)      IP(Vam)      VM(3,4)     VP(3,4)
* real part of I(Vam)          imaginary part of I(Vam)
*                  |            |
.PRINT     AC    IR(Vam)      II(Vam)      VR(3,4)     VI(3,4)
.END
```

The first of these lines asks for the data in polar form as follows

IM(Vam) = current Magnitude through Vam
IP(Vam) = current Phase angle, etc.

and the second line asks for

IR(Vam) = Real part of the current through Vam
II(Vam) = Imaginary part of the current through Vam

The relevant parts of the output file are given in table 14.11.

Table 14.11

```
FREQ          IM(Vam)       IP(Vam)       VM(3,4)       VP(3,4)
   1.592E-01   4.733E-01   4.136E+01    4.733E+00    4.136E+01
FREQ          IR(Vam)       II(Vam)       VR(3,4)       VI(3,4)
   1.592E-01   3.553E-01   3.128E-01    3.553E+00    3.128E+00
```

Worked example 14.6

Use SPICE computer software to calculate the voltage gain of the transistor amplifier equivalent circuit in figure 6.7(a) at a frequency of (a) 1 kHz and (b) 200 kHz (see worked example 6.8.1 on page 139).

Solution

The circuit is fairly straightforward, with the exception that the transistor is simulated by a voltage-controlled current source. The input file is shown in table 14.12, and the appropriate section of the output file in table 14.13. The reader will note that the results agree with those of worked example 6.8.1.

The reader is advised at this point of a difficulty which may arise in some a.c. problems, although it does not occur here, and that is SPICE is organised (for many types of solution) only to deal with angles which lie in the range ±180°. If the angle is (for example) less than −180°, it will be shown as the corresponding positive angle; similarly, positive angles greater than +180° are shown as negative angles. That is, once again, engineering judgement must be applied to the results.

Also, this analysis (and that of worked example 6.8.1) is based on small-signal analysis, and does not allow for practical features such as the effect of 'saturation' of the transistor if too large a voltage is applied to the input.

Table 14.12

```
*** Worked Example 14.6 ***
*** 10 mV a.c. voltage source ***
VS     3    0    AC    10M
*** Circuit components ***
R31    3    1    100
R10    1    0    10K
R12    1    2    100K
R20    2    0    150
C12    1    2    25P
*** Voltage-controlled current source ***
*name of source
*|    (+) output node
*|     |   (-) output node
*|     |    |   (+) controlling node
*|     |    |    |   (-) controlling node
*|     |    |    |    |   transconductance
*|     |    |    |    |    |
G20    2    0    1    0    8
.OPTIONS NOPAGE
.AC        LIN    2    1k    200k
.PRINT     AC    VM(2)     VP(2)
.END
```

Table 14.13

```
      FREQ         VM(2)        VP(2)
   1.000E+03    5.423E+00    1.795E+02
   2.000E+05    2.741E+00    1.204E+02
```

Worked example 14.7

Solve the unbalanced three-phase system in worked example 7.9.1 using SPICE software (see page 153).

Solution

In this case, using the techniques described earlier in the book, we specify a set of unbalanced voltages and unbalanced loads. The input file is given in table 14.14, and the first three lines specify the unbalanced phase voltages as follows

$$V_{an} = 200 \angle 10° \text{ V}$$

$$V_{bn} = 220 \angle -140° \text{ V}$$

$$V_{cn} = 180 \angle 100° \text{ V}$$

where node 0 is the neutral point and lines a, b and c are connected to SPICE nodes 1, 2 and 3, respectively. Three zero-value voltage sources Va, Vb and Vc are used as 'ammeters' to measure the three line currents.

As with worked example 14.5, we use a frequency of 1 rad/s (0.1592 Hz), so that the 'inductance' in phase B is equal to the value of the inductive reactance in phase B, and the capacitance in phase C is equal to $1/X_C$. The '.AC' line specifies that we are using a single frequency of 0.1592 Hz or 1 rad/s.

Finally there is a set of '.PRINT' control lines causing the results obtained to be output; these are listed in table 14.15. Each '.PRINT' line in the input file is seen to produce its own set of results and, in the absence of the '.OPTIONS NOPAGE' line, each will generate a separate page of

Table 14.14

```
*** Worked Example 14.7 ***
*** Unbalanced 3-phase supply ***
Van     1    0     AC     200      10
Vbn     2    0     AC     220      -140
Vcn     3    0     AC     180      100
*** Ammeters in lines a, b and c ***
Va      1    7     AC     0
Vb      2    8     AC     0
Vc      3    9     AC     0
*** Z in phase A = Zas ***
Ras     7    6     10
*** Z in phase B = Zbs ***
Rbs     8    4     14.77
Lbs     4    6     2.605
*** Z in phase C = Zcs ***
Rcs     9    5     4.698
Ccs     5    6     0.5848
.OPTIONS NOPAGE
.AC     LIN    1     0.1592     0.1592
*** V(6) = Vsn ***
.PRINT     AC     VM(6)      VP(6)
*** V(7,6) = Vas, V(8,6) = Vbs ***
.PRINT     AC     VM(7,6)    VP(7,6)     VM(8,6)     VP(8,6)
*** V(9,6) = Vcs ***
.PRINT     AC     VM(9,6)    VP(9,6)
*** I(Va) = Ia, I(Vb) = Ib ***
.PRINT     AC     IM(Va)     IP(Va)      IM(Vb)      IP(Vb)
*** I(Vc) = Ic ***
.PRINT     AC     IM(Vc)     IP(Vc)
.END
```

Table 14.15

```
  FREQ           VM(6)          VP(6)
    1.592E-01     8.224E+01     1.028E+02
  FREQ           VM(7,6)        VP(7,6)        VM(8,6)        VP(8,6)
    1.592E-01     2.200E+02    -1.193E+01     2.678E+02    -1.241E+02
  FREQ           VM(9,6)        VP(9,6)
    1.592E-01     9.794E+01     9.764E+01
  FREQ           IM(Va)         IP(Va)         IM(Vb)         IP(Vb)
    1.592E-01     2.200E+01    -1.193E+01     1.785E+01    -1.341E+02
  FREQ           IM(Vc)         IP(Vc)
    1.592E-01     1.959E+01     1.176E+02
```

paper! When the results are compared with those of worked example 7.9.1, the reader will note that they are the same and will, of course, appreciate that while the computer analysis is straightforward, it does not necessarily give a clear understanding either of the processes involved or of the practicability of the results. Only an experienced 'engineering eye' can give the latter.

Worked example 14.8

Using SPICE software, calculate the value of the parameter $\boldsymbol{y}_{21}$ in the circuit in worked example 8.3.1 (see page 178).

Solution

Strictly speaking, SPICE was not developed for this type of problem, but it is so versatile that it can be used to solve most problems. The only limitation to its use is in our own mind! The reader will recall that

$$\boldsymbol{y}_{21} = \left.\frac{\boldsymbol{I}_2}{\boldsymbol{V}_1}\right|_{\boldsymbol{V}_2 = 0}$$

All we need to do is to apply a 1 V source to the input, and measure $\boldsymbol{I}_2$ with the output short-circuited. This is a simple matter to implement with SPICE.

The input file is shown in table 14.16, in which node 0 is the node common to the input and the output, node 1 being the input node, and node 2 is the output node. In this case, V2 acts as an 'ammeter', with 'positive' current flowing into node 2. The appropriate part of the output file is given in table 14.17.

As with earlier SPICE files, we do not need to specify the type of

Table 14.16

```
*** Worked Example 14.8 ***
*** Calculation of y21 = I2/V1 with V2 = 0 ***
*** Input source ***
V1    1    0    DC    1
*** passive circuit elements ***
R1    1    0    600
R2    1    2    1.5K
R3    2    0    10K
*** Voltage-controlled current source ***
G     2    0    1    0    0.04
*** V2 = 0 is used as an ammeter ***
V2    0    2    DC    0
*** Note: y21 = current in short-circuiting "ammeter" V2 ***
.OPTIONS NOPAGE
.END
```

Table 14.17

```
VOLTAGE SOURCE CURRENTS
NAME          CURRENT
V1           -2.333E-03
V2            3.933E-02
```

analysis to be performed, since we only need the current flowing into the output node, which is 0.03933 A, hence $\boldsymbol{y}_{21} = \boldsymbol{I}_2/\boldsymbol{V}_1 = 39.33$ mS.

Worked example 14.9

Using SPICE software, solve the linear transformer problem in worked example 9.8.1 (see page 212 and figure 9.10(a)) for the inductive load.

Solution

In this case we will drive the circuit with a $1 \angle 0°$ A a.c. source (I1), whose current flows from node 0 to node 1; the magnitude and the phase angle of the source are given following the 'AC' expression. The default phase angle is zero and can be omitted, but is included in the specification of I_1 for completeness.

The reason for selecting a 1 A current source is that the input impedance of the circuit is equal to the voltage at the input terminals. The SPICE primary circuit consists of a 10 Ω resistance R1, connected between nodes 1 and 2, and a 75 mH inductance L1, connected between nodes 2 and 0. The secondary circuit contains the following: a 10 Ω resistance R2 between nodes 4 and 5, a 150 mH inductance L2 between nodes 4 and 3 together

```
****     CIRCUIT DESCRIPTION     ****

* The following is an a.c. CURRENT source of 1 A
I1    0    1    AC    1    0
* See figure 9.10(a) for the circuit elements
R1   1    2    10
R2   4    5    10
* PRIMARY winding
L1   2    0    75M
* SECONDARY winding
L2   4    3    150M
* MAGNETIC COUPLING
* name of coupled circuit
* ¦   name of first inductor
* ¦   ¦   name of second inductor
* ¦   ¦   ¦   magnetic coupling coefficient
* ¦   ¦   ¦   ¦
K1  L1  L2  0.75425
* Extra resistance to 'electrically' link the * primary winding to the
secondary winding.
Rextra 0    3      1MEG
* Inductive load
Lload  5    3      0.4
.OPTIONS NOPAGE
* Calculate values at 500, 1000 and 1500 rad/s
.AC    LIN    3    79.58    238.7
* Print the rectangular form of the input voltage
.PRINT   AC        VR(1)    VI(1)
.END
```

with the load. The latter is a 0.4 H inductance Lload connected between nodes 3 and 5. Coils L1 and L2 are magnetically coupled with a coupling coefficient of

$$k = M/\surd(\text{L1} \times \text{L2}) = 80 \times 10^{-3}/\surd\,(75 \times 10^{-3} \times 150 \times 10^{-3})$$
$$= 0.75425$$

Since SPICE demands that all nodes must have a 'd.c. link' to node 0, a large value of resistance, Rextra, is used to connect the primary and secondary windings together. Since current does not flow in this resistor, it has no significant effect on the result of the analysis.

The '.AC' control line calls for calculations at three frequencies, namely 500, 1000 and 1500 rad/s, whose values are converted into Hz in the .AC line.

In worked example 9.8.1, the results were presented in rectangular form, and this is what we have done here. The .PRINT control line requests the 'real' part of the voltage at node 1, i.e., (VR(1)), and the 'imaginary' part (VI(1)) to be printed. The reader should compare the value of the input impedance at 159.2 Hz (1000 rad/s) with that obtained above. It is, of course, a simple process to repeat the calculation for a capacitive load.

The relevant part of the output file is as follows.

FREQ	VR(1)	VI(1)
7.958E + 01	1.021E + 01	3.169E + 01
1.592E + 02	1.021E + 01	6.337E + 01
2.387E + 02	1.021E + 01	9.505E + 01

Worked example 14.10

Plot graphs of $i_1(t)$ and $i_2(t)$ in figure 10.29(a) – see also worked example 10.13.1 on page 259.

Solution

In this case we will choose to excite the circuit with a PULSE, as described in the V1 line in the input file (see table 14.18). A reason for selecting this form of driving function is that we need to do a transient response (see the '.TRAN' control line) on the output, and, for this purpose it is necessary to energise the system by a time-dependent function, the repetitive PULSE being one example. The PULSE function has the following variables

PULSE(IV PV TD TR TF PW PP)

where

IV is the initial value = 0
PV is the pulsed value = 10 V
TD is the delay time before the pulse is applied = 0
TR is the rise time of the pulse = 1 μs
TF is the fall time of the pulse = 1 μs
PW is the pulse width = 50 ms
PP is the pulse period = 51 ms

In this case, it is merely necessary to make the pulse width longer than the transient period involved in the problem, and the pulse period longer than the sum of TR, TF and PW.

Perhaps, at this time, we should look at the '.TRAN' control line. This calls for SPICE to evaluate the transient solution for 40 ms in steps of 2 ms. The reader will recall that we estimated that the secondary current would reach its steady-state value (zero) in about 40 ms. Consequently, a transient period of 40 ms is allowed for in the input file; a step period of 2 ms provides (40/2) + 1 = 21 results in this period.

Returning to the circuit description, 'Vpri' and 'Vsec' act, respectively, as primary and secondary ammeters. The reader will note the ';' separator

Table 14.18

```
*** Worked Example 14.10 ***
*** The input source is a PULSE ***
*output nodes   pulsed value        fall time
*   |    | initial value | time delay       | pulse width
*   |    |            |    |      | rise time |      |   period
*   |    |            |    |      |      |      |      |    |
V1  1    0   PULSE (0    10    0    1U     1U    50M   51M)
*** Ammeters ***
Vpri     1    2    0 ;primary ammeter
Vsec     4    7    0 ;secondary ammeter
*** Primary circuit resistance ***
Rpri     2    3    20
*** Load ***
Rload    7    6    20
Lload    6    5    0.01
*** Coupled circuit ***
L1       3    0    0.1
L2       4    5    0.1
K        L1   L2   0.5
*** d.c. linking resistor ***
Rlink    0    5    1MEG
.OPTIONS NOPAGE
.TRAN    2M    40M
.PLOT    TRAN    I(Vsec)    (0,0.1)    I(Vpri)    (0,0.5)
.END
```

Table 14.19

```
LEGEND:
*: I(Vsec)
+: I(Vpri)

   TIME        I(Vsec)
(*)------          0.0000E+00   2.5000E-02   5.0000E-02   7.5000E-02   1.0000E-01
(+)------          0.0000E+00   1.2500E-01   2.5000E-01   3.7500E-01   5.0000E-01

 0.000E+00 0.000E+00 X              .              .              .              .
 2.000E-03 7.193E-02 .              .       +      .           *  .              .
 4.000E-03 9.129E-02 .              .              .     +        .        *     .
 6.000E-03 8.735E-02 .              .              .             +.      *       .
 8.000E-03 7.561E-02 .              .              .              *  +           .
 1.000E-02 6.229E-02 .              .              .      *       .      +       .
 1.200E-02 4.996E-02 .              .              *              .        +     .
 1.400E-02 3.951E-02 .              .        *     .              .         +    .
 1.600E-02 3.093E-02 .              .  *           .              .          +   .
 1.800E-02 2.411E-02 .              *              .              .           +  .
 2.000E-02 1.871E-02 .           *  .              .              .           +  .
 2.200E-02 1.452E-02 .        *     .              .              .             +.
 2.400E-02 1.123E-02 .      *       .              .              .             +.
 2.600E-02 8.693E-03 .    *         .              .              .             +.
 2.800E-02 6.717E-03 .  *           .              .              .             +.
 3.000E-02 5.196E-03 .  *           .              .              .              +
 3.200E-02 4.013E-03 . *            .              .              .              +
 3.400E-02 3.103E-03 . *            .              .              .              +
 3.600E-02 2.396E-03 .*             .              .              .              +
 3.800E-02 1.853E-03 .*             .              .              .              +
 4.000E-02 1.430E-03 .*             .              .              .              +
```

used between the circuit element field and the comment field. The electrical circuit and the magnetically coupled circuit are defined in the input file in the manner described earlier in the book. Once again, all nodes must have a 'd.c.' link to node 0, so that a large value of linking resistor, Rlink, connects one point in the primary circuit to one point in the secondary circuit.

Finally, we will take a look at the method of displaying the output data (see also table 14.19). The '.PLOT' control line calls on SPICE to PLOT the current through the voltage source Vsec and through Vpri, in that order. However, the .PLOT control line used here differs from that used hitherto, since we use it to set limiting values to the range over which the current is plotted. As a general rule, if the user does not know the range over which the results will appear, it is advisable to let SPICE itself make the decision. A disadvantage which sometimes occurs with this arrangement is that the resulting graphs do not quite appear in the position that one would like!

In this case, we have a general idea of the values involved, and we can specify them in the .PLOT line. The limits are quoted in parenthesis, with the minimum value first and the maximum value second. That is, the secondary current is to be plotted between values of zero and 0.1 A, and the primary current between zero and 0.5 A. The output from the printer is shown in table 14.19, with each point on the graph of the first named variable, namely I(Vsec), being plotted with a '*', and each of the points on the graph of the second named variable, I(Vpri), being plotted with a '+'.

When I(Vsec) and I(Vpri) have the same value (as they both do initially), the point is marked with an upper-case 'X'.

The table of results on the left of the graphs shows, in the first column the time in ms and, in the second column the value of the first named variable, namely I(Vsec). A print-out of all the values of all variables could have been obtained had we included a '.PRINT' control line in the input file. Once again, we see that the transients have practically reached their final value in about 40 ms.

Worked example 14.11

Plot the frequency response diagram of the circuit in worked example 11.11.1 (see page 285) using SPICE software.

Solution

The circuit is described in the input file in table 14.20 in standard SPICE format. In this case we have chosen to drive the circuit using an a.c. current source of $1 \angle 0°$ A, so that the voltage at node 1 (the input) is equal to the input impedance of the circuit. The '.AC' control line requests the computer to calculate the results in frequency decades, five points per decade, from 0.1592 Hz (1 rad/s) to 1592 Hz (10 000 rad/s), that is, over 4 decades.

The '.PRINT' and '.PLOT' control lines request the computer to output the magnitude in dB and the phase shift (in degrees). The relevant sections of the output file are shown in tables 14.21 and 14.22.

The .PRINT control line causes both the decibel value of the voltage at node 1 and the corresponding phase angle to be printed for each frequency (see also table 14.11).

The .PLOT line does not specify the range over which the results are to be plotted and (see table 14.22) SPICE chose the following

VDB(1): 20 dB to 100 dB, and marks points with a '*'
VP(1): $-50°$ to $150°$, and marks points with a '+'.

However, the .PLOT control line only results in the value of the first named variable in the line to be printed, that is, VDB(1). The first column of results in table 14.22 corresponds to the frequencies at which the calculations occur, and the second column contains the corresponding values of VDB(1).

Table 14.20

```
*** Worked Example 14.11 ***
I    0    1    AC    1
R    1    2    25
L    2    0    0.25
.OPTIONS NOPAGE
.AC     DEC     5     0.1592     1592
* The voltage V(1) is equal to the circuit impedance *
.PRINT    AC    VDB(1)    VP(1)
.PLOT     AC    VDB(1)    VP(1)
.END
```

Table 14.21

FREQ	VDB(1)	VP(1)
1.592E-01	2.796E+01	5.731E-01
2.523E-01	2.796E+01	9.083E-01
3.999E-01	2.796E+01	1.439E+00
6.338E-01	2.797E+01	2.280E+00
1.004E+00	2.798E+01	3.611E+00
1.592E+00	2.800E+01	5.712E+00
2.523E+00	2.807E+01	9.008E+00
3.999E+00	2.822E+01	1.410E+01
6.338E+00	2.860E+01	2.171E+01
1.004E+01	2.941E+01	3.226E+01
1.592E+01	3.097E+01	4.501E+01
2.523E+01	3.342E+01	5.776E+01
3.999E+01	3.660E+01	6.830E+01
6.338E+01	4.023E+01	7.590E+01
1.004E+02	4.407E+01	8.100E+01
1.592E+02	4.800E+01	8.429E+01
2.523E+02	5.198E+01	8.639E+01
3.999E+02	5.597E+01	8.772E+01
6.338E+02	5.996E+01	8.856E+01
1.004E+03	6.396E+01	8.909E+01
1.592E+03	6.796E+01	8.943E+01

Table 14.22

```
LEGEND:
*: VDB(1)
+: VP(1)

(*)--------       2.0000E+01   4.0000E+01   6.0000E+01   8.0000E+01   1.0000E+02
(+)--------      -5.0000E+01   0.0000E+00   5.0000E+01   1.0000E+02   1.5000E+02

 1.592E-01  2.796E+01  .    *       +            .            .            .
 2.523E-01  2.796E+01  .    *       +            .            .            .
 3.999E-01  2.796E+01  .    *       +            .            .            .
 6.338E-01  2.797E+01  .    *       .+           .            .            .
 1.004E+00  2.798E+01  .    *       .+           .            .            .
 1.592E+00  2.800E+01  .    *       .+           .            .            .
 2.523E+00  2.807E+01  .    *       . +          .            .            .
 3.999E+00  2.822E+01  .    *       .   +        .            .            .
 6.338E+00  2.860E+01  .     *      .     +      .            .            .
 1.004E+01  2.941E+01  .     *      .       +    .            .            .
 1.592E+01  3.097E+01  .      *     .           +.            .            .
 2.523E+01  3.342E+01  .        *   .            . +          .            .
 3.999E+01  3.660E+01  .          * .            .    +       .            .
 6.338E+01  4.023E+01  .            *            .      +     .            .
 1.004E+02  4.407E+01  .            .  *         .        +   .            .
 1.592E+02  4.800E+01  .            .    *       .        +   .            .
 2.523E+02  5.198E+01  .            .       *    .        +   .            .
 3.999E+02  5.597E+01  .            .         *  .         +  .            .
 6.338E+02  5.996E+01  .            .            *         +  .            .
 1.004E+03  6.396E+01  .            .            .  *      +  .            .
 1.592E+03  6.796E+01  .            .            .    *    +  .            .
```

Worked example 14.12

Using PSpice software, plot the gain and phase response of the generalised quadratic transfer function

$$\boldsymbol{H}(\mathrm{j}\omega) = 1 + 2\zeta\left[\frac{\mathrm{j}\omega}{\omega_0}\right] + \left[\frac{\mathrm{j}\omega}{\omega_0}\right]^2 \text{ for } \zeta = 0.1 \text{ (see also section 11.13).}$$

Solution

PSpice is a version of SPICE which can handle complex transfer functions which are expressed in the s-domain, that is as a Laplace transformation (see also chapter 10). For the purpose of the program, we allow the complex operator jω to be replaced by s, that is, $s \equiv \mathrm{j}\omega$. The input file in table 14.23 gives a solution to the problem.

PSpice deals with an s-domain transfer function as though it were the 'gain' of a voltage-dependent voltage source, or an 'E' source. This (the s-domain transfer function) has been given the name 'Equadratic' in the program. The output from the source is between nodes 2 and 0, and the input is the voltage at node 1, that is, V(1). In the line defining the transfer function, the expression LAPLACE informs the computer that it must expect a transfer function as a Laplace transform for $\zeta = 0.1$, which it expects to find within a pair of { } brackets.

As with most SPICE programs, the software expects to find a 'd.c.' link between each node and node 0, and a one megohm resistor is connected between the input source nodes (node 1 and 0) and between the output nodes (nodes 2 and 0).

The .PLOT control line requests a plot of the output magnitude response in dB, together with phase shift in degrees to a base of frequency plotted to a decade scale of frequency (10 points per decade) – see also the '.AC' control line. The resulting output is shown in table 14.24. The points

Table 14.23

```
*** Generalised Quadratic Transfer Function ***
* H(s) = 1 + (2*zeta*s/omega O) + (s*s/(omega O*omega O))*
*** where s replaces jw ***
V        1    0    AC    1
*** ZETA = 0.1, omega O = 1 ***
*name of source
*   |     (+) output node
*   |      | (-) output  node
*   |      |  | LAPLACE transform
*   |      |  |       |   signal source
*   |      |  |       |     |            LAPLACE transform expression
*   |      |  |       |     |                       |
Equadratic 2 0 LAPLACE {V(1)} = {1 + (2*0.1*s/1) + (s*s/(1*1)}
* A high value resistor must be connected between node 1 and 0 *
* and between node 2 and 0 *
R1        1    0    1MEG
R2        2    0    1MEG
.OPTIONS NOPAGE
*** Logarithmic plot of frequencies between 0.1 and 10 rad/s ***
*AC sweep
*|   sweep in DECades
*|        |      10 points per decade
*|        |       |  start freq  end freq
*|        |       |       |         |
.AC      DEC     10    15.92M    1.592
.PLOT    AC     VDB(2)    VP(2)
.END
```

corresponding to the gain in dB are plotted using '*', and the phase angles are plotted using '+'. The values on the left of table 14.24 are, firstly, a list of frequencies (in Hz) and, secondly, corresponding gain values (in dB) of the transfer function.

When evaluating phase angles using SPICE, the reader should be aware that some versions output angles for certain solutions only in the range ±180°. That is, if the phase angle associated with a given transfer function slowly changes from, say, −90° to −270°, then SPICE may convert phase angles in excess of −180° into their 'positive' phase angle equivalent, that is, −190° becomes +170°.

The graph in table 14.24 should be compared with those in section 11.13.

Table 14.24

```
LEGEND:
*: VDB(2)
+: VP(2)

  FREQ          VDB(2)
(*)------            -2.0000E+01  0.0000E+00  2.0000E+01  4.0000E+01  6.0000E+01
(+)------             0.0000E+00  5.0000E+01  1.0000E+02  1.5000E+02  2.0000E+02

  1.592E-02 -8.557E-02 +          *          .          .          .
  2.004E-02 -1.360E-01 +          *          .          .          .
  2.523E-02 -2.165E-01 +          *          .          .          .
  3.176E-02 -3.456E-01 .+         *          .          .          .
  3.999E-02 -5.540E-01 .+         *          .          .          .
  5.034E-02 -8.943E-01 .+        *.          .          .          .
  6.338E-02 -1.461E+00 .+        *.          .          .          .
  7.979E-02 -2.437E+00 . +      * .          .          .          .
  1.004E-01 -4.226E+00 .  +    *  .          .          .          .
  1.265E-01 -7.927E+00 .    + *   .          .          .          .
  1.592E-01 -1.398E+01 .  *       .       +  .          .          .
  2.004E-01 -3.909E+00 .       *  .          .          . +        .
  2.523E-01  3.785E+00 .          . *        .          .    +     .
  3.176E-01  9.571E+00 .          .    *     .          .     +    .
  3.999E-01  1.455E+01 .          .       *  .          .     +    .
  5.034E-01  1.911E+01 .          .         *.          .      +   .
  6.338E-01  2.345E+01 .          .          . *        .      +   .
  7.979E-01  2.766E+01 .          .          .   *      .      +   .
  1.004E+00  3.179E+01 .          .          .      *   .      +   .
  1.265E+00  3.587E+01 .          .          .        * .      +   .
  1.592E+00  3.992E+01 .          .          .          *       +  .
```

Worked example 14.13

Using SPICE software, write a program which calculates the impedance of the series *RLC* circuit in worked example 12.3.1 (see page 303), in 10 per cent steps, from 20 per cent below to 20 per cent above resonance.

Solution

Not every implementation of SPICE provides simple facilities for determination of impedance, but we can get around this by driving the circuit with an alternating current of $1\angle 0°$ A. The complex value of voltage between the current source terminals is then equal to the complex input impedance of the circuit. An input file for the circuit in worked example 12.3.1 is given in table 14.25.

To remind the reader, the series circuit contains $R = 4\ \Omega$, $L = 0.1$ mH and $C = 1\ \mu$F. The resonant frequency is 100 000 rad/s, which we take to be 15 920 Hz (to an accuracy of four decimal places).

SPICE detects the capacitor as an open-circuit, and it is necessary to include an additional resistor, 'Rextra', to shunt the current source in order to provide a 'd.c.' path between node 1 and node 0. The value of 'Rextra' is sufficiently high to have no effect on our calculation.

The '.AC' control line causes the frequency to sweep from 12.73 kHz to 19.1 kHz, in five linear steps. This provides us with data at (approx.) 80 krad/s to 100 krad/s in steps of 10 krad/s, which will include the resonant frequency of 100 krad/s.

Table 14.25

```
*** Worked Example 14.13 ***
*** Input impedance calculation ***
*** Circuit is driven by a constant current ***
I    0    1    AC    1
R    1    2    4
L    2    3    0.1M
C    3    0    1U
** Rextra shunts the current source **
Rextra    0    1    1MEG
.OPTIONS NOPAGE
** Calculate values in 10 per cent steps around resonance **
*  Five LINEAR steps between 80 k rad/s to 100 k rad/s  *
.AC    LIN    5    12.73K    19.1K
** Input impedance is equal to V(1) **
.PRINT    AC    VM(1)    VP(1)
.END
```

Table 14.26

FREQ	VM(1)	VP(1)
1.273E+04	6.024E+00	-4.839E+01
1.432E+04	4.524E+00	-2.785E+01
1.592E+04	4.000E+00	-8.898E-03
1.751E+04	4.432E+00	2.552E+01
1.910E+04	5.427E+00	4.252E+01

The '.PRINT' control line requests the magnitude (VM(1)) and the phase (VP(1)) of the voltage across the current source to be output. These values are, incidentally, the magnitude and the phase of impedance of the circuit at the respective frequencies.

The reader will note that the results in table 14.26 (which is the relevant part of the output file) at 1.432E+04 Hz (approx. 80 000 rad/s) and 1.910E+04 Hz (approx. 120 000 rad/s) generally agree with the results obtained in worked example 12.3.3. The impedance at the resonant frequency of 1592 Hz (100 000 rad/s) is seen to be 4 Ω.

Worked example 14.14

Solve the Fourier analysis problem in worked example 13.3.1 (see page 329) using SPICE software.

Solution

Fourier analysis (that is, a '.FOUR' control line appears in the input file in table 14.27) can be performed by SPICE in conjunction with a transient ('.TRAN') analysis of the waveshape; that is, a '.TRAN' control line must appear in the same input file as a '.FOUR' control line.

SPICE performs FOURier analysis on waveforms which exist for $t > 0$, that is, the values exist only in positive time. The rectangular waveform is therefore described in this case for the time interval $0 < t < 1$ s by means of a piece-wise linear stimulus 'Vpwl', which is applied to a resistor R of value 1 Ω. Each point on the waveform is defined by two values, namely by its time in seconds and its value, as follows (see also worked example 14.4)

PWL(0, 1 0.25, 1 0.25001, −1 0.75, −1 0.75001, 1 1, 1)

That is, at $t = 0$ the voltage has a value 1 V, and remains at that value until $t = 0.25$ s. SPICE only accepts a practical waveform, that is it must have a practical value of fall time from 1 V to −1 V. A fall time of 0.01 ms has been assumed; similarly, a rise time of 0.01 ms has been adopted at $t = 0.75$ ms. The points specified on the waveform are joined linearly. That is to say, the waveform is not an ideal square wave, but is trapezoidal.

The transient response is computed by the .TRAN control line, which requests the computer to evaluate the transient response every millisecond for a period of 1 s (this period is also referred to later).

The .FOUR control line causes the computer to evaluate the d.c. component together with the amplitude and phase of the fundamental frequency and all harmonics up to the 9th harmonic. The fundamental frequency is specified by the user, and is the first value in the .FOUR control line, that is, it is 1 Hz. Also written in this line is the output variable on which the Fourier analysis is to be performed, namely V(1). There can be several output variables in this line, but we only need one. The reader is asked to note that the transient analysis period must be at least 1/(fundamental frequency) s long; in our case this is 1/1 = 1 s.

The relevant part of the output file is listed in table 14.28. The reader will note that SPICE reports a small d.c. component, which is due to the fact that the PWL waveform used in the program is not a true square wave, together with any mathematical 'noise' produced by the program. The table produced by the program lists not only each Fourier component and its associated phase angle, but also the 'normalized' values. The latter assumes that the magnitude of the fundamental frequency is 1.000, and that its phase shift is 0.000°.

One needs to look carefully at the table in order to decide what values are significant; in practice we can ignore any harmonic with a magnitude less than about 1 per cent of the normalised fundamental frequency component. Applying this as a general rule, we can ignore all the even harmonics. Looking at the PHASE (DEG) column, we see that the phase shift of the fundamental frequency is 90°, that is, it is a cosine term in the Fourier series. Similarly, the third harmonic is a −cos term, the fifth harmonic is a cos term, etc. The results in table 14.18 are in general agreement with the results of worked example 13.3.1.

Once again, it is pointed out that a computer package may not provide all the answers needed for a particular problem. This example can be used to demonstrate that engineering judgement is often needed to make a final decision. If the .TRAN control line had been written '.TRAN 10M 1', that is, transient values to be calculated at 10 ms intervals rather than the 1 ms period used here, the phase angle values would have differed by many degrees from the values in table 14.28, and the result would not have agreed very well with the theoretical analysis. The result would, of course, be produced much more quickly, but not so accurately!

Table 14.27

```
  Worked Example 14.14
*** Fourier analysis of a square wave ***
*** Voltage is a Piece-Wise Linear (PWL) source ***
Vpwl 1 0 PWL(0,1  0.25,1  0.25001,-1  0.75,-1  0.75001,1  1,1)
R   1   0   1
.OPTIONS NOPAGE
*TRANsient analysis
* |  1 ms steps
* |       |  up to 1 s
* |       |     |
.TRAN    1Ms    1s
*FOURier analysis
* |   1 Hz fundamental frequency
* |       |  analyse the voltage at node 1
* |       |     |
.FOUR     1     V(1)
.END
```

Table 14.28

FOURIER COMPONENTS OF TRANSIENT RESPONSE V(1)

DC COMPONENT = -1.001001E-03

HARMONIC NO	FREQUENCY (HZ)	FOURIER COMPONENT	NORMALIZED COMPONENT	PHASE (DEG)	NORMALIZED PHASE (DEG)
1	1.000E+00	1.273E+00	1.000E+00	9.000E+01	0.000E+00
2	2.000E+00	2.002E-03	1.572E-03	9.000E+01	-1.401E-08
3	3.000E+00	4.244E-01	3.333E-01	-9.000E+01	-1.800E+02
4	4.000E+00	2.002E-03	1.572E-03	-9.000E+01	-1.800E+02
5	5.000E+00	2.547E-01	2.000E-01	9.000E+01	1.868E-08
6	6.000E+00	2.002E-03	1.572E-03	9.000E+01	-3.268E-08
7	7.000E+00	1.819E-01	1.429E-01	-9.000E+01	-1.800E+02
8	8.000E+00	2.002E-03	1.572E-03	-9.000E+01	-1.800E+02
9	9.000E+00	1.415E-01	1.111E-01	9.000E+01	3.735E-08

TOTAL HARMONIC DISTORTION = 4.288100E+01 PERCENT

15

Complex Numbers, Matrices, Determinants and Partial Fractions

15.1 Imaginary numbers

The concept of imaginary numbers was introduced by mathematicians to allow them to express the square root of a negative number. For example, if $x^2 = -9$ the solution is given by saying

$$x = \text{imaginary operator} \times \sqrt{9}$$

The imaginary number is useful to electrical engineers, who gave the imaginary operator the symbol j so that, in the above case

$$x = \mathrm{j}3$$

(Mathematicians use the symbol i to represent the imaginary operator, but as this symbol is used by electrical engineers to represent electrical current, the symbol j is used.)
Thus

$$\mathrm{j} = \sqrt{(-1)}$$
$$\mathrm{j}^2 = (\sqrt{(-1)})^2 = -1$$
$$\mathrm{j}^3 = \mathrm{j} \times \mathrm{j}^2 = -\mathrm{j}$$
$$\mathrm{j}^4 = \mathrm{j}^2 \times \mathrm{j}^2 = 1, \text{ etc.}$$

The reader should note that

$$\text{imaginary number} = \text{imaginary operator (j)} \times \text{real number}$$

15.2 Complex numbers

A *complex number* is the sum of a real number and an imaginary number, and is either written in bold Roman type (as in this book), or has a bar drawn over it (as is often the case in hand-written material). Thus

$$\boldsymbol{V} = a + \mathrm{j}b$$

or

$$\overline{V} = a + \mathrm{j}b$$

It is important to note that both a and b are *real numbers*, but the component $\mathrm{j}b$ is an *imaginary number* since b is multiplied by the imaginary operator j.

15.3 Representation of complex numbers

There are four ways of representing a complex number, namely

rectangular or cartesian form	$\boldsymbol{V} = a + \mathrm{j}b$
polar form	$\boldsymbol{V} = r\angle\theta$
exponential form	$\boldsymbol{V} = r \times \mathrm{e}^{\mathrm{j}\theta}$
trigonometric form	$\boldsymbol{V} = r(\cos\theta + \mathrm{j}\sin\theta)$

The last three forms above are, from Euler's identity, generally the same form. The relationship between them is

$$r = \sqrt{(a^2 + b^2)} \qquad \theta = \tan^{-1}(b/a)$$

The rectangular and polar forms are most widely used in electrical engineering. The seven complex numbers $\boldsymbol{V}_1$ to $\boldsymbol{V}_7$ represented in figure 15.1 are written in polar and rectangular form as follows.

$$\boldsymbol{V}_1 = 3 = 3 + \mathrm{j}0 = 3\angle 0°$$

$$\boldsymbol{V}_2 = 4 + \mathrm{j}2 = 4.47\angle 26.57°$$

$$\boldsymbol{V}_3 = \mathrm{j}3 = 0 + \mathrm{j}3 = 3\angle 90°$$

$$\boldsymbol{V}_4 = -2 + \mathrm{j}4 = 4.47\angle 116.57°$$

$$\boldsymbol{V}_5 = -2 = -2 + \mathrm{j}0 = 2\angle 180° \text{ (or } 2\angle -180°)$$

$$\boldsymbol{V}_6 = -4 - \mathrm{j}3 = 5\angle 216.87° \text{ (or } 5\angle -143.13°)$$

$$\boldsymbol{V}_7 = 1 - \mathrm{j}3 = 3.16\angle -71.57°$$

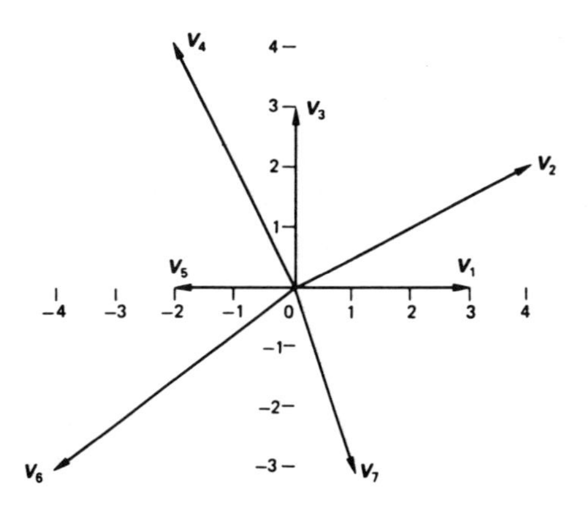

Figure 15.1 *Representation of complex numbers.*

15.4 Conjugate of a complex number

The conjugate, $\boldsymbol{V}^*$, of the complex number $\boldsymbol{V} = a + \mathrm{j}b = r\angle\theta$ is

$$\boldsymbol{V}^* = a - \mathrm{j}b = r\angle -\theta$$

where $r = \sqrt{(a^2 + b^2)}$ and $\theta = \tan^{-1}(b/a)$.

If $\boldsymbol{V} = 3 + \mathrm{j}4 = 5\angle 53.13°$, then $\boldsymbol{V}^* = 3 - \mathrm{j}4 = 5\angle - 53.13°$. Also $(\boldsymbol{V}^*)^* = \boldsymbol{V}$. A complex number and its conjugate are said to form a *conjugate complex pair* of numbers. Other useful properties are

$$(\boldsymbol{V}_1\boldsymbol{V}_2)^* = \boldsymbol{V}_1^*\boldsymbol{V}_2^*$$

$$(\boldsymbol{V}_1 \pm \boldsymbol{V}_2)^* = \boldsymbol{V}_1^* \pm \boldsymbol{V}_2^*$$

$$\left(\frac{\boldsymbol{V}_1}{\boldsymbol{V}_2}\right)^* = \frac{\boldsymbol{V}_1^*}{\boldsymbol{V}_2^*}$$

15.5 Sum and difference of complex numbers

To perform addition (or subtraction) of complex numbers, the numbers must first be converted into rectangular form, and the real parts must be added together (or subtracted from one another), and then the imaginary parts added together (or subtracted from one another) as follows.

If $\boldsymbol{V}_1 = 5 + \mathrm{j}6$ and $\boldsymbol{V}_2 = -3 - \mathrm{j}8$, then

$$\boldsymbol{V}_1 + \boldsymbol{V}_2 = (5 + \mathrm{j}6) + (-3 - \mathrm{j}8) = 2 - \mathrm{j}2$$

$$\boldsymbol{V}_1 - \boldsymbol{V}_2 = (5 + \mathrm{j}6) - (-3 - \mathrm{j}8) = 8 + \mathrm{j}14$$

15.6 Multiplication of complex numbers

When the numbers are expressed in polar form, the multiplication proceeds as follows

$$r_1\angle\theta_1 \times r_2\angle\theta_2 = r_1 r_2\angle(\theta_1 + \theta_2)$$

In general, it is far more convenient to carry out multiplication of complex numbers using the polar form than it is using rectangular form. However, when numbers are expressed in rectangular form, the multiplication can be carried out as follows.

$$(a + \mathrm{j}b)(c + \mathrm{j}d) = (ac + \mathrm{j}^2 bd) + \mathrm{j}(ad + bc)$$
$$= (ac - bd) + \mathrm{j}(ad + bc)$$

If $\boldsymbol{V}_1 = 3 + \mathrm{j}4 = 5\angle 53.13°$ and $\boldsymbol{V}_2 = -5 - \mathrm{j}5 = 7.07\angle -135°$, then

$$\boldsymbol{V}_1 \times \boldsymbol{V}_2 = 5\angle 53.13° \times 7.07\angle -135° = 35.35\angle -81.87°$$

or

$$\boldsymbol{V}_1 \times \boldsymbol{V}_2 = (3 + \mathrm{j}4)(-5 - \mathrm{j}5) = (-15 + 20) + \mathrm{j}(-15 - 20) = 5 - \mathrm{j}35$$

The *product of a conjugate complex pair* of rectangular numbers is

$$(a + \mathrm{j}b)(a - \mathrm{j}b) = a^2 + b^2$$

and the *product of a conjugate complex pair* of polar numbers is

$$r\angle\theta \times r\angle -\theta = r^2\angle(\theta - \theta) = r^2$$

both giving a real number and no imaginary term. For example, if $\boldsymbol{V} = 3 + \mathrm{j}4 = 5\angle 53.13°$, then

$$\boldsymbol{V} \times \boldsymbol{V}^* = (3 + \mathrm{j}4)(3 - \mathrm{j}4) = 9 + 16 = 25$$

or

$$\boldsymbol{V} \times \boldsymbol{V}^* = 5\angle 53.13° \times 5\angle -53.13° = 25\angle 0°$$

15.7 Division of complex numbers

If $\boldsymbol{V}$ and $\boldsymbol{I}$ are complex numbers then, using polar values, division is carried out as follows

$$\frac{\boldsymbol{V}}{\boldsymbol{I}} = \frac{r_1\angle\theta}{r_2\angle\phi} = \frac{r_1}{r_2}\angle(\theta - \phi)$$

In general, it is far more convenient to use polar values rather than rectangular values when performing division.

Using rectangular complex values, division is carried out as follows

$$\frac{\boldsymbol{V}}{\boldsymbol{I}} = \frac{\boldsymbol{V} \times \boldsymbol{I}^*}{\boldsymbol{I} \times \boldsymbol{I}^*}$$

The product $\boldsymbol{I} \times \boldsymbol{I}^*$ gives a real number with no imaginary part (see section 15.6), and the process of dividing by $(\boldsymbol{I} \times \boldsymbol{I}^*)$ is known as *rationalising the denominator*. For example

if $\boldsymbol{V} = a + \mathrm{j}b$ and $\boldsymbol{I} = c + \mathrm{j}d$, then

$$\frac{\boldsymbol{V}}{\boldsymbol{I}} = \frac{(a + \mathrm{j}b)(c - \mathrm{j}d)}{(c + \mathrm{j}d)(c - \mathrm{j}d)} = \frac{(ac + bd) + \mathrm{j}(bc - ad)}{c^2 + d^2}$$

Suppose $\boldsymbol{V} = 4 + \mathrm{j}4 = 5.66\angle 45°$ and $\boldsymbol{I} = 3 + \mathrm{j}4 = 5\angle 53.13°$, then

$$\frac{\boldsymbol{V}}{\boldsymbol{I}} = \frac{5.66\angle 45°}{5\angle 53.13°} = 1.132\angle -8.13°$$

or

$$\frac{\boldsymbol{V}}{\boldsymbol{I}} = \frac{(4 + \mathrm{j}4)(3 - \mathrm{j}4)}{(3 + \mathrm{j}4)(3 - \mathrm{j}4)} = \frac{(12 + 16) + \mathrm{j}(12 - 16)}{9 + 16}$$

$$= \frac{28 - \mathrm{j}4}{25} = 1.12 - \mathrm{j}0.16$$

15.8 Powers and roots of complex numbers

The Nth power of a complex number is calculated as follows

$$(R\angle\theta)^N = R^N\angle N\theta$$

For example

$$(2\angle 60°)^3 = 8\angle 180°$$

There are N roots of the complex number $R\angle\theta$ as follows

$$(R\angle\theta)^{1/N} = R^{1/N} \angle \frac{\theta + [n \times 360]}{N}$$

where n has a value in the range 0, 1, 2, . ., $(N - 1)$.

For example $\sqrt[3]{(8\angle 60°)}$ has the values $2\angle 20°$, $2\angle 140°$ and $2\angle 260°$.

15.9 Matrix representation

Consider the equations

$$V_1 = Z_{11}I_1 + Z_{12}I_2 + Z_{13}I_3$$

$$V_2 = Z_{21}I_1 + Z_{22}I_2 + Z_{23}I_3$$

$$V_3 = Z_{31}I_1 + Z_{32}I_2 + Z_{33}I_3$$

These can be written in a *matrix form* of ordered rows and columns of elements of the same kind as follows

$$\begin{bmatrix} V_1 \\ V_2 \\ V_3 \end{bmatrix} = \begin{bmatrix} Z_{11} & Z_{12} & Z_{13} \\ Z_{21} & Z_{22} & Z_{23} \\ Z_{31} & Z_{32} & Z_{33} \end{bmatrix} \begin{bmatrix} I_1 \\ I_2 \\ I_3 \end{bmatrix}$$

or, alternatively, in the even more ordered form

$$[V] = [Z][I]$$

where $[V]$ is a voltage matrix, $[Z]$ is an impedance matrix and $[I]$ is a current matrix. The formalised matrix representation is well suited to calculator and computer solution of equations.

The double subscript notation is used to identify the position of an element within a matrix. The first subscript denotes the row in which the element is found, and the second denotes the column. Thus Z_{23} is the element in the second row of the third column. A simple mnemonic to remember the order of the subscripts is **R**oman **C**atholic (**R**ows, **C**olumns).

Both the voltage and the current matrix, above, are written in the form of a column, and each is described as a *column matrix* or *vector*. These have only one column, but may contain any number of rows (in the case considered, both have the same length). Since there is only one column in a column matrix, only one subscript is necessary to define the position of an element within it.

The impedance matrix is a *square matrix*, in which the number of rows is equal to the number of columns.

The *major diagonal* of a square matrix is the diagonal line of elements going from the top leftmost element to the bottom rightmost element, that is, the elements Z_{11}, Z_{22}, Z_{33} in the impedance matrix above lie on the major diagonal.

A *diagonal matrix* is a rectangular matrix in which all the elements are zero except those on the major diagonal.

An *identity matrix* or *unit matrix* is a diagonal matrix in which the value of each of the elements on the major diagonal is unity.

Other forms include the following. A *row matrix* has only one row containing number of columns as follows

$$[a_{11} \quad a_{12} \quad a_{13} \quad \ldots \quad a_{1N}]$$

A *rectangular matrix* is one having M rows and N columns ($M \neq N$), and is described as an M *by* N or $M \times N$ matrix, as shown below.

$$\begin{bmatrix} a_{11} & a_{12} & \ldots & a_{1N} \\ a_{21} & a_{22} & \ldots & a_{2N} \\ \ldots & \ldots & \ldots & \ldots \\ a_{M1} & a_{M2} & \ldots & a_{MN} \end{bmatrix}$$

A *null matrix* is one in which every element is zero.

15.10 Matrix addition and subtraction

Two matrices can either be added together or subtracted from one another *if they are of the same order*. If $\boldsymbol{A} = [a_{ij}]$ and $\boldsymbol{B} = [b_{ij}]$ are two $M \times N$ matrices, their sum (or difference) is the matrix $\boldsymbol{C} = [c_{ij}]$, where each element of $\boldsymbol{C}$ is the sum (or difference) of the corresponding elements of $\boldsymbol{A}$ and $\boldsymbol{B}$. That is

$$\boldsymbol{A} \pm \boldsymbol{B} = [a_{ij} \pm b_{ij}]$$

$$\text{If } \boldsymbol{A} = \begin{bmatrix} -3 & 4 & 5 \\ 0 & -6 & 7 \end{bmatrix} \text{ and } \boldsymbol{B} = \begin{bmatrix} 3 & 4 & -5 \\ 6 & -7 & 8 \end{bmatrix} \text{ then}$$

$$\boldsymbol{A} + \boldsymbol{B} = \begin{bmatrix} -3+3 & 4+4 & 5-5 \\ 0+6 & -6-7 & 7+8 \end{bmatrix} = \begin{bmatrix} 0 & 8 & 0 \\ 6 & -13 & 15 \end{bmatrix}$$

$$\boldsymbol{A} - \boldsymbol{B} = \begin{bmatrix} -3-3 & 4-4 & 5-(-5) \\ 0-6 & -6-(-7) & 7-8 \end{bmatrix} = \begin{bmatrix} -6 & 0 & 10 \\ -6 & 1 & -1 \end{bmatrix}$$

15.11 Matrix multiplication

The matrix product $\boldsymbol{AB}$ (which *must be carried out in that order*) can be computed only if the *number of columns in* $\boldsymbol{A}$ is equal to the *number of rows in* $\boldsymbol{B}$. $\boldsymbol{B}$ is not necessarily conformable to $\boldsymbol{A}$ for multiplication, that is, the product $\boldsymbol{BA}$ may not be defined. The following should be observed:

1. $\boldsymbol{AB} \neq \boldsymbol{BA}$ generally.
2. $\boldsymbol{AB} = 0$ does not always imply $\boldsymbol{A} = 0$ or $\boldsymbol{B} = 0$.
3. $\boldsymbol{AB} = \boldsymbol{AC}$ does not always imply $\boldsymbol{B} = \boldsymbol{C}$.

If matrices are conformable, multiplication is performed on a *row by column* basis; each element in a row is multiplied by the corresponding element of a column, and the products are summed.

If $\boldsymbol{A}$ is a $1 \times M$ row matrix, and $\boldsymbol{B}$ is a $M \times 1$ column matrix, then

$$\boldsymbol{AB} = [a_{11} \quad a_{12} \quad \ldots \quad a_{1M}] \begin{bmatrix} b_{11} \\ b_{21} \\ \cdot \\ \cdot \\ \cdot \\ b_{M1} \end{bmatrix}$$

$$= [a_{11}b_{11} + a_{12}b_{21} + \ldots + a_{1M}b_{M1}]$$

$$= \left[\sum_{k=1}^{M} a_{1k}b_{k1} \right]$$

Note: $\boldsymbol{BA}$ is not defined.

Suppose that $\boldsymbol{A} = [-1 \quad 2 \quad 3]$ and $\boldsymbol{B} = \begin{bmatrix} -6 \\ 0 \\ 7 \end{bmatrix}$ then

$$\boldsymbol{AB} = [-1 \quad 2 \quad 3] \begin{bmatrix} -6 \\ 0 \\ 7 \end{bmatrix} = [-1(-6) + 2(0) + 3(7)] = [27]$$

If $\boldsymbol{A} = \begin{bmatrix} a_{11} & a_{12} \\ a_{21} & a_{22} \\ a_{31} & a_{32} \end{bmatrix}$ and $\boldsymbol{B} = \begin{bmatrix} b_{11} & b_{12} \\ b_{21} & b_{22} \end{bmatrix}$ then

$$\boldsymbol{AB} = \begin{bmatrix} a_{11}b_{11} + a_{12}b_{21} & a_{11}b_{12} + a_{12}b_{22} \\ a_{21}b_{11} + a_{22}b_{21} & a_{21}b_{12} + a_{22}b_{22} \\ a_{31}b_{11} + a_{32}b_{21} & a_{31}b_{12} + a_{32}b_{22} \end{bmatrix}$$

Note: $\boldsymbol{BA}$ is not defined.

Suppose that $\boldsymbol{A} = \begin{bmatrix} -1 & 0 \\ 3 & -4 \\ 5 & 6 \end{bmatrix}$ and $\boldsymbol{B} = \begin{bmatrix} 4 & 5 \\ -6 & 7 \end{bmatrix}$ then

$$\boldsymbol{AB} = \begin{bmatrix} -1(4) + 0(-6) & -1(5) + 0(7) \\ 3(4) + (-4)(-6) & 3(5) + (-4)(7) \\ 5(4) + 6(-6) & 5(5) + 6(7) \end{bmatrix} = \begin{bmatrix} -4 & -5 \\ 36 & -13 \\ -16 & 67 \end{bmatrix}$$

Also, if $\boldsymbol{R} = \begin{bmatrix} 5 & -3 & 0 \\ -3 & 12 & -5 \\ 0 & -5 & 11 \end{bmatrix}$ and $\boldsymbol{I} = \begin{bmatrix} \boldsymbol{I}_1 \\ \boldsymbol{I}_2 \\ \boldsymbol{I}_3 \end{bmatrix}$ then

$$\boldsymbol{RI} = \begin{bmatrix} 5I_1 & -3I_2 & -0I_3 \\ -3I_1 & 12I_2 & -5I_3 \\ -0I_1 & -5I_2 & 11I_3 \end{bmatrix}$$

A *matrix may be multiplied by a scalar k* (which should not be confused with the 1 × 1 matrix $[k]$), to give

$$k\boldsymbol{A} = \boldsymbol{A}k = [ka_{ij}]$$

that is, each element in the matrix is multiplied by k.

15.12 The determinant of a square matrix

A matrix is simply an ordered array of elements, and *has no numerical value*. On the other hand, the *determinant* of a square matrix has a numerical value, which is given the symbol Δ, or det $\boldsymbol{A}$ or $|\boldsymbol{A}|$. This value can be used in the computation of the value of unknown variables in the equations represented by the matrix equations.

The value of a determinant of order 2 is calculated as follows.

$$\begin{vmatrix} a_{11} & a_{12} \\ a_{21} & a_{22} \end{vmatrix} = a_{11}a_{22} - a_{12}a_{21}$$

For example

$$\begin{vmatrix} 2 & 4 \\ -5 & -3 \end{vmatrix} = 2(-3) - 4(-5) = 14$$

For a determinant of order 3

$$\begin{vmatrix} a_{11} & a_{12} & a_{13} \\ a_{21} & a_{22} & a_{23} \\ a_{31} & a_{32} & a_{33} \end{vmatrix} = a_{11}a_{22}a_{33} + a_{12}a_{22}a_{31} + a_{13}a_{21}a_{32} - a_{13}a_{22}a_{31} - a_{12}a_{21}a_{33} - a_{11}a_{23}a_{32}$$

For example

$$\begin{vmatrix} 2 & -3 & 4 \\ 5 & 6 & -2 \\ -3 & -4 & 7 \end{vmatrix} = \begin{matrix} 2.6.7 + (-3).(-2).(-3) + 4.5.(-4) \\ -4.6.(-3) - (-3).5.7 - 2.(-2).(-4) = -63 \end{matrix}$$

15.13 Minors and cofactors

The *minor* of the element a_{ij} (row i, column j) of a determinant is obtained by deleting row i and column j of the determinant; the minor of this element is given the symbol M_{ij}. The value of the minor is multiplied by $(-1)^{i+j}$ to give the *cofactor* of a_{ij}; the cofactor is given the symbol Δ_{ij}.

For example, in the determinant of order 3 in section 15.12

$$M_{22} = \begin{vmatrix} a_{11} & a_{13} \\ a_{31} & a_{33} \end{vmatrix}$$

and

$$\Delta_{22} = (-1)^{2+2} \begin{vmatrix} a_{11} & a_{13} \\ a_{31} & a_{33} \end{vmatrix} = + \begin{vmatrix} a_{11} & a_{13} \\ a_{31} & a_{33} \end{vmatrix} = + a_{11}a_{33} - a_{13}a_{31}$$

15.14 Evaluating a determinant

The value of a determinant of order N is the sum of the N products of each element in a selected row (or column) and its cofactor (great care should be taken in ensuring that the cofactor has the correct mathematical sign, see section 15.13).

Consider the following determinant of order 3, which can be evaluated by (for example) selecting the elements in the first row as follows

$$\begin{vmatrix} a_{11} & a_{12} & a_{13} \\ a_{21} & a_{22} & a_{23} \\ a_{31} & a_{32} & a_{33} \end{vmatrix} = a_{11}\Delta_{11} + a_{12}\Delta_{12} + a_{13}\Delta_{13}$$

$$= a_{11} \begin{vmatrix} a_{22} & a_{23} \\ a_{32} & a_{33} \end{vmatrix} - a_{12} \begin{vmatrix} a_{21} & a_{23} \\ a_{31} & a_{33} \end{vmatrix} + a_{13} \begin{vmatrix} a_{21} & a_{22} \\ a_{31} & a_{32} \end{vmatrix}$$

$$= (a_{11}a_{22}a_{33} - a_{11}a_{23}a_{32}) - (a_{12}a_{21}a_{33} - a_{12}a_{23}a_{31}) + (a_{13}a_{21}a_{32} - a_{13}a_{22}a_{31})$$

Alternatively, if the elements in the second column are selected, then the value of the determinant is calculated from

$$a_{12}\Delta_{12} + a_{22}\Delta_{22} + a_{32}\Delta_{32}$$

Both of the above calculations produce the result given earlier in section 15.12.

15.15 The rule of Sarrus

A determinant of order 3 can be evaluated using the rule of Sarrus as follows

$$\begin{vmatrix} a_{11} & a_{12} & a_{13} \\ a_{21} & a_{22} & a_{23} \\ a_{31} & a_{32} & a_{33} \end{vmatrix} \begin{matrix} a_{11} & a_{12} \\ a_{21} & a_{22} \\ a_{31} & a_{32} \end{matrix}$$

The determinant is written down, and the first two columns are repeated to the right of the determinant. Diagonal lines are drawn joining sets of three elements together; the product of the diagonally downwards terms are given a positive sign, and the product of the diagonally upwards terms are given a negative sign. The value of the determinant is the sum of these products. For example, by the rule of Sarrus

$$\begin{vmatrix} 5 & 6 & 7 \\ 2 & -3 & 4 \\ 1 & -2 & 3 \end{vmatrix} \begin{aligned} &= 5.(-3).3 + 6.4.1. + 7.2.(-2) - 7.(-3).1 - 6.2.3 - 5.4.(-2) \\ &= -45 + 24 - 28 + 21 - 36 + 40 = -24 \end{aligned}$$

15.16 Cramer's rule

Linear simultaneous equations can be solved by *Cramer's rule* as follows. Consider the following matrix form of equation

$$\begin{bmatrix} y_1 \\ y_2 \\ \cdot \\ \cdot \\ \cdot \\ y_M \end{bmatrix} = \begin{bmatrix} a_{11} & a_{12} & \cdots & a_{1N} \\ a_{21} & a_{22} & \cdots & a_{2N} \\ \cdot & \cdot & \cdots & \cdot \\ \cdot & \cdot & \cdots & \cdot \\ \cdot & \cdot & \cdots & \cdot \\ a_{M1} & a_{M2} & \cdots & a_{MN} \end{bmatrix} \begin{bmatrix} x_1 \\ x_2 \\ \cdot \\ \cdot \\ \cdot \\ x_N \end{bmatrix}$$

The value of x_K in the Kth row is obtained from the computations

$$x_K = \frac{1}{\det \mathbf{A}} \begin{vmatrix} a_{11} & \cdots & a_{1(K-1)} & y_1 & a_{1(K+1)} & \cdots & a_{1N} \\ a_{21} & \cdots & a_{2(K-1)} & y_2 & a_{2(K+1)} & \cdots & a_{2N} \\ \cdot & \cdots & \cdot & \cdot & \cdot & \cdots & \cdot \\ \cdot & \cdots & \cdot & \cdot & \cdot & \cdots & \cdot \\ \cdot & \cdots & \cdot & \cdot & \cdot & \cdots & \cdot \\ a_{M1} & \cdots & a_{M(K-1)} & y_M & a_{2(K+1)} & \cdots & a_{MN} \end{vmatrix}$$

For example, solve for I_2 in the following

$$5 = 10I_1 - 3I_2 - 5I_3$$

$$10 = -3I_1 + 7I_2 - 4I_3$$

$$-15 = -5I_1 - 4I_2 + 9I_3$$

The matrix form of the equation is

$$\begin{bmatrix} 5 \\ 10 \\ -15 \end{bmatrix} = \begin{bmatrix} 10 & -3 & -5 \\ -3 & 7 & -4 \\ -5 & -4 & 9 \end{bmatrix} \begin{bmatrix} I_1 \\ I_2 \\ I_3 \end{bmatrix}$$

From Cramer's rule

$$I_2 = \begin{vmatrix} 10 & 5 & -5 \\ -3 & 10 & -4 \\ -5 & -15 & 9 \end{vmatrix} \Bigg/ \begin{vmatrix} 10 & -3 & -5 \\ -3 & 7 & -4 \\ -5 & -4 & 9 \end{vmatrix}$$

and using the rule of Sarrus

$$I_2 = \frac{10.10.9 + 5.(-4).(-5) + (-5).(-3).(-15) - (-5).10.(-5) - 5.(-3).9 - 10.(-4).(-15)}{10.7.9 + (-3).(-4).(-5) + (-5).(-3).(-4) - (-5).7.(-5) - (-3).(-3).9 - 10.(-4).(-4)}$$

$$= \frac{60}{94} = 0.638$$

15.17 Matrices and determinants containing complex numbers

Matrices and determinants containing complex numbers can be handled by the methods described above, but the reader must use the methods described in sections 15.5 to 15.8 when dealing with complex numbers.

15.18 Partial fractions

Functions of s can be transformed directly using integral calculus but, in practice, it is easier to arrange the functions so that they fit one or more of the terms in a table of Laplace transforms. One method of doing this is by the use of *partial fractions*.

Analysis of electrical problems by the Laplace transform method generally requires the derivation of the inverse transform from a result which, usually, is in the form of the ratio of two polynomials in s.

If

$$F(s) = \frac{N(s)}{D(s)}$$

where $N(s)$ and $D(s)$ are polynomials in s, and the degree of $N(s)$ is less than the degree of $D(s)$, then:

1. For every linear factor $(As + B)$ in $D(s)$ there is a corresponding partial fraction

$$\frac{1}{As + B}$$

2. For every quadratic factor $(As^2 + Bs + C)^2$ in $D(s)$ which has real roots, there is a corresponding partial fraction

$$\frac{Ps + Q}{As^2 + Bs + C}$$

3. For every repeated factor $(As + B)^2$ in $D(s)$ there is a corresponding partial fraction

$$\frac{P}{As + B} + \frac{Q}{(As + B)^2}$$

4. For every repeated quadratic factor $(As^2 + Bs + C)^2$ in $D(s)$ there is a corresponding partial fraction

$$\frac{Ps + Q}{As^2 + Bs + C} + \frac{Rs + T}{(As^2 + Bs + C)^2}$$

5. For every thrice-repeated character $(As + B)^3$ in $D(s)$ there is a corresponding partial fraction

$$\frac{P}{As + B} + \frac{Q}{(As + B)^2} + \frac{R}{(As + B)^3}$$

6. For every cubic factor $As^3 + Bs^2 + Cs + D$ in $D(s)$ there is a corresponding partial fraction

$$\frac{Ps^2 + Qs + R}{As^3 + Bs^2 + Cs + D}$$

Worked example 15.1

Determine the partial fraction expansion of

$$\frac{200}{s(s + 1)(s + 2)}.$$

Solution

Applying the rules laid down above

$$\frac{200}{s(s+1)(s+2)} = \frac{A}{s} + \frac{B}{s+1} + \frac{C}{s+2}$$

A, B and C in the above expression are known as the *residues*. The residues are quickly evaluated in this case by the *cover-up rule* as follows.

(i) Determine the *value* of s which makes the denominator of that particular term zero.
(ii) Substitute this *value* into the full expression (both numerator and denominator) and, ignoring or 'covering up' the factor in question, the *residue* is the result of the calculation.

The residue A is calculated by letting $s = 0$. Substituting this value in the original equation whilst 'covering up' the factor s gives

$$A = \left.\frac{200}{(s+1)(s+2)}\right|_{s=0} = \frac{200}{((0)+1)((0)+2)} = 100$$

The residue B is calculated by letting $s = -1$, as follows

$$B = \left.\frac{200}{s(s+2)}\right|_{s=-1} = \frac{200}{(-1)((-1)+2)} = -200$$

and the residue C is calculated by letting $s = -2$ as follows

$$C = \left.\frac{200}{s(s+1)}\right|_{s=-2} = \frac{200}{(-2)((-2)+1)} = 100$$

That is

$$\frac{200}{s(s+1)(s+2)} = \frac{100}{s} - \frac{200}{s+1} + \frac{100}{s+2}$$

Worked example 15.2

Determine the partial fraction expansion of

$$\frac{54}{(4s^2 - 5s + 1)^2}$$

Solution

This can be re-written in the form

$$\frac{54}{(4s^2 - 5s + 1)^2} = \frac{54}{(1 - s)^2(1 - 4s)^2}$$

Since there are repeated factors in the denominator, the partial fraction expansion is written as follows

$$\frac{A}{1 - s} + \frac{B}{(1 - s)^2} + \frac{C}{1 - 4s} + \frac{D}{(1 - 4s)^2}$$

Since $(1 - s)^2$ and $(1 - 4s)^2$ appear in the original equation, we can evaluate the residues B and D by the cover-up rule as follows

$$B = \left.\frac{54}{(1 - 4s)^2}\right|_{s = 1} = \frac{54}{(1 - 4(1))^2} = 6$$

$$D = \left.\frac{54}{(1 - s)^2}\right|_{s = 0.25} = \frac{54}{(1 - 0.25)^2} = 96$$

The residues A and C are calculated by substituting the known residues, and multiplying both sides of the equation by the denominator of the polynomial as follows

$$54 = A(1 - s)(1 - 4s)^2 + 6(1 - 4s)^2 + C(1 - s)^2(1 - 4s) + 96(1 - s)^2$$

Equating coefficients of s^3 gives $-16A - 4C = 0$, and equating coefficients of s^2 gives $8A + 3C = -64$. Solving for A and C yields $A = 16$, $C = -64$. That is

$$\frac{54}{(4s^2 - 5s + 1)^2} = \frac{16}{1 - s} + \frac{6}{(1 - s)^2} - \frac{64}{1 - 4s} + \frac{96}{(1 - 4s)^2}$$

Bibliography

There are many excellent texts in the field of circuit analysis, the following being a representative selection. In particular, the attention of the reader is directed towards those marked with a *; these texts contain introductory material on the computer program SPICE.

1. W. Banzhaf, *Computer-Aided Circuit Analysis using SPICE*, Prentice-Hall, 1989*
2. L. S. Bobrow, *Elementary Linear Circuit Analysis*, 2nd edn, Holt, Rinehart and Winston, 1987*
3. S. A. Boctor, *Electric Circuit Analysis*, Prentice-Hall, 1987*
4. R. L. Boylestad, *Introductory Circuit Analysis*, 5th edn, Merrill Publishing Co., 1987
5. M. D. Ciletti, *Introduction to Circuit Analysis and Design*, Holt, Rinehart and Winston, 1988. Supplement available on SPICE
6. T. L. Floyd, *Principles of Electric Circuits*, 2nd edn, Merrill Publishing Co., 1985
7. J. D. Irwin, *Basic Engineering Circuit Analysis*, 2nd edn, Macmillan, 1987*
8. M. C. Kelly and B. Nichols, *Introductory Linear Electrical Circuits and Electronics*, John Wiley, 1988
9. N. M. Morris and F. W. Senior, *Electric Circuits*, Macmillan WORK OUT series, 1991*
10. J. W. Nilsson, *Electric Circuits*, 3rd edn, Addison-Wesley, 1990*
11. P. W. Tuinenga, *SPICE – A Guide to Circuit Simulation & Analysis Using PSpice*, Prentice-Hall, 1988*

PSpice software suitable for personal computers both in full and educational versions are available from ARS Microsystems, Herriard Business Centre, Alton Road, Herriard, Basingstoke, Hampshire. The educational version is available at low cost from many Shareware suppliers.

Index